Biological Effects of Polynucleotides

Biological Effects of Polynucleotides

Proceedings of the Symposium on Molecular Biology Held in New York, June 4–5, 1970 Sponsored by Miles Laboratories, Inc.

Edited by

Roland F. Beers, Jr.

Johns Hopkins University
Baltimore, Maryland

and

Werner Braun

Rutgers University
New Brunswick, New Jersey

SPRINGER-VERLAG NEW YORK · HEIDELBERG · BERLIN
1971

Softcover reprint of the hardcover 1st edition 1971

Library of Congress Catalog Card Number 74-14350

ISBN-13: 978-3-642-85774-4 e-ISBN-13: 978-3-642-85772-0
DOI: 10.1007/978-3-642-85772-0

PREFACE

In recent years our horizons regarding the role of nucleic acids in biological systems have been expanded vastly by the finding that these molecules not only carry and transmit specific information but also can act as less specific triggers of antiviral factors and of immunological responses. The latter properties are of particular interest in terms of possible utilization in human and veterinary medicine and consequently led, in the last few years, to the development of a new research area that combines both fundamental and applied problems in a uniquely attractive way. Furthermore, the importance and the complexity of the problems has attracted investigators from many fields, including molecular biologists, virologists, immunologists, chemists, biophysicists, oncologists, pharmacologists, and clinicians. A discussion of new developments in this area of biological effects of polynucleotides, with particular emphasis on interferon induction and modification of immune responses, therefore, seemed a logical topic for one of the annual symposia that Miles Laboratories, Inc., has sponsored in recent years. The management of Miles accepted the suggestion with enthusiasm and thus once again earned the gratitude of the scientific community for sponsoring a catalytic meeting that was principally concerned with basic research problems and only tangentially with immediate applications. Springer-Verlag agreed to publish the proceedings of the meeting which was held at the Americana Hotel in New York City, June 4–5, 1970 and attracted an unusually large audience.

Many people have helped to make the meeting and this volume possible and we cannot possibly list them all. However, we would like to mention particularly and gratefully Dr. Walter Compton, President of Miles Laboratories, Inc., and Dr. Edward G. Basset of the Research Products Division of Miles who served as the symposium coordinator. We also are very grateful to Mr. Herbert Stillman, Production Manager of Springer-Verlag New York who helped efficiently in getting the present material between covers. Indispensible for the preparation of the discussions and for the numerous details involved in bringing continuity in style of the manuscripts have been the long hours of work by Evelyn Sales of Johns Hopkins University, for which we are most indebted and thankful. Our special thanks, however, must go to the contributors who not only shared their findings and views with us but gave us permission to edit their discussions without seeing the finished product. We hope that by such editing we made the discussions more readable without harming the spirit and accuracy of the discussants' words.

February 1971

ROLAND F. BEERS, JR.

WERNER BRAUN

INTRODUCTORY REMARKS*

Nearly 50 years ago James Cabell made the observation that "The optimist proclaims that we live in the best of all possible worlds; the pessimist fears this is true". We are being constantly warned of impending doom of our civilization and even of our species. Whenever a society becomes aware of a crisis that threatens its survival, there is immediately a search for the cause and cure of this crisis. During the past 10 years we have experienced a variety of thrusts in different directions, many of them labeled some kind of war: war on poverty, war on hunger, war on racial bigotry, and now war on pollution. It would seem that the American character can only respond to a challenge by taking the stance of a warrior going into battle to slay the enemy, an oversimplification of the issues which reduces the objective to a simple "victory". Certainly this posture has dominated our action in Vietnam.

In remarkable contrast to this single purposed approach to some of the major problems of our world today is the multivariant approach used so successfully during the past two decades in the fields of molecular biology. If there are "victories" in molecular biology, they are the creation of self-consistent models or the elimination of regions of ignorance; they are victories of understanding rather than of conquest. Nevertheless, molecular biologists have been under attack because of their "reductionist" approach to their subject, an attitude that has led to a framework of concepts that has become almost a dogma, replete with myths and the equivalent of sacred cows. It is a transformation of nineteenth century attitudes of physics into twentieth century biology. Such a condition, if it exists (and I think there is some evidence for believing that it does), provides the basis for the kinds of scientific revolution described by Thomas Kuhn. This can be very threatening to those whose careers and accomplishments are measured on the basis of a particular paradigm as well entrenched as that of molecular biology. But this is the inevitable price for the evolution of man's understanding of

* Presented at the banquet preceding the symposium.

his universe and of himself. It is clear that at some stage in the evolution of a science the reductionist's approach reaches a limit of detail and understanding or of suitable model constructs. At this point the holistic approach must become operational. Indeed, it is fair to conclude that the degree of maturity of a scientific field is often times reflected in the extent to which the model constructs reflect the holistic perspective.

Related to but not identical with this is the increasing concern over the relationship of the models of molecular biology to man, his society, and his concerns. The phrase, "social relevance" included, for example, in the NSF program for Interdisciplinary Research Relevant to the Problems of Society (IRRPOS) reflects this concern. The present fiscal draught in the biomedical research field, especially in "nonmission-oriented areas", may be the consequence of misplaced priorities resulting from a misunderstanding of the significance of such research in the larger framework of man's world, but if so it is an inconsistent and illogical conclusion in the face of such priorities as the Vietnam adventure, the Apollo program, and the SST. But a more reasonable and, perhaps, modest explanation is an apparent irrelevancy of the goals of some of the fields of biomedical research with those of society. When the research does not appear to contribute to the solution or to alleviate the crises facing society, society ceases to support it or even tolerate it, especially if there is any suggestion that the research may contribute to the magnitude and complexity of these crises. One serious consequence, seen prominently in our rebellious younger generation, is the rejection of the rational methods of understanding and the growth of anti-intellectualism and "quickie" solutions. One of the major responsibilities of scientists and engineers today is to fight this trend by educating the public to the indispensible need for rational understanding, and by applying the methods of science and technology to the problems of man's troubled world.

It is true, although some would deny it, that science and technology in the present condition of mankind can only buy time, that is, delay the predicted catastrophes while man comes to grips with himself and his institutions. But time is a resource we can ill afford to squander or ignore. This Symposium, like its predecessors, constitutes a part of a sequence that reflects an attempt to provide a particular and timely perspective toward molecular biology and related biomedical areas of research; interdisciplinary in its content to emphasize a more comprehensive view of the topics and somewhat mission-oriented to indicate the relevancy of the topics to man and his society. The relevancy consists of the contributions of the fields of study toward and understanding of man and his world and the application of the information and concepts of these fields of study in the technologies employed in finding solutions to the problems of man.

It will become apparent in the following articles that there are several "socially relevant" areas but I would stress one in particular. One of the technologies that is answering the need for time is agriculture. Although I agree with the Paddocks and the Erlichs regarding the ultimate inability of agricultural technology to meet the demands for food as long as the world population increases exponentially, nevertheless, without the advances of agriculture the plight of the world would already be disasterous.

However, the problems of adequate food production, especially of protein, extend beyond high yields per acre through improved genetic strains and application of chemical fertilizers. Equally important are the competitors of man for the same food resources. Clearly the problem of pesticides and their impact on the ecology of the earth points out the fallacy of single-purposed approaches to problems of this nature. But similar complex problems looms on the horizon in the production of animal protein. As the world population of mammals and fowls increase at a rate faster than that of man, and their density and highly inbred character increases, the prospects for massive epidemics increase alarmingly. Contributing to this hazard are the modern methods for fast transportation which makes isolation increasingly difficult.

To put these remarks in perspective we could examine the economic cost. The recent foot-and-mouth disease outbreak in Great Britain cost that nation over $ 300,000,000 just to eradicate it. It has been estimated that the annual loss in South America from this same disease is equivalent to the annual U.S. economic aid given there. Comparable losses occur in Africa and in Asia. Similar statistics can be cited for the poultry industry abroad and in the United States. Israel, which used to be an exporter of poultry products, is now an importer. These chronic losses are increasing at a time when the need for increased protein production is becoming critical.

What these considerations say, of course, is that the prevention of viral epidemics by methods other than those now employed (eradication by destruction of the diseased and exposed animals and isolation by quarantine) must be replaced by methods that directly attack the viral-host interaction. That challenge makes the topics of this Symposium particularly significant and relevant, a fact that has already been emphasized by Philip Handler in his defense for continued Federal support for basic biomedical research. If molecular biology is to develop beyond the model-building stage it appears to this observer that some of the best opportunities lie in the research studies represented by this assembled group of investigators.

June 3, 1970

ROLAND F. BEERS, JR.
The Johns Hopkins University
Baltimore, Maryland

LIST OF AUTHORS, COAUTHORS AND DISCUSSANTS

AARONSON, J. N., State University of New York (SUNY) Albany, N.Y.
ADAMSON, R., National Institutes of Health, Bethesda, Md.
AMIEL, J. L., Hopital Paul-Brousse, Paris, France
BARON, S., National Institutes of Health, Bethesda, Md.
BEERS, R. F., JR., The Johns Hopkins University, Baltimore, Md.
BISHOP, D. C., Rutgers University, New Brunswick, N. J.
BRAUN, W., Rutgers University, New Brunswick, N. J.
BROOM, A., University of Utah, Salt Lake City, Utah.
BUCKLER, C. E., National Institutes of Health, Bethesda, Md.
CABRERA, E., University of Illinois, Urbana, Ill.
CARBONE, P., National Institutes of Health, Bethesda, Md.
CARTER, W. A., The Johns Hopkins University, Baltimore, Md.
CELERCQ, E. D., Stanford University, Stanford, Calif.
CHAMBERLIN, M. J., University of California, San Diego, La Jolla, Calif.
CHOAY, J., Hopital Paul-Brousse, Paris, France
CLARK, D. D., Fordham University, New York, N.Y.
COLBY, C., University of California, San Diego, La Jolla, Calif.
CONE, R. E., University of Michigan, Ann Arbor, Mich.
DE VITA, V., National Institutes of Health, Bethesda, Md.
DOVTHART, R., Eli Lilly & Co., Indianapolis, Indiana
DU BUY, H., National Institutes of Health, Bethesda, Md.
D'ANTONIA, L., University of Illinois, Urbana, Ill.
DUESBERG, P. H., University of California, San Diego, La Jolla, Calif.
EBEL, J. P., University of Strasbourg, Strasbourg, France
ECKSTEIN, F., Stanford University, Stanford, Calif.
FAUVE, R. M., Institut Pasteur, Garches 92, France
FIELD, A. K., Merck Institute for Therapeutic Research, West Point, Pa.
FLEISHER, M., Sloan Kettering Institute for Cancer Research, New York, N.Y.
FORLANO, A., Ferris State College, Big Rapids, Mich.

FRAENKEL-CONRAT, H., University of California, Berkeley, Calif.
FRIEDMAN, H. M., University of Michigan, Ann Arbor, Mich.
FRIEDMAN, S., Laval University, Quebec, Canada
GALLAGHER, J. G., University of Vermont, Burlington, Vt.
GALLO, R. C., National Institutes of Health, Bethesda, Md.
GAZDAR, A., National Institutes of Health, Bethesda, Md.
GOTTLIEB, A. A., Rutgers University, New Brunswick, N. J.
HAMILTON, L. D., Brookhaven National Lab., Upton, N. Y.
HAN, I. H., University of Michigan, Ann Arbor, Mich.
HAYAT, M., Hopital Paul-Brousse, Paris, France
HECHT, S., University of Wisconsin, Madison, Wisc.
HILLEMAN, M. R., Merck, Sharp and Dohme, West Point, Pa.
HOMAN, E. R., National Institutes of Health, Bethesda, Md.
ISHIZUKA, M., Rutgers University, New Brunswick, N. J.
JASMIN, C., Hopital Paul-Brousse, Paris, France
JOHNSON, A. G., University of Michigan, Ann Arbor, Mich.
JOHNSON, M., National Institutes of Health, Bethesda, Md.
KARKAS, J., Columbia University, New York, N. Y.
KRAKOFF, I. H., Sloan-Kettering Institute, New York, N. Y.
LAMPSON, G. P., Merck Institute for Therapeutic Research, West Point, Pa.
LEVY, H. B., National Institutes of Health, Bethesda, Md.
LUDLUM, D., University of Maryland, College Park, Md.
MATHÉ, G., Hopital Paul-Brousse, Paris, France
MERIGAN, T. C., Stanford University, Stanford, Calif.
MIRO-QUESDA, O., National Institutes of Health, Lima, Peru
MITTELMAN, A., Roswell Park Memorial Institute, Buffalo, N. Y.
MORRELL, R. M., The Upjohn Company, Kalamazzo, Mich.
MOZES, E., Weizmann Institute of Science, Rehovot, Israel
NEMES, M. M., Merck Institute for Therapeutic Research, West Point, Pa.
NIBLACK, J., Charles Pfizer, Inc., Groton, Conn.
PHILIPS, F. S., Sloan Kettering Institute for Cancer Research, New York, N. Y.
PITHA, P., The Johns Hopkins University, Baltimore, Md.
PLESCIA, O. J., Rutgers University, New Brunswick, N. J.
REGELSON, W., Medical College of Virginia, Charlottesville, Va.
RICHMOND, J. Y., U. S. Dept. of Agriculture, Greenport, N. Y.
RHIM, J., National Institutes of Health, Bethesda, Md.
RILEY, F., National Institutes of Health, Bethesda, Md.
ROELANTS, G. E., National Institute for Medical Research, London, England
ROSENFELD, C., Hopital Paul-Brousse, Paris, France

SAKOUHI, M., Hopital Paul-Brousse, Paris, France
SCHMIDTKE, J. R., University of Michigan, Ann Arbor, Mich.
SCHNEIDER, M., Hopital Paul-Brousse, Paris, France
SCHWARTZ, M., Sloan Kettering Institute for Cancer Research, New York, N. Y.
SCHWARZENBERG, L., Hopital Paul-Brousse, Paris, France
SEEGER, R. C., National Institutes of Health, Bethesda, Md.
SELA, M., The Weizmann Institute of Science, Rehovot, Israel
SHEARER, G., The Weizmann Institute of Science, Rehovot, Israel
SHEFFMAN, G., University of Pennsylvania, Philadelphia, Pa.
SIMON, M. I., University of California, San Diego, La Jolla, Calif.
SINGER, B., University of California, Berkeley, Calif.
STEINBERG, A. D., National Institutes of Health, Bethesda, Md.
STERNBERG, S., Sloan-Kettering Institute, New York, N. Y.
STINEBRING, W. R., University of Vermont, Burlington, Vt.
STOUT, R. D., University of Michigan, Ann Arbor, Mich.
TALAL, N., National Institutes of Health, Bethesda, Md.
TYTELL, A. A., Merck Institute for Therapeutic Research, West Point, Pa.
DE VASSAL, F., Hopital Paul-Brousse, Paris, France
VILCEK, J., New York University, New York, N. Y.
WEBB, D., Rutgers University, New Brunswick, N. J.
WEINSTEIN, A., National Institutes of Health, Bethesda, Md.
WEINSTEIN, I. B., Columbia University, New York, N. Y.
WELLS, R. D., Stanford University, Stanford, Calif.
WHANG-PENG, J., National Institutes of Health, Bethesda, Md.
WINCHURCH, R., Rutgers University, New Brunswick, N. J.
WORTHINGTON, M., National Institutes of Health, Bethesda, Md.
YAJIMA, Y., Rutgers University, New Brunswick, N. J.
YOUNG, C. W., Sloan Kettering Institute for Cancer Research, New York, N. Y.

CONTENTS

Part I

Template Effects and Their Modification

QUALITATIVE AND QUANTITATIVE ASPECTS OF POLYNUCLEOTIDE MODIFICATION*

B. SINGER and H. FRAENKEL-CONRAT

Space Sciences Laboratory
and
Department of Molecular Biology and Virus Laboratory
University of California, Berkeley, California

A discussion of the chemistry of polynucleotide modification involves not only the reactivity of the individual units but also the role of secondary structure or conformation of the polymer. As will be discussed later, the reactivity of bases in a polymer with particular reagents can be quite dissimilar to that in a monomer, depending on whether the base is hydrogen bonded in a double-stranded structure, whether there are stacking interactions, or whether it is protein encased. Although the chemistry of base modification is often determined using bases, nucleosides, or nucleotides, this is only a general guide to the sites and extent of modification by the same reagents acting on polynucleotides.

The reagents that we have used for modification of polynucleotides are several of those known or believed to be mutagenic in biological systems. As model systems we have used homo- or heteropolymers—both single- and double-stranded. As biological materials we have used TMV-RNA and TMV to correlate chemical modification with mutation.

The best known and most effective chemical mutagen is nitrous acid. Adenine, guanine, and cytosine can be deaminated to hypoxanthine, xanthine, and uracil. However, the rate of deamination of poly A or poly C is extremely slow compared to that of the monomers (Singer and Fraenkel-Conrat, 1970), and poly AT is deaminated only above 60°, a temperature at which the helical structure is melted out (Kotaka and Baldwin, 1964). In the case of TMV-RNA, all three bases react at similar rates, but when

* This investigation was supported by research grants NsG 479 from the National Aeronautics and Space Administration and GB 6209 from the National Science Foundation.

TMV is treated with nitrous acid, adenine and cytosine, but not guanine, become deaminated (Schuster and Wilhelm, 1963). This illustrates the differences in reactivity of monomer, single- and double-stranded polymer, and protein-encased nucleic acid.

Hydroxylamine reacts primarily with the pyrimidine bases. Uridine is split into ribosyl-urea and 5-isoxazolone at *p*H 9, but does not appreciably react at *p*H 6. The reaction with cytidine which is favored at *p*H 6 is shown in Figure 1. Both II (6-hydroxylamino-5, 6-dihydrocytidine) and III (N4-hy-

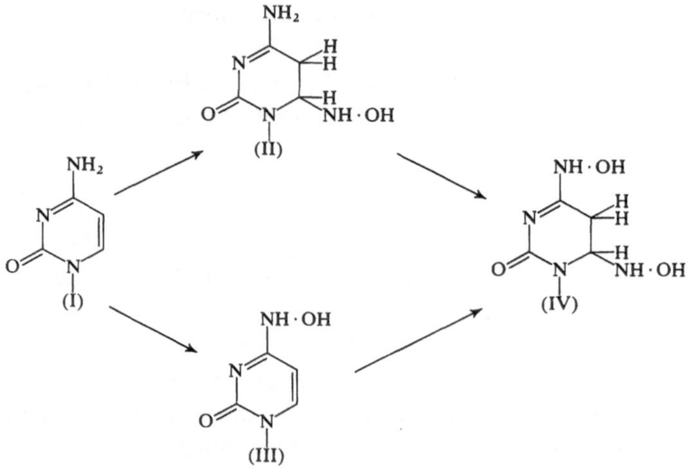

Fig. 1. Products formed in reaction of cytidine with hydroxylamine. II, 6-hydroxyamino-5,6-dihydrocytosine is an intermediate of very limited stability. III, N4-hydroxycytosine is referred to as the "Mono"-product in the text; and IV, N-4 hydroxy, 6-hydroxyamino-5,6-dihydrocytosine as the doubly modified product, or "Bis".

droxycytidine) are formed directly (Lawley, 1967; Kochetkov *et al.*, 1967). The relative amounts of the final products (III and IV), formed in poly C as a function of time, are shown in Table 1. Both III and IV are found in the early stages of reaction. However, III is generally the predominant product particularly at *p*H 4.5. As with nitrous acid, the reaction of polymers is slower than that of the monomer. Similar methods have given data on the reactivity of TMV-RNA with hydroxylamine. The relative proportion of III is greater than that found in the homopolymer.

O-methyl hydroxylamine (methoxyamine) reacts only with cytidine and at a much slower rate than hydroxylamine (Figure 2) (Phillips *et al.*, 1966; Kochetkov *et al.*, 1963). The mechanism is believed to be analogous to that in Figure *a*. With very little modification N4-methoxy cytidine is the only product detectable in both, poly C and TMV-RNA (Table 1). However, as

Table 1. Products of Reaction of Poly C with Hydroxylamine and Methoxyamine[a]

Reagent	*p*H	Hours	Temperature	Mono[b] %	Bis[c] %	Total %
Hydroxylamine	6.0	1	~25° C	2.8	1.0	3.8
		3		3.5	1.5	5.0
		24		35	23	58
	4.5	18		2.8	n.d.	2.8
Methoxyamine	5.5	1	37° C	0.08	n.d.	0.08
		3		0.7	n.d.	0.7
		18		3.4	0.1	3.5
		72		9.2	0.2	9.4
	4.0	18		6.8	0.7	7.5

[a] Modified ^{14}C-labeled poly C is digested with snake venom diesterase and *E. coli* phosphatase, and the nucleosides are separated chromatographically, using largely the solvents advocated by Small and Gordon (1968); however, the separation of the two products obtained with methoxyamine requires other systems, such as 80% saturated ammonium sulfate containing 2% isopropanol.

[b] N4-hydroxy-(or methoxy) cytidine.

[c] N4-hydroxy, 6-hydroxyamino 5,6 dihydrocytidine or the corresponding methoxy compounds.

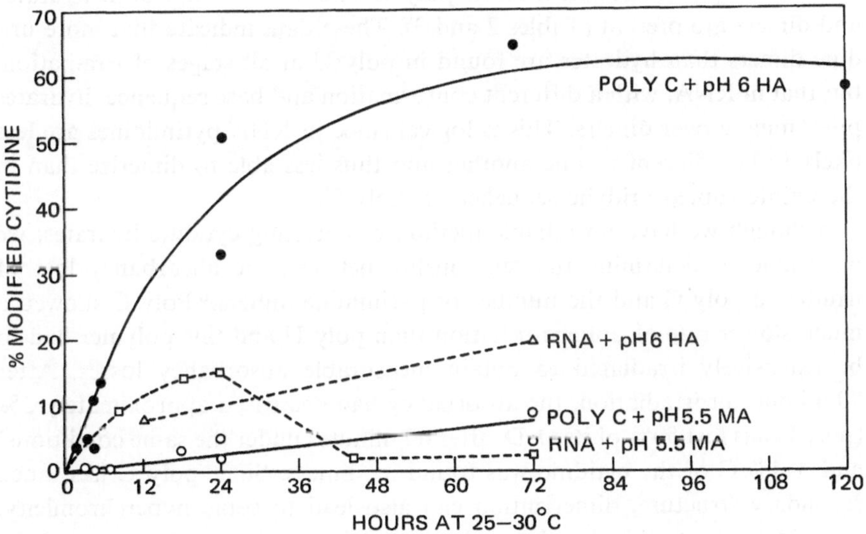

Fig. 2. Rate of modification (Products III + IV of Figure 1) of cytidine in poly C and TMV-RNA. HA stands for hydroxylamine and MA for methoxyamine.

the reaction proceeds, the doubly substituted product is found in an equivalent amount.

N4-hydroxycytidine (III) has been found to prefer the tautomeric form,

NOH
HN—C⟨
CH
| ||

, corresponding to uracil oxime, by a factor of 10 (Brown *et al.*, 1968). The same is presumed to be true for the *O*-methyl compound. There are no data on the relative amounts of IV and III formed after reacting double-stranded polymers with hydroxylamine or methoxyamine.

The effect of ultraviolet irradiation on bases has been studied very extensively and is reviewed by Setlow (1966). Both cytidine and uridine can form hydrates and dimers but quantitation has been difficult, particularly of cytidine hydrate since it easily reverts to cytidine. We have investigated the effect of UV irradiation on poly U, poly C, and TMV-RNA and found that the rate of modification is dependent on the conformation, as well as the composition of the polymer. For poly U and TMV-RNA, irradiated in water, we find that the decrease in ultraviolet adsorption is approximately equivalent to the number of uridine hydrates and pyrimidine dimers found upon analysis. In 10^{-3} *M* Mg^{++}, TMV-RNA is hypochromed and the absorbancy loss is much less than the amounts of products found. However, in such an H-bonded conformation, the RNA is also much less reactive than as a random coil. The rate of formation of UV products is about $^1/_3$ of that in water; while the O.D. decrease is about $^1/_5$.

At all stages of modification of poly U and TMV-RNA, both hydrates and dimers are present (Tables 2 and 3). These data indicate that more uridine dimers than hydrates are found in poly U at all stages of irradiation, but that in RNA, with a different conformation and base sequence, hydrates predominate over dimers. This is logical since in RNA pyrimidines are less likely to be adjacent to one another and thus less able to dimerize than in the uninterrupted uridylic sequences of poly U.

Although we have no reliable method of detecting cytidine hydrates, we were able to determine the relationship between the absorbancy loss of irradiated poly C and the number of pyrimidine dimers.[1] Poly C showed a much slower rate of dimer formation than poly U and this polymer had to be extensively irradiated to obtain measurable absorbancy losses. After 90 minutes of irradiation, the absorbancy had decreased approximately 2% (poly U has lost 50% of its O.D. after 60 minutes under the same conditions) and 3–4% C of the cytidine was found as dimers. Since poly C has much secondary structure, dimerization can also lead to some hyperchromicity, compensating for the absorbancy loss upon irradiation, and thus the O.D.

[1] Cytidine dimers tend to become deaminated.

Table 2. Effect of Ultraviolet Irradiation on Poly U: The Relationship between O.D. Loss and the Formation of Uridine Hydrates and Dimers

Experiment	O.D. loss %	U hydrates %	U dimers %	Total hydrates + dimers %
1	7	1.5	3.6	5.1
	20	6	16	22
2	(15)	6	12	18
	27	7	20	27
	50	17	27	44

Table 3. Effect of Ultraviolet Irradiation on TMV-RNA: The Relationship between O.D. Loss and the Formation of Uridine Hydrates and Pyrimidine Dimers per Mole RNA

Experiment	O.D. loss[a] %	Calc. No. bases losing O.D.	U hydrates	U dimers (including C dimers)	Undigested "dimer"	Total UV products
1	(0.22)	(14)	14	4	...	18
	1.1	70	59	26	...	85
	3.3	211	128	68	8	204
2	(0.25)	(16)	6	6	...	12
	0.6	38	26	12	...	38
	4.4	280	125	61	75	261
	6.5	415	254	118	192	446
3	12.8	820	330	312	30	672

[a] Figures in parentheses were obtained by extrapolating from the O.D. loss found on more extensive treatment.

data represent minimum figures. Therefore, it appears that formation of cytidine dimers is a significant event but that in a heteropolymer, such as TMV-RNA, the contribution of the cytidine reaction is small.

There are several recent reports on the effects of UV irradiation on protein-encapsulated RNA. Irradiation of both R17 phage and TMV has been found to produce only uridine hydrates (Tao *et al.*, 1969; Carpenter and Kleczkowski, 1969; Remsen *et al.*, 1970). The absence of any pyrimidine dimers confirms the belief that the RNA in these viruses is surrounded by protein so as to prevent two bases from being in the proper relationship for dimerization. In contrast, in double-stranded polymers, hydration is suppressed and dimers are the major products of irradiation (Setlow, 1966).

Last year we published a comparative study of the effects of various alkylating agents, including nitrosoguanidine, on polynucleotides of various conformations (Singer and Fraenkel-Conrat, 1969a; Singer and Fraenkel-Conrat, 1969b). The results of these studies will now be briefly reviewed.

Table 4. The Reaction of MNNG and DMS with Monomeric and Homopolymeric Nucleotides in Aqueous Solution

	MNNG						DMS					
	Monomer			Polymer			Monomer			Polymer		
	A	G	C	A	G	C	A	G	C	A	G	C
	Percent of original base											
Methyl A	<1.5			38			60			40		
Methyl G		5			15			25			12	
Methyl C			20			<1.5			20			50

Many reagents can introduce alkyl groups into nucleotides and polynucleotides in neutral aqueous solution. The possible sites of alkylation are the N-7 of guanine, the N-1, N-3, and N-7 of adenine, and the N-3 of cytosine. Typical alkylating agents, such as dimethyl sulfate (DMS) or methyl methanesulfonate (MMS) alkylate adenosine, guanosine and cytidine to about equal extents in both monomeric and homopolymeric form. Another type of alkylating agent, nitrosoguanidine (MNNG), reacts with cytidylic acid, but poorly with adenylic acid or guanylic acid. But when poly C is treated with MNNG, little methylation occurs, while both poly A and poly G are methylated. This is shown in Table 4. Further experiments established that MNNG acted preferentially on guanosine and adenosine when they were stacked, as in a polymer, but that stacking interactions interfered with the methylation of cytidine. Thus, in a dispersing solvent such as dimethyl formamide (DMF), poly C was methylated to a much greater extent than in water. The reverse was found for poly G, where in DMF the guanosine was not methylated and behaved as the monomer.

In aqueous solution, MNNG, DMS, and MMS all showed the same pattern of methylation of TMV-RNA (Table 5). Performing the reaction in DMF depressed the guanosine methylation and increased the cytidine methylation with MNNG, but not with the other reagents.

The effect of hydrogen bonding in a double-stranded polymer is quite different from the variations observed with stacking interactions. As an illustration, Table 6 shows the relative amounts and extents of methylation of poly A and poly A:U. In poly A:U the N-1 position is base paired and its reactivity is extremely small while the methylation of N-3 and N-7 is not

Table 5. Relative Amounts of Methylated Bases Formed Upon Reaction of TMV-RNA with Alkylating Agents in H_2O and DMF

	MNNG		DMS	NMS
	H_2O	DMF	H_2O	H_2O
	Percent of total alkylation			
1-Methyl A	9.2	12.2	11.7	15.6
3-Methyl A	1.3	1.8	1.8	1.3
7-Methyl A	6.3	5.1	3.7	3.5
7-Methyl G	76	59	77	73
3-Methyl C	7.1	22	5.8	6.7

Table 6. The Effect of Double-Strandedness on Methylation of Poly A by Dimethyl Sulfate and Nitrosoguanidine

	MNNG		DMS	
	Poly A % of total alkylation	Poly A · U % of poly A alkylation	Poly A % of total alkylation	Poly A · U % of poly A alkylation
1-Methyl A	92	1.2	81	8
3-Methyl A	2	45	5	142
7-Methyl A	7	65	13	44

appreciably affected. The same behavior is found with both DMS and MNNG, although double-strandedness depresses the reaction of the N-1 of adenosine with MNNG more then that with DMS.

In poly C:G only the alkylation of the cytidine is depressed, in agreement with expectation. The alkylation of guanosine occurs normally with both types of reagents. The effect of encasing a single-stranded RNA in protein is well illustrated with TMV. The rate and extent of alkylation are very low compared to TMV-RNA but the most dramatic and important result is given in Table 7. With DMS and MMS 7-methyl guanosine is greatly

Table 7. Relative Yields of Methylated Bases after Reacting TMV with Dimethyl Sulfate, Methyl Methanesulfonate, or Nitrosoguanidine

	Dimethyl sulfate	Methyl methane-sulfonate	Nitroso-guanidine
	Percentage of total methylation		
1-Methyladenine	1	0.7	6.3
3-Methyladenine	0	0	4.9
7-Methyladenine	2	0.4	6
7-Methylguanine	97	97	48
3-Methylcytosine	0.2	1.7	35

Table 8. Principal Reactions of Nucleotides in Various Conformations

Reagent	Monomer	Single-stranded polymer	Double-stranded polymer	Protein-encased single-stranded RNA (TMV)
HNO_2	deamination A, G, C	deamination A, G, C	none	deamination *A, C*
HA, MA[a] (*p*H 6)	mono, bis C[a]	*mono*, bis C	?	?
UV	U, (C) hydrates	U, (C) hydrates, dimers	U, C dimers	U, (C) hydrates
DMS, etc.	7 mG, 3 mC, 1 mA[b] (3 mA, 7 mA)	7 mG, 3 mC, 1 mA (3 mA, 7 mA)	7 mG (3 mA, 7 mA)	7 mG
MNNG	3 mC	7 mG, 1 mA (3 mA, 7 mA)	7 mG (3 mA, 7 mA)	7 mG, *3 mC* (1 mA, 3 mA, 7 mA)

[a] See Table 1. HA and MA stand for hydroxylamine and methoxyamine.
[b] m stands for methyl.

predominant (97% of the methyl groups introduced are found as 7-methyl guanosine), but with MNNG methylation of guanosine and cytosine is almost equal (7-methyl guanosine 48% and 3-methyl cytidine 35% of the total methylation). This reaction of TMV with MNNG is the only instance of alkylation of another RNA base approaching in magnitude the alkylation of guanosine. These observations can be correlated with the fact that the bases in TMV are not stacked in terms of base interactions and thus behave in a characteristically different manner with MNNG compared to other alkylating agents.

The effects of the various reagents discussed in this paper on monomeric and polymeric nucleotides, including viral RNA and viruses, is summarized in Table 8. Reactions which are regarded as highly mutagenic are underlined (see following pages).

References

Brown, D. M., Hewlins, M. J. E., and Shell, P. (1968). The tautomeric state of *N*(4)-hydroxy- and of *N*(4)-amino-cytosine derivatives. J. Chem. Soc. (C): 1925–1929.

Carpenter, J. M. and Kleczkowski, A. (1969). The absence of photoreversible pyrimidine dimers in the RNA of ultraviolet-irradiated tobacco mosaic virus. Virology **39**: 542–547.

Kochetkov, N. K., Budowsky, E. I., and Shibaeva, R. P. (1963). The selective reaction of *O*-methylhydroxylamine with the cytidine nucleus. Biochim. Biophys. Acta **68**: 493–496.

Kochetkov, N. K., Budowsky, E. I., Sverdlov, E. D., Shibaeva, R. P., Shibaev, V. N., and Monastirskaya, G. S. (1967). The mechanism of the reaction of hydroxylamine and *O*-methylhydroxylamine with cytidine. Tetrahedron Letters **34**: 3253–3257.

Kotaka, T. and Baldwin, R. L. (1964). Effects of nitrous acid on the dAT copolymer as a template for DNA polymerase. J. Mol. Biol. **9**: 323–339.

Lawley, P. D. (1967). Reaction of hydroxylamine at high concentration with deoxycytidine or with polycytidylic acid: Evidence that substitution of amino groups in cytosine residues by hydroxylamino is a primary reaction, and the possible relevance to hydroxylamine mutagenesis. J. Mol. Biol. **24**: 75–81.

Phillips, J. H., Brown, D. M., and Grossman, L. (1966). The efficiency of induction of mutations by hydroxylamine. J. Mol. Biol. **21**: 405–419.

Remsen, J. F., Miller, N., and Cerutti, P. A. (1970). Photohydration of uridine in the RNA of coliphage R 17. II. The relationship between ultraviolet inactivation and uridine photohydration. Proc. Natl. Acad. Sci. **65**: 460–466.

Schuster, H. and Wilhelm, R. C. (1963). Reaction differences between tobacco mosaic virus and its free ribonucleic acid with nitrous acid. Biochim. Biophys. Acta **68**: 554–560.

Setlow, J. K. (1966). The molecular basis of biological effects of ultraviolet radiation and photoreactivation. Current Topics Radiation Res. **2**: 195–248.

Singer, B. and Fraenkel-Conrat, H. (1969a). Chemical modification of viral ribonucleic acid. VII. The action of methylating agents and nitrosoguanidine on polynucleotides including tobacco mosaic virus ribonucleic acid. Biochemistry **8**: 3260 3266.

Singer, B. and Fraenkel-Conrat, H. (1969b). Chemical modification of viral ribonucleic acid. VIII. The chemical and biological effects of methylating agents and nitrosoguanidine on tobacco mosaic virus. Biochemistry **8:** 3266–3269.

——— (1970). The messenger and template activites of chemically modified polynucleotides. Biochemistry **9:** 3694–3701.

Small, G. D. and Gordon, M. P. (1968). Reaction of hydroxylamine and methoxyamine with the ultraviolet-induced hydrate of cytidine. J. Mol. Biol. **34:** 281–291.

Tao, M., Small, G. D., and Gordon, M. P. (1969). Photochemical alterations in ribonucleic acid isolated from ultraviolet-irradiated tobacco mosaic virus. Virology **39:** 534–541.

TEMPLATE AND MESSENGER ACTIVITIES OF MUTAGEN-TREATED POLYNUCLEOTIDES*

H. Fraenkel-Conrat and B. Singer

Department of Molecular Biology and Virus Laboratory
and
Space Sciences Laboratory
University of California, Berkeley, California

The elucidation of the structure of DNA by Watson and Crick (1953) led to an immediate realization of the nature of point mutations: replacement of one purine or pyrimidine by the other or a base resembling the other in its preferred tautomeric state and thus in its hydrogen bonding pattern. Much of the data obtained with mutagenic agents acting on DNA, RNA, or polynucleotides in subsequent years supported this concept. However, other observations seemed not to be interpretable on this basis, and a feeling of uncertainty has arisen in this regard among workers in the area of chemical mutagenesis during the past decade. We believe that these seeming discrepancies and doubts were due to three principle causes: Several of the discordant results were experimentally unsound, and were not repeatable at later times or in different laboratories. Other data were misinterpreted in terms of incomplete knowledge of the chemistry of the mutagenic reaction. Finally, the properties of very extensively modified polynucleotides were studied in the assumption that these would reflect the nature of the mutagenic event, without due consideration of the infrequency of actual mutagenic reactions.

In recent years we have obtained and reported comparative data on the efficiency of various mutagens acting on the RNA in the tobacco mosaic virus (TMV) particle and on isolated TMV-RNA (Singer and Fraenkel-Conrat, 1969c, d). Both are effectively mutated by nitrous acid, hydroxyl-

* This investigation was supported by research grants GB 6209 from the National Science Foundation and NsG 479 from the National Aeronautics and Space Administration.

amine, methoxyamine, and bromine; less by alkylating agents, and not at all by UV light. In contrast to typical alkylating agents, nitrosoguanidine was a very effective mutagen, but only when acting on intact TMV, and we were able to show that under these conditions nitrosoguanidine caused appreciable methylation of cytosine (Singer and Fraenkel-Conrat, 1969a, b). When acting on the isolated RNA, however, nitrosoguanidine methylated, like typical alkylating agents, predominantly the 7 position of guanine. These

Table 1. Triphosphate Incorporation with Modified Templates

Polynucleotide (% modification in parentheses)	GTP[a]	ATP[a]	UTP[a]	CTP[a]
		cpm × 10^{-3}		
Poly C, unmodified	49	0.3	0.2	0.3
Poly C, hydroxylamine-treated (3.5)	12	*0.8*	0.2	0.2
Poly C, methoxyamine-treated (3.5)	15	*0.9*	0.2	0.2
Poly C, methylated (5)	8	*0.8*	*0.6*	*0.6*
Poly C, nitrous acid-treated (7% U)	19	*1.6*	0.3	0.2
Poly (C, N4-hydroxy C) (4.1)	26	*1.1*	0.3	0.2
Poly (C, N4-methoxy C) (6.1)	29	*2.2*	0.3	0.4
Poly (C, 3-methyl C) (4.0)	27	*1.4*	*0.7*	*0.9*
Poly (C, U) (1.5% U)	41	*1.5*	0.5	0.5

[a] Radioactive triphosphate; unlabeled GTP is present together with labeled ATP, UTP, and CTP.

results led us to suggest that methylation of cytosine might actually be the most mutagenic event resulting from alkylation.

We have now compared the *in vitro* mutagenicity of these same reagents by studying the effects of modification of polynucleotides on their messenger and template activities in cell-free systems for the incorporation of either amino acids or nucleoside triphosphates. These studies, with particular reference to poly C, and in conjunction with the studies of the nature and extent of the various modification reactions discussed in the preceding paper, have led to the following conclusions:

Concerning the template activities of modified polynucleotides, we have found hydroxylamine and methoxyamine-treated poly C to incorporate adenylic acid into the poly G, in amounts increasing with increasing extent of reaction (Table 1). This is confirmatory of earlier studies (Phillips *et al.*, 1965; Wilson and Caicuts, 1966). We now know that hydroxylamine causes both the formation of N4-hydroxycytosine and of its C5,6 · NH_2OH addition product, while methoxyamine initially causes only the formation of N4-methoxycytosine in polynucleotides. We also know, or presume by analogy, that the N4-substituted cytosines exist predominantly in the uracil oxime form (>C=NOR) (Brown and Hewlins, 1968; Brown *et al.*, 1968). Therefore, these derivatives are expected to behave like point mutations in polynucleotide linkage. Thus, through the recent chemical data on the mode and extent of action of these reagents with polynucleotides (Singer and Fraenkel-Conrat, 1971), and through confirmation of older incorporation data, the Watson-Crick concept of mutagenesis appears upheld.

Concerning the mutagenicity and oncogenicity of alkylating agents, it has until recently been assumed that the alkylation of guanine in the 7 position represented the mutagenic event. However, no evidence for this arose from *in vitro* incorporation experiments (Wilhelm and Ludlum, 1966; Ludlum, 1970). On the other hand, the presence of 3-methylcytosine in poly C was definitely shown in two laboratories to represent a mutagenic event when this polynucleotide was used as template. Ludlum and Wilhelm (1968) reported that UTP was incorporated, while we now report that the incorporation of ATP, UTP, and CTP is stimulated by the introduction of methyl groups into poly C (Table 1). This suggests that the result of the change from

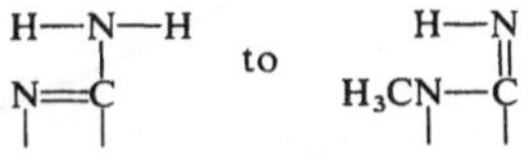

(a pseudotautomeric shift with replacement of H by CH_3) gives the resultant base a nonspecific tendency to bind at least 3 typical bases. It appears probable that this new binding capability represents the basis of the high mutagenicity of nitrosoguanidine acting on TMV. Quite possibly the action on cytidine residues may also contribute to the high mutagenic and oncogenic activity of many other alkylating agents acting on DNA.

The modified polynucleotides heretofore discussed have been prepared not only by chemical treatments of the polymer but also, in collaboration with Dr. G. Means, by copolymerization of the modified with the unmodified diphosphates using polynucleotide phosphorylase. N4-hydroxycytidylic containing polymers have been previously prepared by Janion and Shugar (1968). Our copolymers, however, were made to contain smaller amounts of either the N4-hydroxy- or N4-methoxycytidylic acid, for the purpose of serving as models for mutagenic events. The advantage of the enzymatic method over that of chemically modifying the poly C is that the doubly substituted cytidine diphosphate is not utilized by the polynucleotide phosphorylase and the resulting polymers thus consist of only cytidylic and N4-substituted cytidylic acid residues. The incorporating activities of the two types of polynucleotides, the treated poly C and the copolymerized materials, were the same in that both showed ATP and decreased GTP incorporation. Copolymers of cytidylic with 3-methylcytidylic acid also behaved like methylated poly C in causing all four nucleotides to become incorporated (Table 1).

Negative results in terms of mutagenesis were obtained with UV irradiated poly C. In contrast to earlier reports by Ono *et al.* (1965), we were unable to detect any ATP incorporation under our experimental conditions as well as under conditions similar to those used by those authors (15° or 20° C,

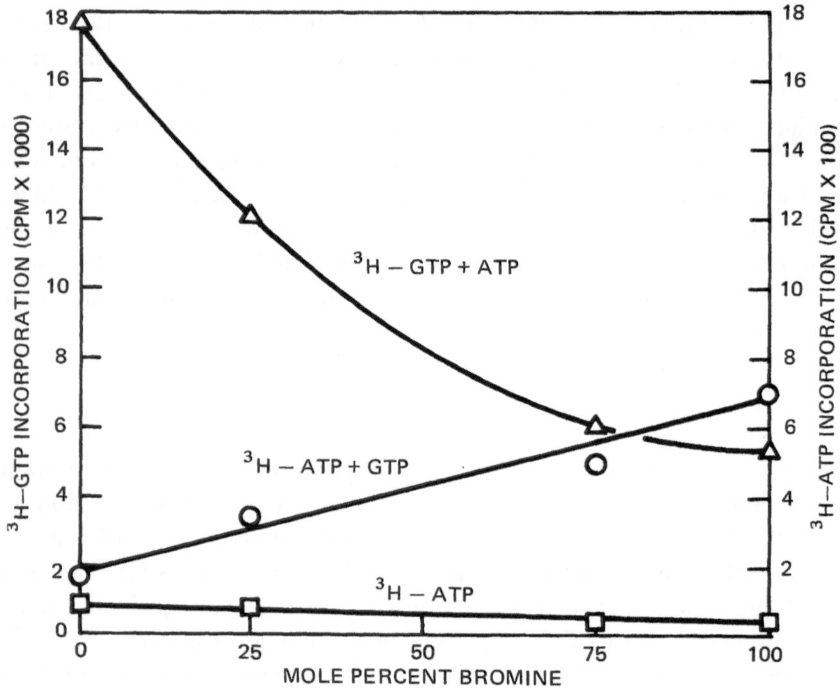

Fig. 1. Incorporation of triphosphates by RNA polymerase using poly C treated with increasing amounts of bromine as template. The brominated samples have been incubated at 37 °C for 4 hours at *p*H 5. The mole percent of added bromine then closely approximates the 5-bromocytidine content of the polymer.

*p*H 6.5). This is in line with the nonmutagenicity upon UV irradiation of viral RNA. As shown in the preceding article (Singer and Fraenkel-Conrat, 1971), the only detectable effect of UV on poly C is the formation of dimers, and this reaction probably accounts for the inactivating action of irradiation on poly C, in terms of its GTP incorporating activity.

Finally, we wish to present some data obtained by Dr. G. Means on the effect of bromination on the *in vitro* activities of poly C. The addition of bromine water to poly C leads to the addition of BrOH to the double bond.

Subsequent exposure to *p*H 4–5 causes loss of water and formation of 5-bromocytosine residues in the polymer. Only very small amounts of other reaction products were detected. When such copolymers of C and BrC were used as templates, ATP became incorporated in amounts proportional to the bromocytosine content and amounting per bromocytosine residue to about 12% of the GTP incorporation (Figure 1). Thus, it would appear that bromination of C causes a sufficient shift in its tautomeric equilibrium

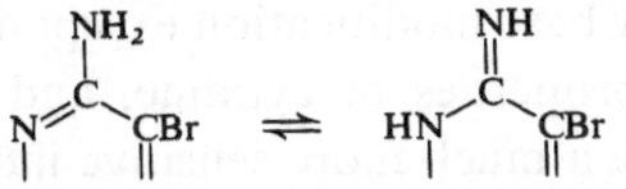

to effect appreciable mutation. However, the copolymerization of CDP and 5-bromo-CDP with polynucleotide phosphorylase yielded polymers which did not stimulate ATP incorporation (see Table 2). Several potential causes for this singular discrepancy were investigated unsuccessfully. The difference in the template properties of these two presumably identical types of polynucleotides remains unexplained (Means and Fraenkel-Conrat, Biochem., in prepn.).

With regard to the messenger activity of poly (C, 5 BrC) we were able to confirm Grunberg-Manago and Michelson (1964) that threonine was in-

Table 2. Messenger and Template Activities of Cytidylic-Bromocytidylic Copolymers

Amino acid	Codon	Messenger activity		
		Poly C	Brominated poly C	Poly (C, Br C)
Proline	CCX	high	decreased	increased
Threonine	ACX	none	very small	small
Histidine	CAPy	none	none	small
Asparagine	AAPy	none	none	small
Glutamine	CAPu	none	none	small

Triphosphate labeled (unlabeled)	Template activity		
	Poly C	Brominated poly C	Poly (C, Br C)
GTP (± ATP)	high	decreased	high
ATP (+GTP)	none	~12% of GTP	none
CTP (+GTP)	none	none	none
UTP (+GTP)	none	none	none

corporated by the copolymer. However, in contrast to these authors, we obtained positive results also with histidine. The properties of the brominated poly C, presumably identical after *p*H 5 treatment to the copolymer, were qualitatively similar, although the bromine treated polymer was less active than the copolymer in all regards, including the typical polymerization of proline. Thus, it appears that bromination makes cytidine code like adenine in *in vitro* protein synthesis. This is particularly surprising for the two reasons, that (1) no other base modification except deamination with nitrous acid alters the coding properties of cytidine, and (2) we have found the template properties to be a much more sensitive indicator of mutation than the messenger properties for a variety of reactions, whereas bromination makes cytidine act like adenine only in the messenger test.

In conclusion, we have obtained new evidence that hydroxyl- and methoxyamine are good mutagens, *in vivo* and *in vitro*, most probably through their effect on the N4 position of cytidine and the ensuant tautomeric shift.

We have newly demonstrated that alkylation may be mutagenic (and possibly oncogenic) through its action on the 3 position of cytidine, and that methylation in this position leads to nonspecific template action.[1]

We have found UV to be nonmutagenic on polyribonucleotides *in vitro*. With regard to bromine containing poly C, we have confirmed earlier observations of the presence of A-like coding qualities, in terms of threonine and histidine incorporation. This is true for both the copolymer of C and 5 bromo C, and for the presumably identical product resulting from bromination and *p*H 5 treatment of poly C. However, these two copolymers differed in their template activities, the brominated polymer showing a proportional ATP incorporation activity which is not shown by the copolymer of cytidylic and 5-bromocytidylic acid, and the nature of which remains to be established.

References

Brown, D. M. and Hewlins, M. J. E. (1968). The reaction between hydroxylamine and cytosine derivatives. J. Chem. Soc. (C) 1922–1924.

——— and Shell, P. (1968). The tautomeric state of *N*(4)-hydroxy- and of *N*(4)-amino-cytosine derivatives. J. Chem. Soc. (C) 1925–1929.

Grunberg-Manago, M. and Michelson, A. M. (1964). Polynucleotide analogues. II. Stimulation of amino acid incorporation by polynucleotide analogues. Biochim. Biophys. Acta **55:** 431–440.

[1] Some of these results have been recently reported in detail (Singer and Fraenkel-Conrat, 1970).

Janion, C. and Shugar, D. (1968). Studies on possible mechanisms of hydroxylamine mutagenesis. Acta Biochim. Polon. **15:** 107–121.

Ludlum, D. B. (1970). The properties of 7-methylguanine-containing templates for ribonucleic acid polymerase. J. Biol. Chem. **245:** 477–482.

——— and Wilhelm, R. C. (1968). Ribonucleic acid polymerase reactions with methylated polycytidylic acid templates. J. Biol. Chem. **243:** 2750–2753.

Ono, J., Wilson, R. G., and Grossman, L. (1965). Effects of ultraviolet light on the template properties of polycytidylic acid. J. Mol. Biol. **11:** 600–612.

Phillips, J. H., Brown, D. M., Adam, R., and Grossman, L. (1965). The effects of hydroxylamine on polynucleotide templates for RNA polymerase. J. Mol. Biol. **12:** 816–828.

Singer, B. and Fraenkel-Conrat, H. (1969a). Chemical modification of viral ribonucleic acid. VII. The action of methylating agents and nitrosoguanidine on polynucleotides including tobacco mosaic virus ribonucleic acid. Biochemistry **8:** 3260–3266.

——— (1969b). Chemical modification of viral ribonucleic acid. VIII. The chemical and biological effects of methylating agents and nitrosoguanidine on tobacco mosaic virus. Biochemistry **8:** 3266–3269.

——— (1969c). Mutagenicity of alkyl and nitroso-alkyl compounds acting on tobacco mosaic virus and its RNA. Virology **39:** 395–399.

——— (1969d). The role of conformation in chemical mutagenesis. *In Progress in Nucleic Acid Research.* Vol. **9:** 1–29. Ed. by Davidson, J. N. and Cohn, E. Academic Press, New York.

——— (1970). Messenger and template activities of chemically modified polynucleotides. Biochemistry **9:** 3694–3701.

——— (1971). Qualitative and quantitative aspects of polynucleotide modification. This volume. pp. 3–12.

Watson, J. D. and Crick, F. H. C. (1953). A structure of deoxyribonucleic acid. Nature **171:** 737–738; Genetical implications of the structure of deoxyribonucleic acid. Nature **171:** 964–967.

Wilhelm, R. C. and Ludlum, D. B. (1966). Coding properties of 7-methylguanine. Science **153:** 1403–1405.

Wilson, Robert G. and Caicuts, Mary J. (1966). The effects of hydroxylamine on the template properties of polycytidylic acid. J. Biol. Chem. **241:** 1725–1731.

DISCUSSION

DR. R.F. BEERS: I would like to address a question to Dr. Fraenkel-Conrat and to Dr. Singer regarding the specificity of the template assay method. In introducing mutagenic changes in the templates, are you indeed measuring a template rather than an initiating phase of the system?

DR. H. FRAENKEL-CONRAT: I would have thought that initiation factors play a minor role in the case of polynucleotides and that it is mainly a template action that we measure, since initiation can start anywhere non-specifically. Am I wrong?

DR. R. F. BEERS: In a general sense I am sure you are right, but I still wonder whether you have proven that the actual increase of nonspecific activity itself is truly a result of template activity or involves some other factor.

DR. H. FRAENKEL-CONRAT: I should have stressed that we have been able to use only poly C successfully for template work. In the case of poly A and poly U, or any combination of A and U, there is much reiteration and one obtains false results. A paper we have in *Biochemistry* contains much more data and shows that only poly C seems to behave true in a template sense.

DR. J. KARKAS: I would like to ask Dr. Singer whether, in the use of poly C treated with nitrozoguanidine, you ever tried to use just one of the precursors such as GTP as a precursor rather than mixture of all precursors, and see which one was appropriate?

DR. B. SINGER: We have used single-labeled triphosphates, but ordinarily incorporation is too low to give meaningful results.

DR. A. BROOM: Dr. Singer or Dr. Fraenkel-Conrat, would it be fair to suggest that the mutagenic effects you observed with methylated poly C are due to the methyl C residues acting merely as a kind of spacer, as it were, an indiscriminate one, and that the force necessary for the orientation of the nucleotide is just that factor?

DR. H. FRAENKEL-CONRAT: We think not, because many reactions, or all reactions we study cause loss of activity in terms of incorporation of GTP and that, we think, is an expression of the absence of functional nucleotides. Since methylation is the only reaction in which we do find all other bases incorporated, we believe it must be a semispecific action.

DR. D. LUDLUM: I would like to make a couple of comments about the first two papers. First of all, in response to Dr. Beers' initial question, I think that it is template activity that is being measured rather than initiation, because in all cases where we have done a nearest-neighbor analysis it has come out properly.

The second thing I would like to add is that I think there is a little problem about the specificity of the misincorporation which is induced by the alkylation of cytidine. I think both our groups have shown that UTP and ATP can be introduced by alkylation. The question of whether or not CTP is introduced is complicated and is tied up with the possibility of reiteration and the formation of granddaughter strands in the case of methyl poly C.

DR. S. FRIEDMAN: I would like to ask Dr. Singer a question on the action of hydroxylamine on poly C. If my memory serves me right, Professor Shugar of Warsaw University reported that the monosubstituted analog can be prepared by the action of hydroxylamine, followed by dialysis against, I believe, 0.1 N HCl. Have you confirmed this?

DR. B. SINGER: One obtains hydroxyamino cytidine from poly C directly without using any acid. Professor Shugar originally stated that you can convert most of the reaction products to the mono by using low *p*H. But this is not necessary.

DR. S. FRIEDMAN: You had both mono- and di-substituted analogs in your modified poly C?

DR. H. FRAENKEL-CONRAT: Yes, but there is mostly mono formed when we use polynucleotide phosphorylase to synthesize polymers. Shugar has shown, and we have confirmed, that if you polymerize cytidine diphosphate both with the mono- and di-compounds present, the enzyme uses only the mono-substituted compound so that the polymerized copolymer contains only cytidylic and hydroxylamino cytidylic acid.

DR. A. MITTLEMAN: Dr. Singer, have you used X-irradiation as a method of mutagenesis in the poly C and in the TMV experiments, and if so, have you observed any effects on template activity?

DR. B. SINGER: No, we have not. We have only used irradiation at 257 mμ.

DR. A. MITTELMAN: Dr. Fraenkel-Conrat, what evidence do you have for the formation of the 6-hydroxycytosine during the bromination reaction?

DR. H. FRAENKEL-CONRAT: We have only very preliminary data on this but we get a product that has a chromatographic mobility quite different from 5-bromo cytidine and with the pK of both an amino acid and an OH group, as determined spectrophotometrically. The amount of this material is quite low so that it is difficult to see how it could account for the relatively large ATP incorporation that we get, but we are trying to copolymerize cytidine with 5- and/or 6-hydroxcytidine and see if we get similar effects. But these ideas are quite hypothetical.

Part II

Interferon Stimulation

DOUBLE-STRANDED RNA'S IN RELATION TO INTERFERON INDUCTION AND ADJUVANT ACTIVITY

MAURICE R. HILLEMAN, GEORGE P. LAMPSON, ALFRED A. TYTELL, A. KIRK FIELD, MARJORIE M. NEMES, IRWIN H. KRAKOFF, CHARLES W. YOUNG

Division of Virus and Cell Biology Research
Merck Institute for Therapeutic Research
West Point, Pennsylvania
and
the Memorial Hospital and
Sloan-Kettering Institute for Cancer Research
New York, New York

INTRODUCTION

The conduct of research in the medical sciences finds ample justification and support when it is targeted at an applied objective—that of producing something useful. New phenomena are eagerly sought that can be developed toward such an end. Hoskins' (1935) discovery of the interference phenomenon is one such example, and its exciting possibilities were increasingly realized as its broad-spectrum antiviral nature was revealed. The epic discovery by Isaacs and Lindenmann (1957) of the protein mediator of interference, that is interferon, raised immediate hope for direct application to man but this has been subdued to date by problems of production and economic considerations (Hilleman, 1965; Ho and Postic, 1967; Merigan, 1967; Hilleman, 1968). The logical development, therefore, was the search for suitable inducers whereby the body could be stimulated to produce and distribute its own interferon. Early attack on the problem provided an abundance of inducers (Finter, 1966; Merigan and Regelson, 1967; Regelson, 1967) but none was of practical value because of infectiousness, extreme toxicity, antigenicity, and the like. Studies in our laboratories (Merck Institute) (Field *et al.*, 1967a and b; Lampson *et al.*, 1967; Tytell *et al.*, 1967; Field *et al.*, 1968; Hilleman, 1968; Lampson *et al.*, 1969; Nemes *et al.*, 1969a and b; Hilleman, 1970), which sought to learn how viruses induce interferon in nature, ultimately led to discovery of the requirement of double-strandedness of RNA for interferon induction and provided a working hypothesis for the phenomenon. The double-stranded RNA's have given the greatest hope, to date, for the evolution of a practical inducer. Since the time of this discovery, our work has centered on the double-stranded RNA's with one centripetal emphasis—that of application to man and his domestic animals.

THE DOUBLE-STRANDED RNA INDUCERS

The early studies of RNA by our group (Field *et al.*, 1967a and b; Lampson *et al.*, 1967; Tytell *et al.*, 1967; Field *et al.*, 1968; Hilleman, 1968; Lampson *et al.*, 1969; Nemes *et al.*, 1969a and b; Hilleman, 1970) provided an abundance of candidates for interferon induction. As shown in Table 1, all the double-stranded RNA's induced interferon in rabbits. These included synthetic poly I:C (polyriboinosinic:polyribocytidylic acid complex, $rI_n:rC_n$), poly A:U, virion RNA from reovirus 3, and the probable mycovirus of *Penicillium funiculosum*, replicative form RNA from MS2 and MU9 coliphage, rice dwarf and polyhedrosis virus RNA's, self-complexed alternating ribocopolymer poly IC, and hybrid DNA:RNA from F1 bacteriophage. To this can now be added (Colby and Duesberg, 1969; Falcoff and Falcoff, 1969) double-stranded RNA from mengovirus, influenza, and vaccinia viruses. Single-stranded RNA's of natural or synthetic origin were not active. The relatively low level activity reported (Baron *et al.*, 1969) for single homopolymer poly I or poly C may be due to double-stranding through self-complexing (De Clercq and Merigan, 1969) promoted by excess magnesium ion or low *p*H and is of no evident practical importance.

Table 2 shows that the same double-stranded RNA's that were found to be active in inducing interferon in rabbits also produced resistance to vesicular stomatitis virus when added to primary rabbit kidney cell cultures in gamma amount. In each instance, the inducer was added to the culture 18–24 hours prior to challenge with 10 $TCID_{50}$ (50% tissue culture infectious doses) of virus.

Similarly, Table 3 shows that all the double-stranded RNA's prevented death in mice caused by pneumonia virus of mice (PVM). In the tests, the RNA's were given intranasally to the mice 3 hours prior to administration of PVM by the same route.

Table 1. Interferon Induction in Rabbits by Double-Stranded RNA

Nucleic acid: Source or kind	Nucleic acid: Chemical nature	Intravenous dose per rabbit (μg)	Interferon titers of individual rabbit sera	
Synthetic poly I:C ($rI:rC$; $rI_n:rC_n$)	double-stranded RNA	2	> 640,	> 640
Synthetic poly A:U ($rA:rU$; $rA_n:rU_n$)	double-stranded RNA	25	20,	40
Penicillium funiculosum (viral?)	double-stranded RNA	8	> 640,	80
Reovirus 3 virion	double-stranded RNA	8	> 640,	640
MS2 coliphage (replicative)	double-stranded RNA	8	160,	40
MU9 mutant coliphage (replicative)	double-stranded RNA	2	40	
Rice dwarf virus virion	double-stranded RNA	20	> 640	
Cytoplasmic polyhedrosis virion	double-stranded RNA	22	160	
Synthetic rIC (alternating ribocopolymer, self-complexed)	double-stranded RNA	10	320,	> 640
DNA-RNA F1 phage hybrid	double-stranded DNA-RNA	5	40	
Calf thymus	double-stranded DNA	200	0	
Synthetic poly I	single-stranded RNA	25	0	
Synthetic poly C	single-stranded RNA	20	0	
MS2 coliphage virion	single-stranded RNA	8	0,	0
Escherichia coli	single-stranded RNA	100	0,	0
Newcastle disease virus virion	single-stranded RNA	10	0	
Influenza virus virion	single-stranded RNA	10	0	
Tobacco mosaic virus virion	single-stranded RNA	40	0	
Yeast ribosome	single-stranded RNA	1000	0	
Yeast soluble	single-stranded RNA	200	0	
Yeast core	single-stranded RNA	100	0	
Ribosomal mouse liver	single-stranded RNA	200	0	
Bovine liver soluble	single-stranded RNA	200	0	

Table 2. Induction of Resistance by Double-Stranded RNA's against Vesicular Stomatitis Virus in Rabbit Kidney Cell Culture[a]

Source of kind of double-stranded RNA	Least amount of RNA required to protect (μg)
Poly I:C (rI:rC; rI_n:rC_n)	0.002
Penicillium funiculosum (viral ?)	0.30
Reovirus 3 virion	0.04
MS2 coliphage (replicative)	0.04
MU9 mutant coliphage (replicative)	0.125–1.0
Rice dwarf virus virion	0.015
Cytoplasmic polyhedrosis virus virion	0.04
Synthetic poly IC (self-complexed alternating ribocopolymer; rIC)	0.003
DNA-RNA F1 phage hybrid	0.60

[a] Inducer was added 18–24 hours prior to 10 $TCID_{50}$ of vesicular stomatitis virus.

Table 3. Prophylactic Activity of Double-Stranded RNA's against PVM Infection in Mice[a]

Double-stranded RNA	Intranasal dose (μg)	Survival of mice (14 days)		
Source or kind		No. survived / No. tested	Survival %	Excess survival compared with controls (%)
Poly I:C	4	192/211	91	90
Penicillium funiculosum	20	18/20	90	89
Reovirus 3 virion	16	16/16	100	75
MS2 coliphage (replicative)	9	9/12	75	65
MU9 mutant coliphage (replicative)	15	11/15	73	65
Rice dwarf virus virion	3	13/17	76	89
Synthetic poly IC (alternating ribocopolymer self-complexed; rIC)	3	15/15	100	95
Cytoplasmic polyhedrosis virus virion	3	15/15	100	83

[a] RNA's given nasally 3 hours prior to virus given nasally.

PRECLINICAL TESTS IN ANIMALS FOR ANTIVIRAL EFFECTS

Intentions to evaluate the utility of any drug in man must necessarily be preceded by tests in animal model systems whereby regimens can be evolved, efficacy judged, and safety assessment made.

Poly I:C and MU9 double-stranded RNA's have been evaluated for prophylactic efficacy against a variety of RNA and DNA viruses in mouse and chick model systems. In the tests, the inducer was always given prior to

Table 4. Interpretive Summary of Prophylactic Efficacy of Double-Stranded RNA's against Viral Infections in Mice or Chicks (Rous, Marek's)

System		Route of administration		Appraisal of activity
Virus	Polynucleotide	Virus	Polynucleotide	
PVM	poly I:C	intranasal	intranasal	excellent
PVM	MU9 coli-phage (replicative)	intranasal	intranasal	excellent
Columbia SK	poly I:C	intranasal	intranasal	very good
Columbia SK	poly I:C	intranasal	intraperitoneal	very good
Columbia SK	poly I:C	subcutaneous	intraperitoneal	very good
Columbia SK	MU9 coli-phage (replicative)	intranasal	intranasal	excellent
Columbia SK	MU9 coli-phage (replicative)	intranasal	intraperitoneal	excellent
Vaccinia	poly I:C	intravenous	intraperitoneal	very good
Vaccinia	poly I:C	intravenous	subcutaneous	very good
Vaccinia	MU9 coli-phage (replicative)	intravenous	intraperitoneal	very good
Parainfluenza 1 (Sendai)	poly I:C	intranasal	intranasal	good
Parainfluenza 1 (Sendai)	poly I:C	intranasal	intranasal	good
Rabies	poly I:C	intraplantar	intraperitoneal	weak
Rabies	poly I:C	intracerebral	intraperitoneal	weak
Rous sarcoma	poly I:C	intraperitoneal	intraperitoneal	weak
Influenza B	poly I:C	intranasal	intranasal	weak
Influenza A and A2	poly I:C	intranasal	intranasal	none
Yellow fever	poly I:C	intracerebral	intraperitoneal	none
Marek's agent	poly I:C	intraperitoneal	intraperitoneal	none
Marek's agent	poly I:C	contact	intraperitoneal	none

virus and various route combinations were tested. Table 4 presents a brief interpretive summary of the results obtained to date. Both poly I:C and MU9 coliphage RNA were active. Greatest benefit was shown against PVM, Columbia SK, vaccinia, and parainfluenza 1 (Sendai) viruses. Weak activity was found against rabies, Rous sarcoma, and influenza B viruses. There was no activity against influenza A, A2, yellow fever, or Marek's (avian neural lymphomatosis) agent which is a herpesvirus. In addition, minor beneficial effects have been found for neoplasia induced in animals by SV_{40}, adenovirus 12, and Friend leukemia viruses (Larson *et al.*, 1969a and b; 1970). These model systems reflect the effects of a single dose of inducer and a severe virus challenge. Hence, any lack of effect may reflect more the severity of the particular system than the potential for beneficial effects under conditions to be found in nature. A case in point is the recent demonstration by Janis and Habel (1970) and by Postic and Fenje (1970) of high level activity of poly I:C against rabies in rabbits.

A principal disadvantage of the interferon system is its relatively short duration of action. The positive effect that is obtained is generally regarded as prophylactic only for the individual cell, although this may not be true for the composite of cells in the metazoan host in which successive infection of cells may take place and in which therapeutic as well as prophylactic effects can sometimes be demonstrated. In the PVM system, for example (Nemes *et al.*, 1969a), protection at high level was afforded against PVM virus for 6 days after the drug was given, with a decline to low level by day 7. The drug provided near total protection against death when given therapeutically as late as 2 days after virus, and nearly half the mice were saved when it was given on the third day after virus. The known refractoriness (Finter, 1966) of cells to reinduction of interferon prior to the approximate time that the interferon effect is lost might seem an overwhelming obstacle. However, it appears that this might be overcome by administering less than the dose required to stimulate all cells. By such adjustment, it might be possible to keep a substantial portion of cells capable of induction at any particular time; and repeated administration of poly I:C in such reduced amounts might still be capable of providing continuous immunity to viral infection (Hilleman, 1970).

SAFETY ASSESSMENT

Further tests in our laboratory have centered on poly I:C as the prime candidate for human use. Poly I:C is not a chemically defined substance in the ordinary use of the term and it is therefore important to define the properties of the material under test. It has been our practice to limit use of poly I:C to those lots in which the physical and chemical properties fall within sharply defined limits, thereby permitting reproducibility. The individual lots of poly I and of poly C conformed to rigid ultraviolet absorption specifications and revealed approximate sedimentation coefficients of 7.7 and 8.5 for the 2 homopolymers, respectively. The Tm range for the poly I:C was 60–64 °C. Relative viscosity was 3.1–3.9 at 1 mg/ml solution. Hypochromicity on mixing was 36.5% and hyperchromicity on melting was 65%. All poly I:C lots were active in microgram amounts in inducing interferon and resistance to viral infection in *in vivo* and *in vitro* test systems.

Poly I:C was carried in continued passage with human diploid cell[1] strains in culture and did not cause significant alteration of growth rate, plating efficiency, or morphology and did not bring about delay in natural occurrence of senescence in the cultures. The karyotype of the cells was not altered and neoplastic transformation did not take place. The substance was not found to be carcinogenic when tested in newborn hamsters and it did not promote HEp-II tumor growth in the cheek pouch of hamsters. It did not cause anaphylactic sensitization and neither sensitized nor provoked a Shwartzman reaction in rabbits. The poly I:C complex was pyrogenic for rabbits and caused a leukopenia, but the individual homopolymers did not. Very similar results were obtained with double-stranded RNA from MU9 coliphage.

Very extensive assessment for safety[2] of poly I:C was carried out in mice, rats, dogs, monkeys, and hamsters, varying dose, number of doses, and route. Observations included physical signs, hematology, blood chemistry, gross pathology, and histopathology. Toxic effects were noted that had not been apparent in the animals employed in the tests for antiviral activity. The dog appeared to be the most susceptible of the several species to the toxic effect of repeated intravenous administration of poly I:C. The most significant toxicologic changes that were noted appeared to have a vascular, hepatic, or hematologic basis. Physical signs included retching, emesis, diarrhea (sometimes with frank or occult blood), and tremors and convulsions, which were commonly seen especially at high and frequent dosage. Pathology included necrotic changes in several organs, including the liver, bone marrow, bone, spleen, and other organs. Liver function tests, especially alkaline phosphatase, were positive and leukopenia was commonly noted. Most of the other species were far less reactive than the dog, especially the monkey, which showed nothing more than an elevation in alkaline phos-

[1] The *in vitro* safety assessment tests were carried out by Dr. C. L. Baugh and associates in these laboratories.

[2] These investigations were carried out by Drs. R. E. Zwickey, H. M. Peck, W. R. Brown, M. Hite, A. H. Mosher, E. Piperno, and G. E. Dagle.

phatase within 2 days after the first of 3 weekly doses of 1.0 mg/kg of poly I:C given intravenously. The route of administration was very important. Toxicity was negligible or nonexistent at comparable dose levels given by the respiratory or subcutaneous routes. The respiratory findings are especially important in view of the potential for using drug in relation to the common cold.

CLINICAL TESTS IN MAN

Overall consideration of the available data indicated that interferon might be induced in man employing doses of poly I:C which would be less than the significantly toxic level. Therefore, studies were initiated in human

Table 5. First Interferon Response among 13 Patients Given Poly I:C Intravenously

Patient Number	Age (yrs.)	Clinical diagnosis	Poly I:C dose µg/kg	Interferon titer of serum at hour 0	2	12	24	48	72
614230	61	Reticulum cell sarcoma	2	0[a]	8	0	0	...	...
254292	28	Hodgkin's disease	25	0	...	...	32	...	...
275260	31	Nasopharyngeal carcinoma	50	0	0	8	8	0	0
253197	56	Carcinoma of breast	50	0	...	4	8	8	0
272979	57	Mesothelioma	100	0	8	16	8	8	8
271745	30	Hodgkin's disease	1000	0	0	...	8	...	...
614634	58	Carcinoma of bladder	25	4	...	8	16	...	...
613294	18	Embryonal carcinoma	100	4	...	...	16	4	8
672043	24	Leukosarcoma	1000	4	4	8	16	...	...
273384	45	Malignant melanoma	10	4	4	0	0	...	...
613039	52	Laryngeal carcinoma	25	0	0	0	0	0	0
273050	22	Hodgkin's disease	25	0	0	...	0	...	...
273282	40	Cervical carcinoma	50	0	0	0	0	0	0

[a] 0 titer = < 1:4.

subjects by Krakoff and Young at the Memorial Hospital and the Sloan-Kettering Institute for Cancer Research. The purpose was threefold: to evaluate toxicity, to measure the clinical effect in cancer, and to measure interferon induction following intravenous administration of poly I:C.

Table 5 presents the first responses to poly I:C given intravenously to 13 patients with far-advanced cancer. Nine of the 13 patients responded with the development of antiviral substance; 6 from zero initial titer, and

3 from a pre-existing titer of 1:4 to 1:8 or to 1:32. There were 4 failures of response. The source for the pre-existing levels of antiviral activity were not known with certainty but were most likely due to interferon stimulated by intercurrent viral infection or another inducer such as endotoxin. All the patients had far-advanced cancer and were subject to a variety of concurrent infections. The responses were maximal in 1 of the subjects 2 hours after poly I:C and 12 hours after drug in most of the others. Interferon levels persisted for as long as 72 hours in certain patients and for less than 12 hours in 1 of the subjects. There was no certain relationship between poly I:C dose and pattern of response in this small number of patients. The interferon titers were generally low and reflected, possibly, a reduced capability to produce interferon because of overlying neoplastic disease. The interferon nature of the antiviral activity was demonstrated based on narrow host species specificity, broad antiviral spectrum, low molecular weight, and stability at low and high *p*H (*p*H 2–10).

The effect of repeated administration of poly I:C in 1 of the patients (No. 614634) is shown in Figure 1. It is seen that interferon was readily reinduced after a "rest period" of about 10 days to 2 weeks. Daily administration usually failed to reinduce although the titers appeared to be retained at a higher level and for a longer period than expected following a single dose of poly I:C.

Toxic manifestations were not a prominent feature although fever was commonly noted to follow administration of the drug. No hematologic effect was detected and there were no significant alterations in the values found in liver function tests. There were no definite effects on tumor.

The findings in the clinical tests to date have been encouraging and prompt continued studies of prophylaxis and therapy of viral infections in man, especially in relation to the common cold caused by rhinoviruses. Hill *et al.* (1970) have already shown a limited but significant clinical effect against rhinovirus infection in volunteers given poly I:C by nasal drops and Niblack *et al.* (1970) recorded the appearance of serum interferon in human subjects given poly I:C by deep inhalation.

MOLECULE ALTERATION

The inevitable sequel to demonstrated useful biological activity of a chemical substance is the alteration of the molecule to make it even more useful. With the double-stranded RNA's, the object will be to improve the toxicity:activity ratio. Such alteration has already been started. De Clercq *et al.* (1969) have demonstrated a very substantial increase in the interferon inducing capacity of the very weakly active alternating ribocopolymer, poly

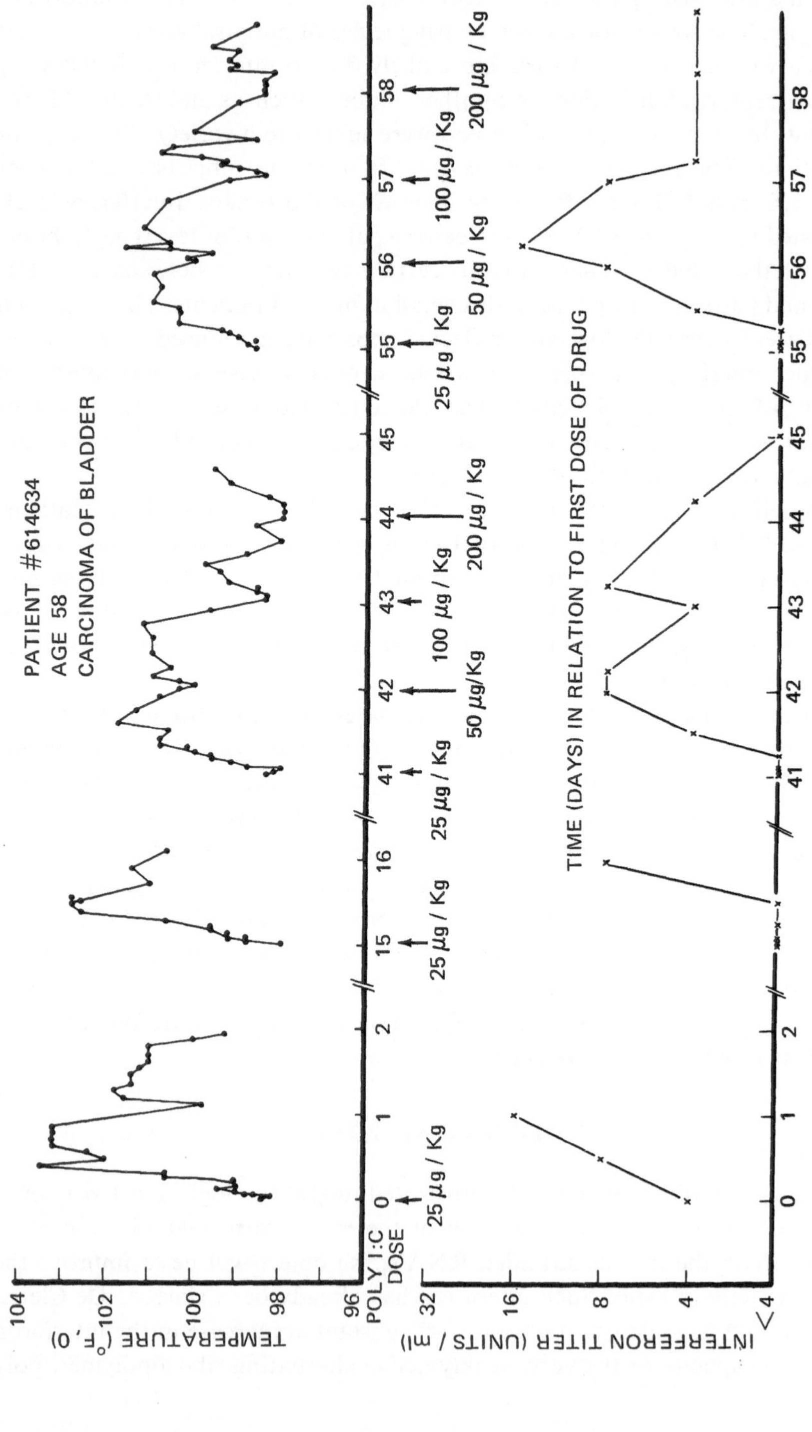

Fig. 1. Febrile and interferon responses in a human subject to repeat intravenous administration of poly I:C.

AU (rAU). Toxicity was not determined and hence no assessment of usefulness in the practical sense can be made at this time. The change could be adverse as well as beneficial.

Systematic investigation of the factors that influence the capacity of polynucleotides to induce interferon and host resistance is currently being investigated by our group (Merck Institute). One such study has been

Table 6. Capacity of Fractions of Sonicated Poly I:C from Sepharose Chromatography to Induce Rabbit Serum Interferon and Interference in Primary Rabbit Kidney Cell Cultures

Approximate molecular weight of poly I:C	Dose of poly I:C per rabbit (μg)	Assay	
		Rabbit serum interferon titer	Interference in cell culture. % of activity relative to untreated fraction
$>1 \times 10^6$	2	80, 160	100
6.4×10^5	2	640, 160	...
4.4×10^5	2	320, 0[a]	200
2.8×10^5	2	20, 80	100
2.0×10^5	2	20, 20	100
1.2×10^5	2	<5, 160	...
1.0×10^5	...	...	50
7.0×10^4	2.8	0, 0	...
6.0×10^4	...	...	12.5
Nil (control)	...	0, 0	...

[a] 0 = titer < 1:5.

devoted to the problem of measuring the influence of molecular weight of the complexed poly I:C on its physical properties and on its capacity to induce interferon and resistance to viral infections *in vivo* and *in vitro* (Lampson *et al.*, 1970). Decrease in the average molecular weight of poly I:C by sonic radiation from 7.8×10^6 to 2.8×10^5 was accompanied by an exponential decrease in viscosity of the preparation, by an increase in the sensitivity of the polynucleotide complex to degradation by pancreatic ribonuclease, by a decrease in the melting point, and by a decrease in the hyperchromicity observed during thermal transition. The effect of such lowering of molecular weight on biological activity is shown in Table 6. Poly I:C fractions that had an approximate average molecular weight of 1.2×10^5 or greater were roughly as active in inducing interferon in rabbits and resistance to viral infection as was the untreated complex. Fractions with lower molecular weight displayed significantly reduced activity.

Protective activity against PVM infection in mice required higher molecular weight. Table 7 shows that reduction of molecular weight to 4.6×10^5 caused marked loss in capacity to protect mice against PVM.

Table 7. Capacity of Sonicated Poly I:C to Induce Resistance against Pneumonia Virus of Mice

Approximate molecular weight of poly I:C	Total dose/ mouse (μg)	Excess survival of treated mice compared with controls: % survival	Excess survival of treated mice compared with controls: Days of survival
7.8×10^6	4	94	>7
	1	80	>7
	0.25	60	>7
4.2×10^6	4	90	>7
	1	75	>7
	0.25	60	>7
2.3×10^6	4	47	>7
	1	55	>7
	0.25	20	3
1.2×10^6	4	75	>7
	1	42	6
	0.25	20	2
7.4×10^5	4	50	>6
	1	45	>6
	0.25	5	1
4.6×10^5	4	30	5
	1	15	3
	0.25	5	1
2.8×10^5	4	24	3
	1	12	2
	0.25	20	2

Precise toxicity:activity ratios for the sonicated fractions have not as yet been worked out. However, it appears that significant improvement will be afforded by reducing the size of the molecule.

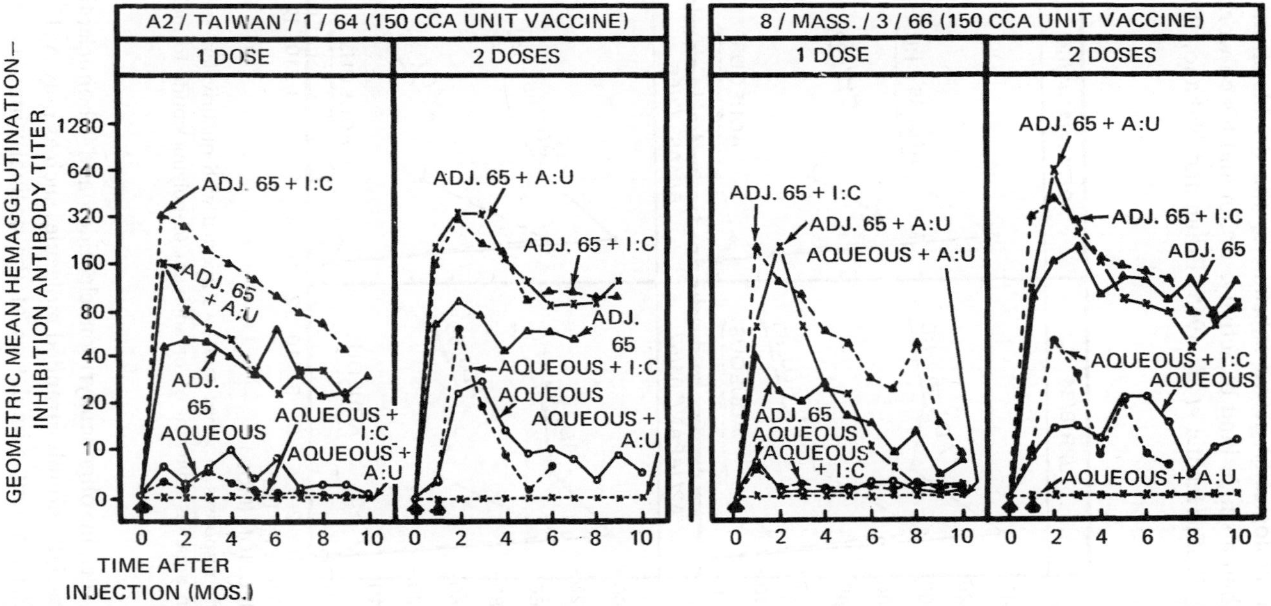

Fig. 2. Guinea pig antibody responses to polyvalent influenza virus vaccine, aqueous and adjuvant 65 types and with or without polynucleotides, given in 1 or 2 dose schedules.

ADJUVANT ACTIVITY OF POLY I:C AND POLY A:U

Like most biologically active substances, the double-stranded RNA's show more than a single kind of activity. Braun and his co-workers (Braun and Nakano, 1965; Hechtel *et al.*, 1965; Braun and Nakano, 1967) demon-

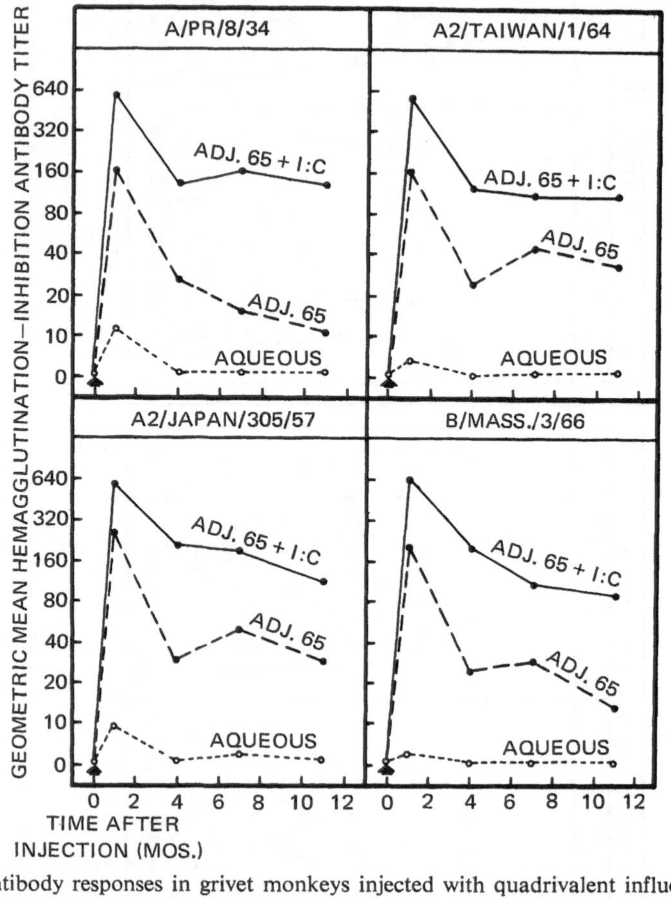

Fig. 3. Antibody responses in grivet monkeys injected with quadrivalent influenza virus vaccine in aqueous adjuvant 65 or adjuvant 65-polynucleotide formulation.

strated that certain oligodeoxyribonucleotides and ribonucleotide homopolymers, especially double-stranded molecules such as poly A:U (rA_n:rU_n), may function as potent stimulators of the ordinary immune mechanism. This may be due to enhancement in the early rate of increase, in the presence of antigen, in numbers of certain antibody forming cells, and may be associated with the stimulation of nucleotide kinase in antibody-forming cells. This action appears separate and distinct from interferon induction.

Our laboratories (Merck Institute) have developed an emulsified peanut oil adjuvant (adjuvant 65) whereby the antibody stimulating capability of ordinary aqueous influenza vaccine is increased 4- to 16-fold even though the amount of antigen may be reduced to $^{1}/_{4}$ the amount as in the aqueous vaccine (Hilleman, 1966; 1967). Studies were carried out by Woodhour *et al.* (1969) in our group in which poly I:C and poly A:U were tested for their capacity to potentiate antibody response in animals when added to ordinary aqueous influenza vaccine compared with the same vaccine in metabolizable adjuvant 65. Figure 2 shows the antibody responses in guinea pigs to 1 or 2 doses of bivalent influenza vaccine given as aqueous material with or without poly I:C or poly A:U and given in adjuvant 65 with or without added poly I:C or poly A:U. Poly I:C or poly A:U administered with aqueous vaccine caused comparatively weak potentiation of antibody response to these antigens. In certain instances, poly A:U was actually inhibitory. However, immunologic adjuvant 65 combined with poly I:C or poly A:U acted synergistically to cause a hyperpotentiation of antibody response which greatly exceeded the additive effects of adjuvant 65 and the complexed polynucleotides given singly. The similar findings in tests in monkeys with poly I:C and adjuvant 65 are given in Figure 3. The combination of poly I:C or poly A:U with adjuvant 65 shows promise of contributing a new magnitude of potential benefit for control of viral infections by means of vaccines.

SUMMARY

The possibility for practicable broad-spectrum control of viral diseases by administration of double-stranded RNA, such as poly I:C, has excited the interest of those who would seek to limit the rate and severity of virus illnesses of man and animals. The toxicity of poly I:C is clearly a problem that will impose limitations upon the usability of the drug, although these would appear to be minimal for administration by the respiratory route. Further, alteration of the molecular entity may substantially improve the toxicity:activity ratio and provide a more favorable position. The second effect of double-stranded RNA to enhance immune responses may well provide a second benefit of considerable magnitude in its inclusion with immunologic adjuvant 65. The great potential benefit of poly I:C and the promise that it has shown in clinical studies to date clearly justify continued vigorous exploration into its potential usefulness in human and animal medicine.

References

Baron, S., Bogomolova, N. N., Billiau, A., Levy, H. B., Buckler, C. E., Stern, R., and Naylor, R. (1969). Induction of interferon by preparations of synthetic single-stranded RNA. Proc. Natl. Acad. Sci. **64:** 67.

Braun, W. and Nakano, M. (1965). Influence of oligodeoxyribonucleotides on early events in antibody formation. Proc. Soc. Exp. Biol. **119:** 701.

——— (1967). Antibody formation: Stimulation by polyadenylic and polycytidylic acids. Science **157:** 819.

Colby, C. and Duesberg, P. H. (1969). Double-stranded RNA in vaccinia virus infected cells. Nature **222:** 940.

De Clercq, E., Eckstein, F., and Merigan, T. C. (1969). Interferon induction increased through chemical modification of a synthetic polyribonucleotide. Science **165:** 1137.

De Clercq, E. and Merigan, T. C. (1969). Requirement of a stable secondary structure for the antiviral activity of polynucleotides. Nature **222:** 1148.

Falcoff, R. and Falcoff, E. T. (1969). Induction de la synthèse d'interféron par des RNA bicaténaires. I. Application a l'étude du cycle de multiplication du virus mengo. Biochim. Biophys. Acta **182:** 501.

Field, A. K., Lampson, G. P., Tytell, A. A., Nemes, M. M., and Hilleman, M. R. (1967a). Inducers of interferon and host resistance. IV. Double-stranded replicative form RNA (MS2-RF-RNA) from *E. coli* infected with MS2 coliphage. Proc. Natl. Acad. Sci. **58:** 2102.

Field, A. K., Tytell, A. A., Lampson, G. P., and Hilleman, M. R. (1967b). Inducers of interferon and host resistance. II. Multistranded synthetic polynucleotide complexes. Proc. Natl. Acad. Sci. **58**, 1004.

——— (1968). Inducers of interferon and host resistance. V. *In vitro* studies. Proc. Natl. Acad. Sci **61:** 340.

Finter, N. B. (Ed.) (1966). *Interferons*. W. B. Saunders Co., Philadelphia.

Hechtel, M., Dishon, T., and Braun, W. (1965). Influence of oligodeoxyribonucleotides on the immune response of newborn AKR mice. Proc. Soc. Exp. Biol. **119:** 991.

Hill, D. A., Baron, S., Levy, H. B., Bellanti, J., Buckler, C. E., Cannellos, G., Carbone, P., Chanock, R. M., De Vita, V., Guggenheim, M. A., Homan, E., Kapikian, A. Z., Kirschstein, R. L., Mills, J., Perkins, J. C., Van Kirk, J. E., and Worthington, M.: Personal communication, February, 1970.

Hilleman, M. R. (1965). Immunologic, chemotherapeutic and interferon approaches to control of viral disease. Amer. J. Med. **38:** 751.

——— (1966). Critical appraisal of emulsified oil adjuvants applied to viral vaccines. Progr. Med. Virol. **8:** 131.

——— (1967). Considerations for safety and application of emulsified oil adjuvants to viral vaccines. Symp. Series Immunobiol. Standard. 13–26 Karger, Basel/New York.

——— (1968). Interferon induction and utilization. J. Cell. Physiol. **71:** 43.

Hilleman, M. R. (1970). Prospects for the use of double-stranded ribonucleic acid (poly I:C) inducers in man. J. Infect. Dis. **121:** 196.

Ho, M. and Postic, B. (1967). Prospects for applying interferon to man. *In* First International Conference on Vaccines Against Viral and Rickettsial Diseases of Man, 632–649. Washington. Pan American Health Organization Scientific Publication No. 147.

Hoskins, M. (1935). A protective action of neurotropic against viscerotropic yellow fever virus in macacus rhesus. Amer. J. Trop. Med. **15:** 675.

Isaacs, A. and Lindenmann, J. (1957). Virus interference. I. The interferon. Proc. Roy. Soc. B **147:** 258.

Janis, B. and Habel, K. (1970). Polyriboinosinic and polyribocytidylic acid polymer (poly I:C) in rabies prophylaxis. Fed. Proc. **29:** 636.

Lampson, G. P., Field, A. K., Tytell, A. A., Nemes, M. M., and Hilleman, M. R. (1970). Relationship of molecular size of $rI_n{:}rC_n$ (poly I:C) to induction of interferon and host resistance. In Press. Proc. Soc. Exp. Biol. Med.

——— (1967). Inducers of interferon and host resistance. I. Double-stranded RNA from extracts of *Penicillium funiculosum.* Proc. Natl. Acad. Sci. **58:** 782.

——— (1969). Influence of polyamines on induction of interferon and resistance to viruses by synthetic polynucleotides. Proc. Soc. Exp. Biol. **132:** 212.

Larson, V. M., Clark, W. R., Dagle, G. E., and Hilleman, M. R. (1969a). Influence of synthetic double-stranded ribonucleic acid, poly I:C, on Friend leukemia in mice. Proc. Soc. Exp. Biol. **132:** 602.

Larson, V. M., Clark, W. R., and Hilleman, M. R. (1969b). Influence of synthetic (poly I:C) and viral double-stranded ribonucleic acids on adenovirus 12 oncogenesis in hamsters. Proc. Soc. Exp. Biol. **131:** 1002.

Larson, V. M., Panteleakis, P. N., and Hilleman, M. R. (1970). Influence of synthetic double-stranded ribonucleic acid (poly I:C) on SV_{40} viral oncogenesis and transplant tumor in hamsters. Proc. Soc. Exp. Biol. **133:** 14.

Merigan Jr., T. C. (1967). Interferon's promise in clinical medicine. Fact or fancy? Amer. J. Med. **43:** 817.

——— and Regelson, W. (1967). Interferon induction in man by a synthetic polyanion of defined composition. New Engl. J. Med. **277:** 1283.

Nemes, M. M., Tytell, A. A., Lampson, G. P., Field, A. K., and Hilleman, M. R. (1969a). Inducers of interferon and host resistance. VI. Antiviral efficacy of poly I:C in animal models. Proc. Soc. Exp. Biol. **132:** 776.

——— (1969b). Inducers of interferon and host resistance. VII. Antiviral efficacy of double-stranded RNA of natural origin. Proc. Soc. Exp. Biol. **132,** 784.

Niblack, J. F., Knirsch, A. K., and Vora, K. R. M.: Personal communication, March, 1970.

Postic, B. and Fenje, P. (1970). Prophylaxis of rabies in rabbits by poly I:C. Bact. Proc. 155.

Regelson, W. (1967). Prevention and treatment of Friend leukemia virus (FLV) infection by interferon-inducing synthetic polyanions. *In* The Reticuloendothelial System and Atherosclerosis. Ed. by N. R. DiLuzio and R. Paoletti, 315–332. Plenum Press, New York.

Tytell, A. A., Lampson, G. P., Field, A. K., and Hilleman, M. R. (1967). Inducers of interferon and host resistance. III. Double-stranded RNA from reovirus type 3 virions (Reo 3-RNA). Proc. Natl. Acad. Sci. **58:** 1719.

Woodhour, A. F., Friedman, A., Tytell, A. A., and Hilleman, M. R. (1969). Hyperpotentiation by synthetic double-stranded RNA of antibody responses to influenza virus vaccine in adjuvant 65. Proc. Soc. Exp. Biol. **131:** 809.

FACTORS AFFECTING THE INTERFERON RESPONSE OF MICE TO POLYNUCLEOTIDES

SAMUEL BARON, HERMAN DUBUY, CHARLES E. BUCKLER, MARTIN JOHNSON, and MICHAEL WORTHINGTON

Laboratory of Viral Diseases
National Institute of Allergy & Infectious Diseases
National Institutes of Health, Bethesda, Maryland

Natural and synthetic RNA's such as polyinosinic·polycytidylic acid (poly I·poly C) stimulate the production of interferon in mice and other animals (Isaacs *et al.*, 1961; Field *et al.*, 1967). The mouse is an important model system for studying the role of interferon in the control of viral infections. The understanding of this model system requires knowledge of the interferon response to RNA stimulators. The present study was made tc help determine the dose-response kinetics of interferon production following single and multiple injections of the synthetic RNA, poly I·poly C.

METHODS

Mice: NIH Swiss mice weighing 20 to 25 g were used in all experiments. *RNA:* The synthetic homopolymers poly I and poly C were prepared and annealed as previously described (Field *et al.*, 1967). *Interferon Induction:* Solutions of poly I·poly C were injected intraperitoneally (i.p.) in mice at the appropriate concentration as indicated in the results section. The interferon titer was determined as the reciprocal of the highest dilution of serum which inhibited the hemaglutinin yield of GD-7 virus in mouse L cells by 0.5 $\log_{10}$.

RESULTS

Response to a single injection of varying amounts of poly I·poly C. Figure 1 shows the results of two representative experiments on the interferon response in groups of 5 mice following a single i.p. injection of increasing amounts of poly I·poly C. A single injection into mice (i.p.) of the synthetic double-stranded RNA, poly I·poly C, produces a serum interferon response with peak titer and total duration of production directly proportional to the amount of inducer injected. The interferon response was still increasing at the highest (and toxic) dose used.

Response to multiple injections of poly I·poly C. In order to determine whether the initial response would be influenced by multiple injections

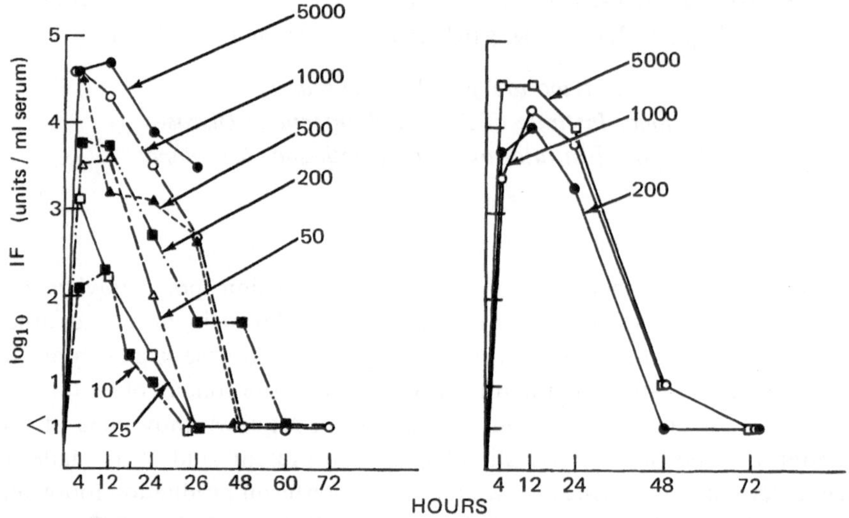

Fig. 1. *Serum interferon response of mice to varying amounts of poly I·poly C.* A single i.p. injection of the indicated concentration/mouse (in μg) was administered at 0 hours. Two experiments are shown.

during the first 48 hours, groups of mice were injected with 200 μg of poly I·poly C every 6, 12, or 24 hours. Their serum interferon response was compared to a group of mice that received only a single dose of 200 μg. The results are shown in Figure 2. Repeated injection within the first 48 hours after an initial induction only increases the total time of production without increasing the peak titer.

Serum interferon following repeated injections of poly I·poly C. The ability to maintain high and sustained levels of interferon may be hampered by hyporesponsiveness to restimulation and by the rapid elimination of the inducer (Cantell and Paucker, 1963; Park *et al.*, 1969). The experiments were undertaken to help determine whether the hyporesponsive state in the mouse could be altered or overcome by varying the dosage schedule of poly I·poly C.

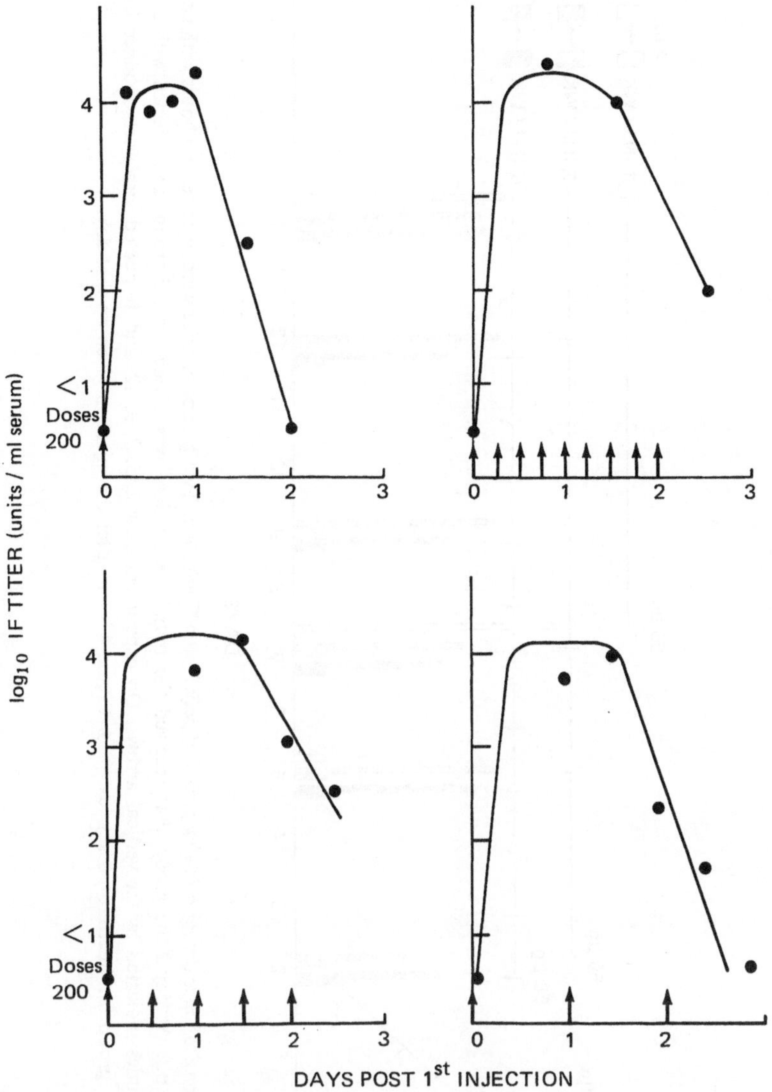

Fig. 2. *Serum response of mice to a single or to multiple injections of poly I·poly C.* Mice were injected with 200 μg/mouse, i.p. at times indicated by vertical arrows. Serum was obtained and interferon titers determined at times indicated by solid circles.

Groups of mice were initially stimulated with 50 μg or 200 μg of poly I·poly C given either intravenously (i.v.) or intraperitoneally (i.p.). The groups that received 200 μg were divided into subgroups which were restimulated with 50 μg once weekly, three times weekly, or seven times

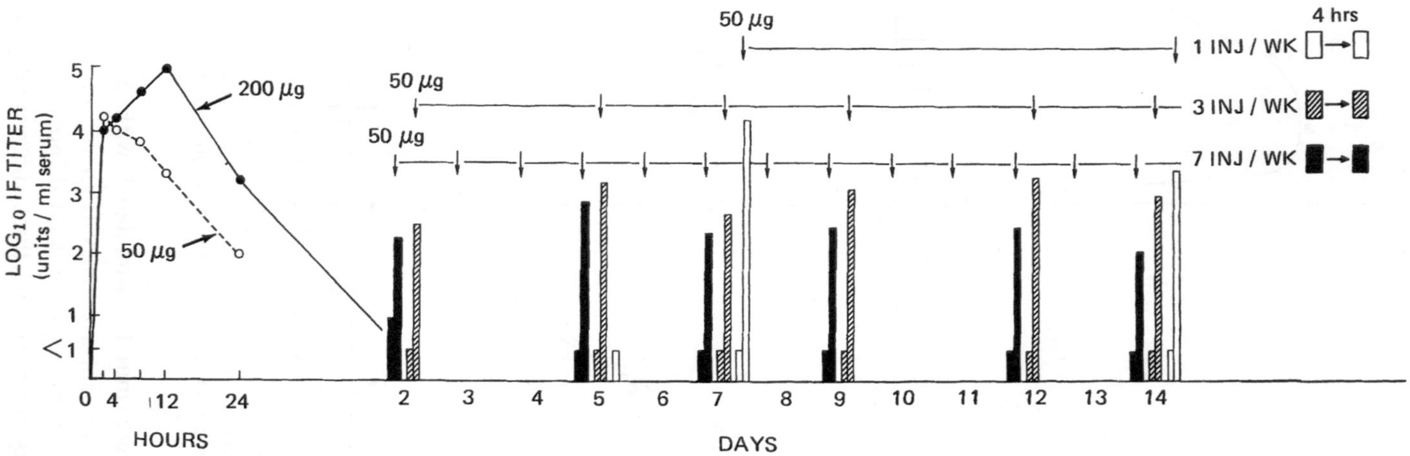

Fig. 3. *Serum interferon response of mice during a 14-day period of poly I·poly C injection.* Two groups of mice were injected at 0 hours with either 50 or 200 μg poly I·poly C (i.p.). At day 2 the group that received the initial 200 μg dose was divided into 3 groups and injected with 50 μg poly I·poly C (i.p.) at the times indicated by the vertical arrows. On experiment days 2, 5, 7, 9, 12, and 14 pooled sera were obtained from each group for interferon assay just prior to and 4 hours after the injection given on that day.

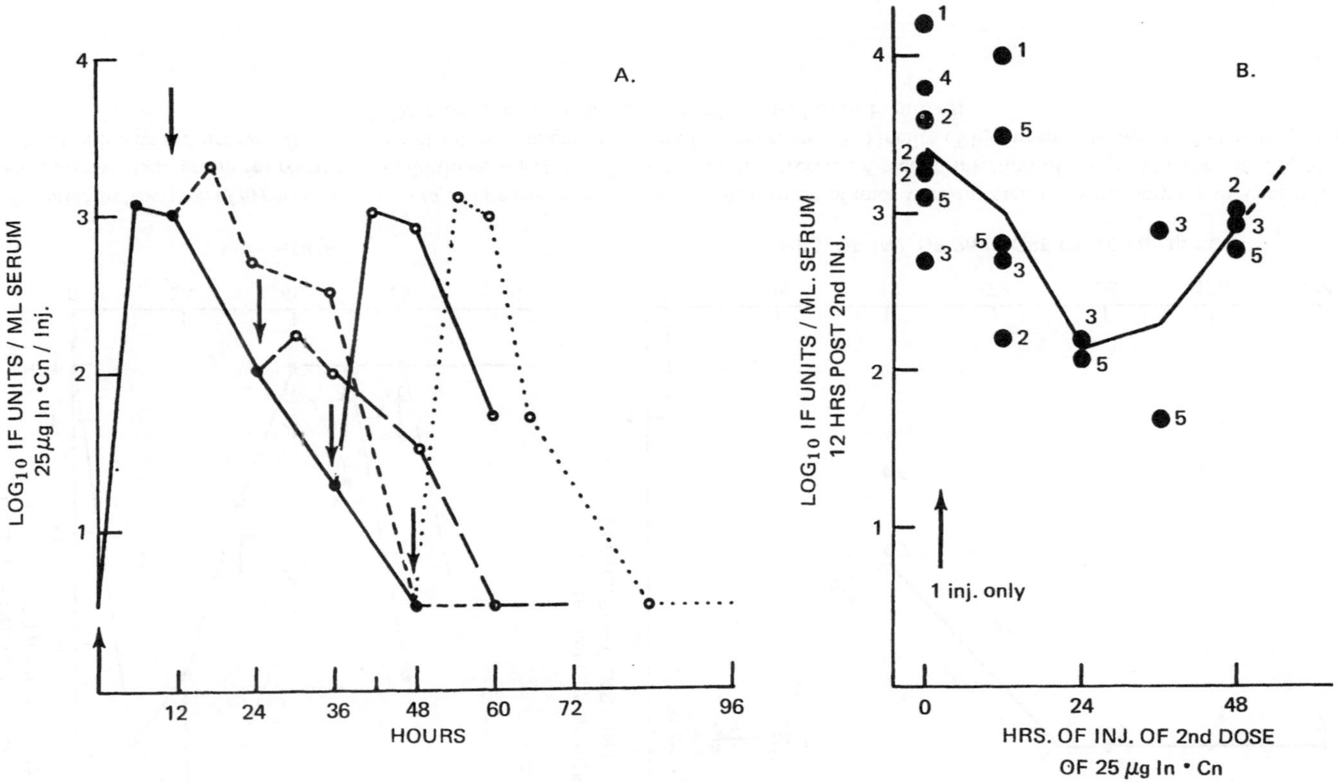

Fig. 4. *Onset and duration of hyporesponsiveness following 25 μg poly I·poly C given i.p. to mice.* A. Groups of mice were injected i.p. with 25 μg poly I·poly C at 0 hours (upward arrow). One group (•) received no additional injections. The other groups received a second injection of 25 μg at either 12, 24, 36, or 48 hours. Downward arrows indicate the times at which the second injections were given. Serum interferon production resulting from the second dose is plotted. B. Summary of all experiments performed to determine the kinetics of hyporesponsiveness to 25 μg poly I·poly C. The serum interferon response determined 12 hours after the second dose of poly I·poly C is plotted against the time of injection of the second dose. Points obtained from the same experiment are designated by a common number. In some experiments multiple determinations were made at a given time.

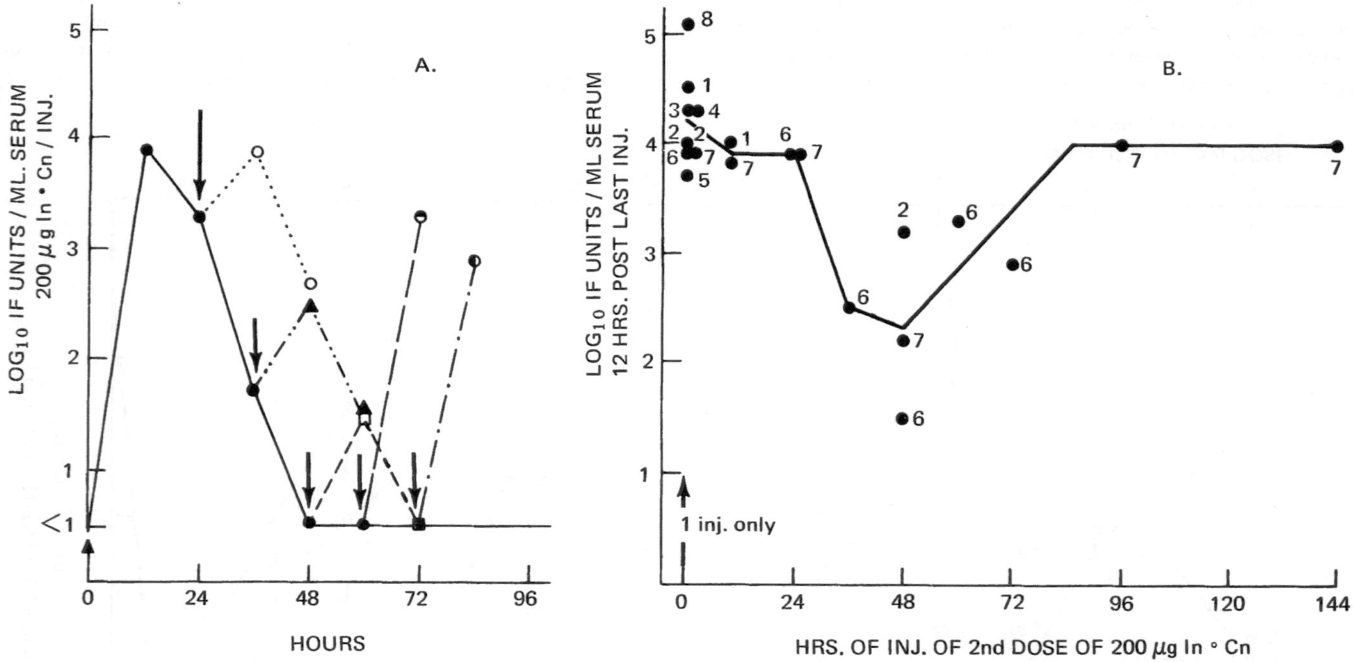

Fig. 5. *Onset and duration of hyporesponsiveness to 200 μg poly I·poly C in mice.* A. Groups of mice were injected i.p. with poly I·poly C at 0 hours (upward arrow). One group (•) received no additional injections. The other groups received a second injection of 200 μg at either 24, 36, 48, 60, or 72 hours (downward arrow). B. Summary of all experiments performed to determine the kinetics of hyporesponsiveness to 200 μg poly I·poly C. Data derived and superscript numbers used as in Figure 4B.

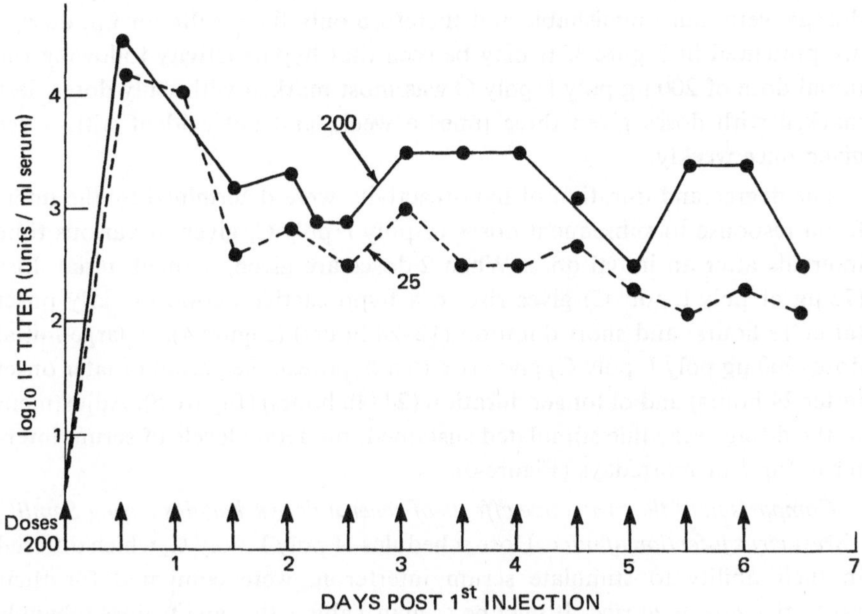

Fig. 6. *Serum interferon response in mice following multiple injections, i.p., of poly I·poly C.* Mice were injected every 12 hours (verticle arrows) with either 200 μg (●——●) or 25 μg (●---●) per mouse.

Table 1. Comparison of the Protective Effects of Different Dose Schedules of Poly I·Poly C on SFV Infection of Mice

Exp. no.	Dose schedule	% Protection relative to day therapy started			
		Day[c] 1	2	3	4
1	100 μg 1/day[a]	100	69	85	77
	100 μg 2/day	77	46	8	8
2	100 μg 1/day[a]	57	51	0	0
	200 μg 1/day	67	8	8	13
3	100 μg 1/day[a]	100	72	65	79
	100 μg 1 dose	37	44	0	10
4	100 μg 1/day[a]	89	67	56	33
	100 μg 1/2 days[b]	89	28	22	22

[a] 100 μg of poly I · poly C injected i.p. 1 time per day.
[b] 100 μg of poly I · poly C injected i.p. 1 time per 2 days (every other day).
[c] Treatment was begun on day indicated for each group and continued until the tenth day after infection.

weekly beginning on the second day. The findings with i.p. dosage or i.v. dosage were indistinguishable and therefore only the results for i.p. dosage are presented in Figure 3. It may be seen that hyporeactivity following the initial dose of 200 μg poly I·poly C was most marked with daily doses, less marked with doses given three times a week, and not evident with doses given once weekly.

The degree and duration of hyporeactivity were determined by the interferon response to subsequent doses of poly I·poly C, given at various time intervals after an initial dose. When 2 doses are given, a small initial dose (25 μg of poly I·poly C) gives rise to a hyporeactive period of early onset (after 12 hours) and short duration (12–24 hours) (Figure 4). A large initial dose (200 μg poly I·poly C) gives rise to a hyporeactive period of later onset (after 24 hours) and of longer duration (24–48 hours) (Figure 5). Adjustment of the dosage schedule stimulated sustained, moderate levels of serum interferon for 7 or more days (Figure 6).

Comparison of the protective effects of several dose schedules during Semliki Forest virus infection of mice. Dose schedules of poly I·poly C, which differed in their ability to stimulate serum interferon, were compared for their protective effects *in vivo*. It may be seen in Table 1 that an i.p. dose schedule of 100 μg of poly I·poly C given once a day gave significantly greater protection than did the same dose given twice a day, every other day, or as 1 dose. No greater protection was achieved by increasing the daily dose to 200 μg. The most protective dose schedule (100 μg given once a day) probably elicited less interferon and certainly elicited no more interferon than did the less protective dose schedule of 100 μg given twice a day and 200 μg given once a day.

DISCUSSION

The results of these studies help define some of the factors governing the circulating interferon response of mice to poly I·poly C. The level of serum interferon observed is related to the amount of a single dose of RNA injected up to 5000 μg. It could not be determined whether doses above 5000 μg would continue to induce increasing amounts of interferon because such doses were toxic, inducing 90 to 100% lethality. Therefore, it is possible that the maximum observed response of $10^{5.0}$ units of interferon per ml of serum is not the maximum amount that the mouse can produce.

A repeated dose(s) of poly I·poly C given during the 24-hour period after the first dose extends the time of maximum interferon production from 24 to 36 hours. Further peak production is prevented by the development of the hyporesponsive period at 36 hours.

The present findings indicate that the dose of poly I·poly C and the dosage schedule exert a strong influence on the hyporeactive period. A small initial dose gives rise to a hyporeactive period of relatively early onset and relatively short duration. A large initial dose gives rise to a hyporeactive period of later onset and of longer duration. These findings have several important implications. These include (1) the potential for prolonging the initial interferon response by employing multiple doses of poly I·poly C before hyporeactivity begins; (2) the possibility of employing doses of poly I·poly C of varying concentration to achieve a significant interferon responsiveness over a more prolonged period; (3) alternating treatment with poly I·poly C and interferon or other interferon inducers to overcome hyporesponsiveness; and (4) the implication that neither interferon nor the antiviral state are the direct causes of hyporesponsiveness because hyporesponsiveness frequently did not occur until long after interferon was produced and long after resistance is known to develop under these conditions of interferon production.

It is noteworthy that an increase of the poly I·poly C dose from 100 µg/day to 2 doses of 100 µg/day resulted in a significantly decreased protection despite the fact that 2 doses of 100 µg a day or 1 dose of 200 µg a day gives rise to an equal amount or more serum interferon. This finding could indicate a disassociation between amount of interferon stimulated and the protective effect. However, we favor the interpretation that since the higher dose approaches the toxic level of poly I·poly C, its decreased protective effect is more likely due to secondary effects of the stimulator.

SUMMARY

A single injection into mice (i.p.) of the synthetic double-stranded RNA poly I·poly C produces a serum interferon response with peak titer and total duration of production directly proportional to the amount of inducer injected. The interferon response was still increasing at the highest (and toxic) dose used. Repeated injection within the first 48 hours after an initial induction only increases the total time of production without increasing the peak titer. The ability to maintain high and sustained levels of interferon is hampered by hyporesponsiveness to restimulation and by the rapid elimination of the inducer.

The degree and duration of hyporeactivity were determined by the interferon response to subsequent doses of poly I·poly C, given at various time intervals after an initial dose. When giving 2 doses, a relatively small initial dose gives rise to a hyporeactive period of early onset (after 12 hours) and short duration. A relatively large initial dose gives rise to a hyporeactive period of later onset and of longer duration.

Adjustment of the dosage schedule can stimulate sustained, moderate levels of serum interferon for 7 or more days. Although serum interferon levels were a good indicator of antiviral activity, this parallel did not always hold true.

References

Cantell, K. and Paucker, K. (1963). Quantitative studies on viral interference in suspended L cells. IV. Production and assay of interferon. Virology **21**: 11–21.

Field, A. K., Tytell, A. A., Lampson, G. P., and Hilleman, M. (1967). Inducers of interferon and host resistance. II. Multistranded synthetic polynucleotide complexes. Proc. Natl. Acad. Sci., U.S. **58**: 1004–1010.

Isaacs, A., Baron, S., and Allison, A. C., as referred to in Isaacs, A. (1961). Interferon. Sci. Am. **204**: 51–57.

Park, J. H., Galin, M. A., Billiau, A., and Baron, S. (1969). Prophylaxis of herpetic keratoconjunctivitis with interferon inducers. Arch. Ophthal. **81**: 840–842.

STUDIES ON THE ANTI-TUMOR ACTION OF POLY I: POLY C

H. B. LEVY,[1] R. ADAMSON,[2] P. CARBONE,[2] V. DEVITA,[2]
A. GAZDAR,[2] J. RHIM,[3] A. WEINSTEIN,[2] and F. RILEY[1]

[1]*National Institute of Allergy & Infectious Diseases*
[2]*National Cancer Institute*
National Institutes of Health, Bethesda, Maryland,
and
[3]*Microbiological Associates, Bethesda, Maryland*

During the past year there has been a good deal of interest in various aspects of the antitumor action of poly I:poly C, ranging from basic studies of the mechanism of its action to the beginning of therapeutic trials in humans with malignant disease. This discussion will touch on both of these areas of interest.

When we first described the antitumor action of poly I:poly C, we thought that not much, if any, of the antitumor activity was attributable to the interferon induced by the compound (Levy *et al.*, 1969 b). Subsequent observations by Gresser and by us suggest that this may be an overstatement. Gresser *et al.* (1967) pointed out that interferon treatment of mice bearing certain viral-induced tumors exerts a strong protective action in these mice. In subsequent tests with Dr. Adamson we also found that interferon exerts some protective activity, although not as much as that found by Gresser. The protection was less than that given by poly I:poly C. We have looked at the effect of daily interferon treatment on the survival time of mice bearing a reticulum cell sarcoma. In many experiments that we have done, poly I:poly C has often doubled and sometimes tripled such survival time. One can look at two measures of the effectiveness of interferon on tumor inhibition. Figure 1 shows the comparative tumor size 28 days after transplant in untreated animals, in animals treated daily with two levels of interferon, and in animals treated with poly I:poly C. It can be seen that although interferon treatment definitely leads to a decreased rate of growth of the tumor,

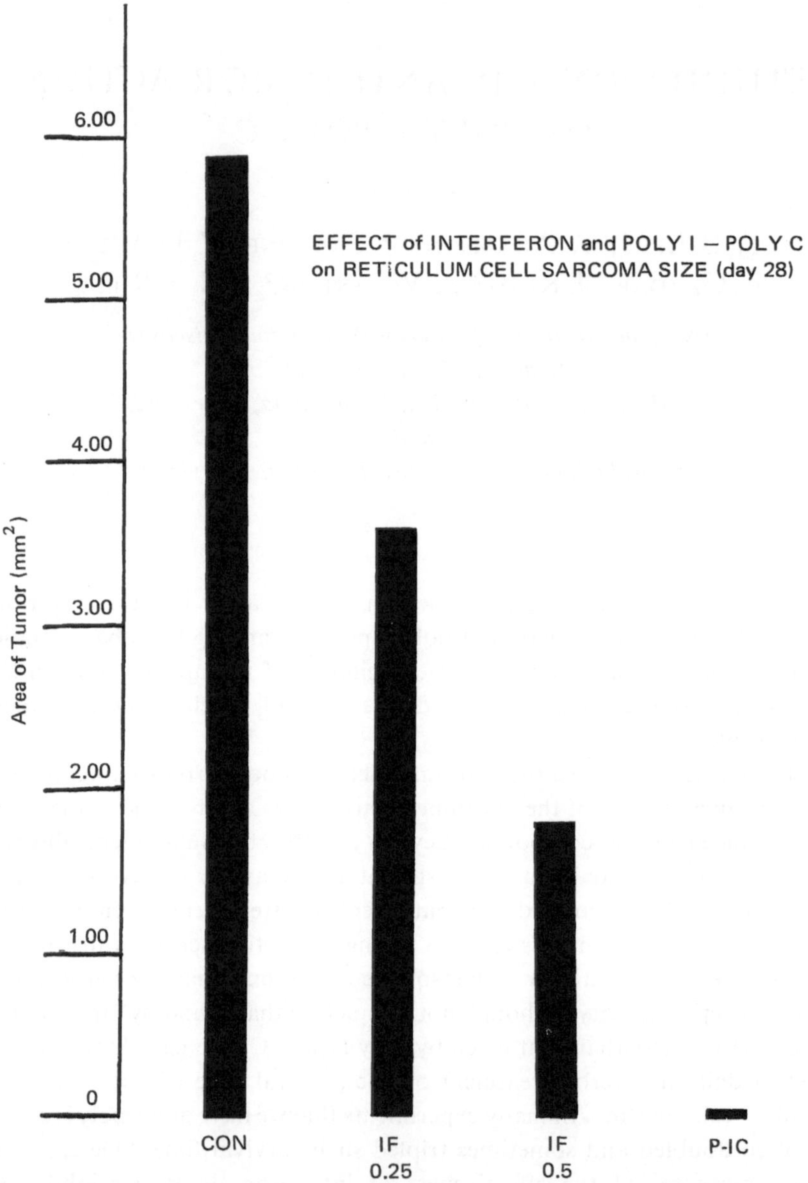

Fig. 1. *Effect of interferon on the growth rate of a reticulum cell sarcoma.* CDF_1 mice were implanted with the A reticulum cell sarcoma, and treated intraperitoneally, with 0.25 ml (5000 units) or 0.50 ml (10,000 units) of interferon daily or 150 µg of poly I:poly C every other day. On day 28 the tumor was measured in two dimensions, and the area calculated.

poly I:poly C does so to a much greater extent. The difference between poly I:poly C and interferon in effects on survival time, although present, is not quite so dramatic, as shown in Figure 2. In other experiments, the protective action of poly I:poly C was a good deal greater than that shown here. It might be pointed out that very often tumor-bearing mice that are

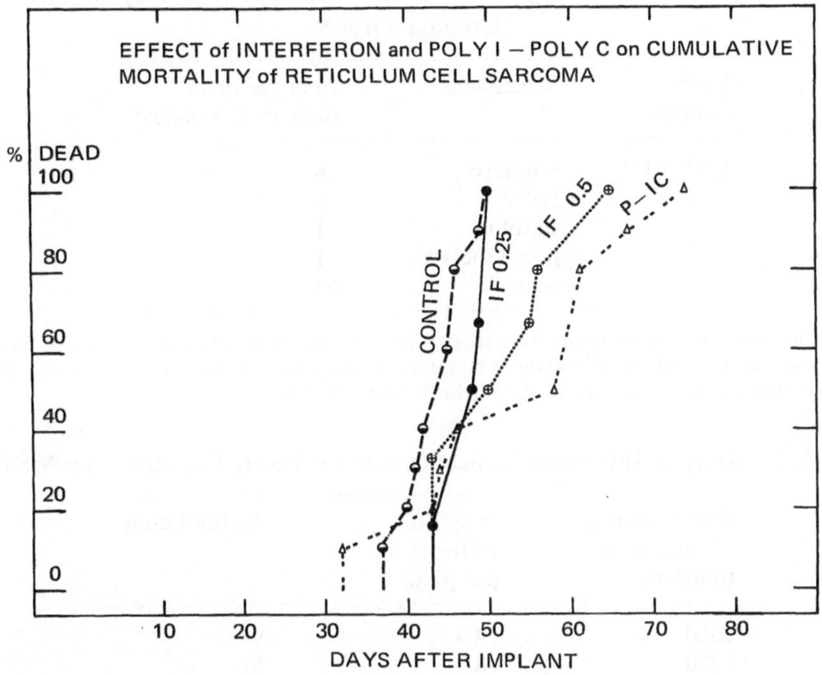

Fig. 2. *Effect of interferon on survival time of mice bearing a reticulum cell sarcoma.* As in Figure 1, except that survival time was measured.

being treated with poly I:poly C die even when their tumors are quite small, hardly the size that one would think sufficient to be responsible for deaths. That is, animals treated with poly I:poly C may be dying not because of massive tumors but because of a synergistic toxicity between the drug and the tumor. With another tumor, the Maloney sarcoma, induced either by the virus or by transplantation of tumor cells to recipient animals, one can use two different criteria for an effect of the development of malignancy. Depending on which criterion we used, we found different results. In tissue culture, using the ability of MSV to form foci of transformed cells, we found that interferon strongly inhibits focus formation, as shown in Table 1. Newcastle disease virus, statolon, and poly I:poly C also show this ability, essentially to a comparable extent. Table 2 shows that higher concentrations of interferon lead to greater inhibition.

In *in vivo* tests the survival time in animals bearing the tumor was used as a criterion of effectiveness. Five thousand units of interferon were given daily

Table 1. Comparative Inhibitory Effect of Interferon, Newcastle Disease Virus, Statolon, and Poly I·Poly C on Focus Formation by Moloney Sarcoma Virus[a]

Virus dilution	Treatment	Foci per plate (avg. of 2–3 plates)
$1/2 \times 10^{-1}$	interferon	18
	NDV	5
	statolon	1
	poly I·poly C	11
	none	85

[a] Primary mouse embryo cells were treated for 24 hours with either 1300 units of interferon, 10^8 PFU of NDV, 100 μg of statolon or 100 μg of poly I:poly C, all per ml. The inhibitors were then washed off, and MSV added as indicated.

Table 2. Effect of Interferon Concentration on Focus Formation by MSV[a]

Concentrations of interferon (unit/ml)	Avg. no. of foci per plate	% Inhibition
2500	4.3	93
1250	8.3	87
625	26.0	59
None	63.7	...

[a] Secondary NIH mouse embryo cultures were pretreated for 24 hours with the indicated concentration of interferon. The cultures were then washed and infected with virus. Foci were counted 5 to 7 days later.

to mice that had received Maloney sarcoma virus. There was very little effect of interferon, as shown in Figure 3. Poly I:Poly C, under these conditions, strongly inhibits tumor growth. Therefore, *in vivo*, with this virus, interferon was not at all comparable to poly I:poly C even though the *in vitro* tests showed that it was a moderately good inhibitor.

Further evidence that interferon may not be playing a major role in the antitumor activity of poly I:poly C comes from the following experiments. NIH Swiss mice and C_3H mice produce the same amount of interferon in

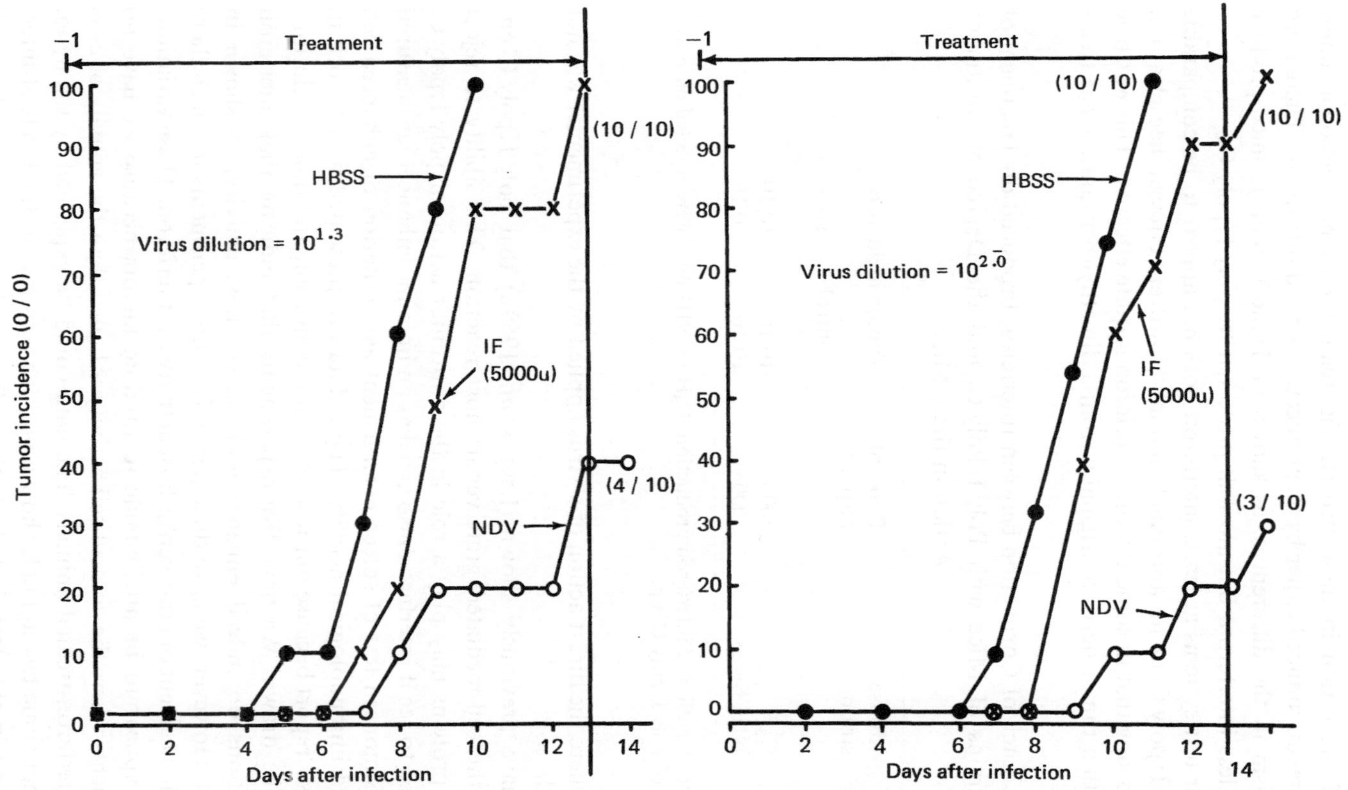

Fig. 3. *Effect of interferon on survival time of mice bearing MSV-induced tumors.* NIH Swiss mice were infected with two different levels of Moloney sarcoma virus, and treated daily with either Hanks balanced salt solution (HBSS), 5000 units of interferon, or were injected once with 10^8 PFU of NDV. The numbers in parentheses are the number of animals bearing tumors per number of animals tested.

response to poly I:poly C. However, poly I:poly C exerts very different degrees of protection in these two strains against Maloney sarcoma tumor development, induced either by virus injection or cell transplant. There is no parallelism in the differences, as shown in Table 3. Balb C mice produce slightly less interferon, but are well protected by poly I:poly C.

In our test systems at least, interferon does not appear to be comparable to poly I:poly C in its antitumor action. However, Gresser has clear-cut data showing that in some systems interferon is quite effective. This was true even with a tumor derived originally from a dimethyl-benzanthracene treatment, where the direct action of a virus applied by the experimenter was not involved.

Table 3. Lack of Correlation between Interferon Produced on Treatment of Two Strains of Mice with Poly I·Poly C, and the Degree of Antitumor Action in these Mice[a]

Mouse strain	Peak inf. response	Tumor incidence	
		Control	Treated
C_3H	3000	10/10	10/10
NIH Swiss	3000	10/10	0/10

[a] For determination of interferon production, C_3H or NIH mice were injected once with 200 μg of poly I:poly C, i.p.

We have previously shown (Levy *et al.*, 1969a) that poly I:poly C enhances the cell-mediated graft versus host reaction. This ability to reject foreign proteins may play a role in the antitumor action of poly I:poly C. We tested to see if interferon also acted as an immune enhancer, as measured by the Simonsen test of spleen enlargement as a criterion of enhanced cell mediated immunological activity. These data are presented as a "work in progress" report because our results are not consistent and final conclusions cannot be drawn. We have five experiments that indicate that interferon preparations do indeed enhance graft versus host activity, as shown in Figure 4. However, we have three perfectly good experiments in which there is no enhancement of graft versus host activity by interferon. These variations do not appear to be attributable to random factors because we have no experiments where the control cell activity is higher than the activity of cells from interferon-treated animals. The data could be explained by the hypothesis that some test animals show a positive response to the levels of interferon used in this test, but that others do not. At the moment the small

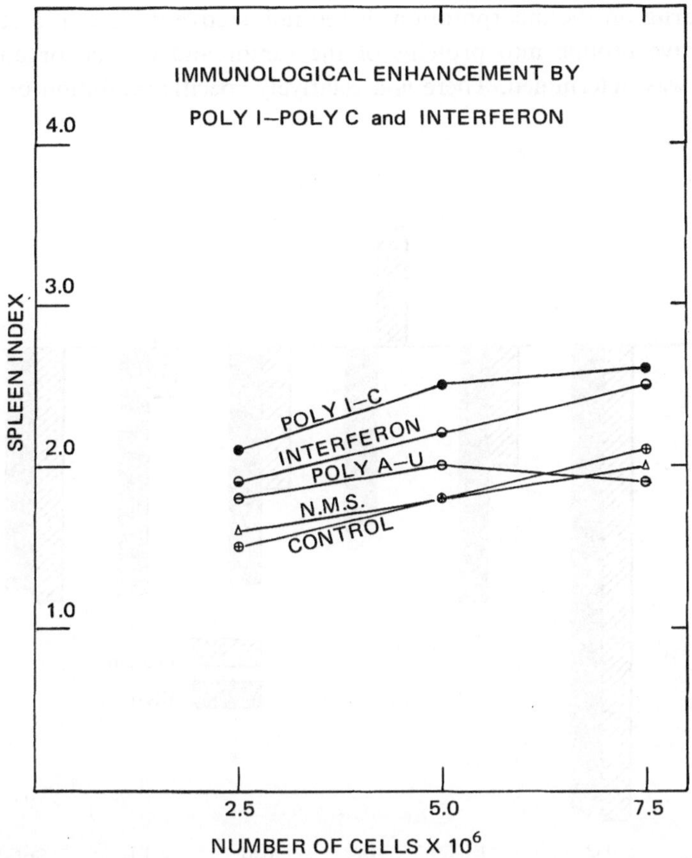

Fig. 4. *Effect of interferon on cell-mediated graft versus host reaction.* Adult Balb C mice were injected i.p. with 7000 units of mouse interferon. Two days later their spleens were removed, minced with scissors, and the cells dispersed with the aid of a pipette. The indicated number of white cells were injected i.p. into 5-day-old or younger F_1 hybrids of Balb C × C57 black mice. After 9 days the ratio of spleen to body weight was determined. The spleen index is defined as:

$$\frac{\text{spleen weight/body weight of baby mice that had received cells from treated donors}}{\text{spleen weight/body weight of baby mice that had received cells from untreated donors}}$$

antitumor action we see and the strong one seen by Gresser is still unexplained and the extent of the role of interferon in the antitumor action of poly I:poly C is also undetermined.

We also have examined further a phenomenon that might be considered a direct chemotherapeutic effect of poly I:poly C on tumors. Animals bearing one of several different types of tumors were exposed to poly I:poly C overnight and then given ^3H uridine and ^{14}C proline for a few hours. The effect of the drug on the incorporation of the radioactive uridine into RNA and radioactive proline into proteins of the tumor and normal organs of the animal was determined. There is a relatively specific inhibition of such in-

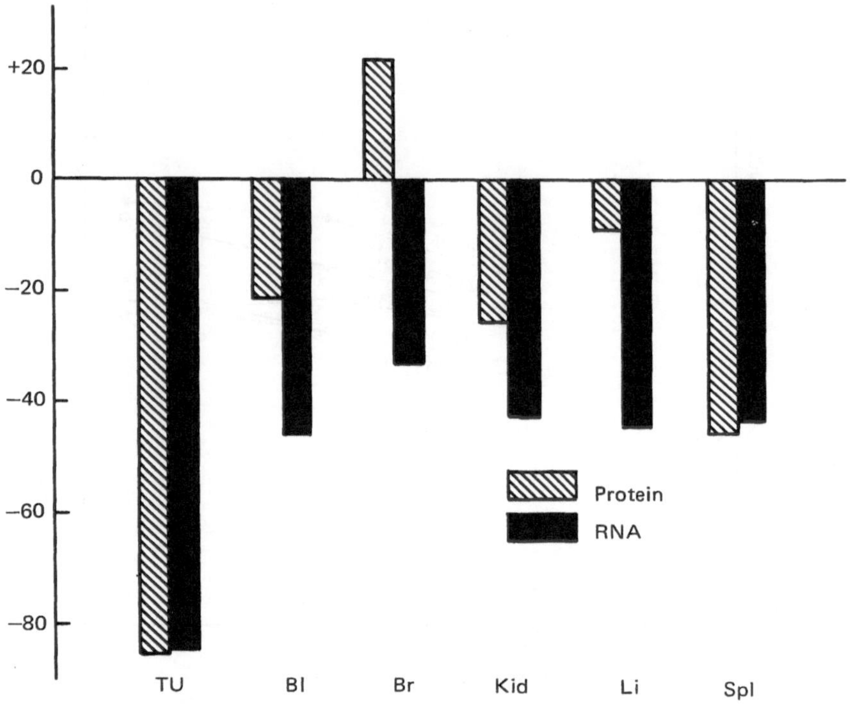

Fig. 5. *The effect of poly I:poly C on RNA and protein synthesis in CDF_1 mice bearing a reticulum cell sarcoma.* Three mice bearing the A reticulum cell sarcoma were injected i.p. with 150 µg of poly I:poly C, and 3 were injected with saline solution. Seventeen hours later, each animal received 10 µC ^{14}C proline, and 100 µC ^3H uridine. After 2 additional hours, the specific radioactivities of the acid-insoluble fractions of organs of the treated and untreated animals were determined and compared.

corporation into the tumor as contrasted with normal organs. Figure 5 shows an example of this. The exact pattern seen varied with the strain of animals but the tumor was always more strongly inhibited than any of the other organs. We have examined in detail the possibility that the differences in specific radioactivities in the acid-insoluble fractions of the organs are merely reflections of the changes in the specific radioactivities of the acid-soluble precursor pools of the cells in the different organs. The data are too complicated to present here in detail. Poly I:poly C does indeed induce changes in the radioactivity and composition of the precursor pools, causing an increase in specific activity in some organs, and a decrease in others. Analysis of both the magnitude and the direction of these changes reveal that they are not responsible for the changes in radioactivity in RNA and protein. It is surprising, with this degree of inhibition of macromolecule synthesis in both the tumors and in the normal organs, that there is not a faster destruction of the tumor, and that there is not greater toxicity in the animals receiving the drug. The reason for the lack of greater effect on the growth of the tumor and the health of the host is suggested by the following experiment. Tumor-bearing animals were treated in the following way. Some animals received 1, some 2, and some 3 injections of poly I:poly C on alternate days and some received no poly I:poly C. As before, all the animals were given radioactive uridine and proline, at the same time. The specific activities of the acid-soluble and acid-insoluble fractions of the several organs were determined. For simplicity of presentation, the data of Figure 6 are restricted to just the acid-insoluble fractions of three organs. The inhibitory effect of poly I:poly C is largely transient, showing its maximum effect after the first treatment and decreasing with subsequent treatments. The other organs, including blood cells, show a comparable decline in the amount of inhibition. Again there is no effect on acid-soluble pools that could explain this decreased inhibition with repeated injection.

The last area that will be mentioned deals with the data obtained to date in man. On the basis of the antitumor action that we saw in mice, the National Cancer Institute undertook extensive toxicological studies of poly I:poly C in small and large mammals and then began a study of the effects of intravenous administration of the compound in man. The goals of this study were: (1) To be alerted to any toxicity in man not suggested by the studies in large animals. Initial treatment therefore was with very small doses. Small stepwise increases were instituted when no toxicity appeared at the previous lower dose. (2) To test man's responsiveness to the compound, as measured by circulating interferon production. (3) To determine what a proper dose for chemotherapeutic treatment of malignancy might be.

Man does make interferon in response to poly I:poly C. The levels found so far have been low, perhaps 100 units/ml, but higher doses of the drug appear to give slightly higher amounts of circulating interferon. In comparison with the amount of drug that has been used in mice, the levels administered to man so far have been quite low. In mice there is a direct relationship between the amount of interferon produced and the amount of poly I:poly C given. The levels of poly I:poly C that would yield 100 units of serum interferon, as found in man, would not exert any antitumor action. Whether sufficient drug can be given to man to elicit really high levels of

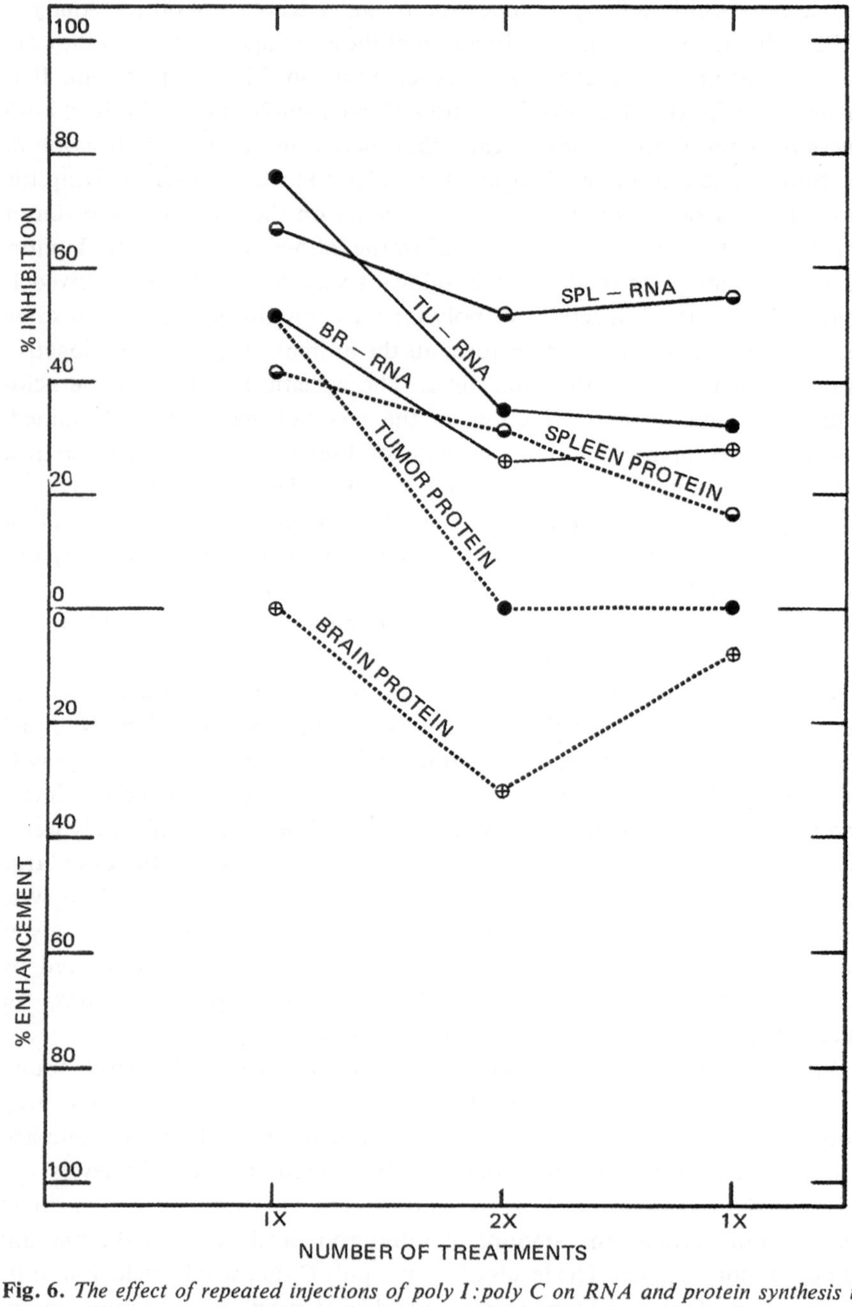

Fig. 6. *The effect of repeated injections of poly I:poly C on RNA and protein synthesis in CDF_1 mice bearing a reticulum cell sarcoma.* Conditions as for the data of Figure 5, except for the modifications described in the text.

interferon remains to be determined. Relevant to this question is the observation reported by Nordlund, Wolff, and Levy (1970) that human serum contains an enzyme that rather rapidly destroys poly I:poly C and renders it inactive. Mice contain very much less of this enzymatic activity.

Thus, the mode of action of poly I:poly C appears complex, and still poorly understood. Although three types of action can be seen, the relative importance of these components is not known. Our trials in man are too early to permit either optimism or pessimism. Man does show a hyporeactive state in a manner analogous to other animals. That is, man responds maximally to the first dose of the compound whereas subsequent doses, unless spaced at least 1 week apart, give a small response. With the doses used thus far, there appears to be a requirement for a recovery period of about a week before circulating interferon can be found again. This should not be interpreted to mean that the compound is not having an effect during this period of time. The problem of hyporeactivity is under study in many laboratories.

References

Gresser, I., Coppey, J., Fontaine-Brouty-Boye, D., Falcoff, R., Falcoff, E., and Zajdela, A. (1967). Interferon and murine leukaemia. III: Efficacy of interferon preparations administered after inoculation of Friend virus. Nature **215:** 174.

Levy, H. B., Asofsky, R., Riley, F., Garapin, A., Cantor, H., and Adamson, R. (1969). The antitumor action of polyinosinic-polycytidylic acid. Ann. N.Y. Acad. Sci., in press.

Levy, H. B., Law, L. W., and Rabson, A. S. (1969). Inhibition of tumor growth by polyinosinic-polycytidylic acid. Proc. Natl. Acad. Sci. U.S. **62:** 357.

Nordlund, J. J., Wolff, S. M., and Levy, H. B. (1970). Inhibition of biologic activity of poly I:poly C by human plasma. Proc. Soc. Exptl. Biol. Med. **133:** 439.

MOLECULAR REQUIREMENTS FOR SYNTHETIC RNA TO ACT IN INTERFERON STIMULATION[1]

THOMAS C. MERIGAN, ERIK DE CLERCQ[2], FRITZ ECKSTEIN, and ROBERT D. WELLS

Division of Infectious Diseases, Department of Medicine, Stanford University Medical School, Stanford, California, Max-Planck-Institut für Experimentelle Medizin, Abteilung Chemie, Göttingen, West Germany
and
Department of Biochemistry, University of Wisconsin, Madison, Wisconsin

INTRODUCTION

Three varieties of polyanions have been associated with interferon production; that is, plastics, polysaccharides, and polynucleotides (reviewed by De Clercq and Merigan, 1970). Only with the latter do we have any knowledge of three-dimensional structure, due to the work of Watson and Crick (1953), Wilkins *et al.* (1953), and others as reviewed by Davies (1967). Knowledge of the genetic role of nucleic acids in the last 2 decades has led to an immense amount of chemical information on their structure and synthesis. However, proteins are the biologically important macromolecules about which most is known as to mechanisms on the organic chemical level which underly their activities. Here, it is quite clear that in contrast to the limited knowledge provided by information on primary sequence, major insights were afforded only by precise descriptions of the three-dimensional

[1] Supported by United States Public Health Service Grant AI-05629, the Deutsche Forschungsgemeinschaft and the National Science Foundation Grant GB-8786.

[2] Fellow of the Damon Runyon Memorial Fund for Cancer Research and "Aangesteld Navorser" of the Belgian N.F.W.O. (Nationaal Fonds voor Wetenschappelijk Onderzoek).

structure of proteins (Barnard, 1969). Therefore, we have been attracted to study the structural requirements and fate of nucleotides acting as interferon inducers in the cell, because of the potential for meaningful molecular level understanding of this phenomenon. As this represents a new function for nucleic acids, it is possible that the structural basis of this activity may even have implications for nucleic acid structure and genetic function.

It seems most likely that polynucleotides are the critical interferon stimulating factors during the process of natural virus infection. In addition, events occurring within the cell during the process of interferon production by polynucleotides may lead to insights into what occurs during natural virus replication. Synthetic polynucleotides have the obvious advantage of precise characterization. Studies in our laboratory have been carried out with these model compounds as well as with certain related analogs. It seems reasonable to extrapolate most of the conclusions to naturally occurring polynucleotides as well.

STRUCTURAL REQUIREMENTS

Initially our studies focused on the importance of a stable secondary structure in the antiviral and interferon stimulating ability of polynucleotides. From work carried out in our and other laboratories with a large number of synthetic polynucleotides (De Clercq and Merigan, 1969; Colby and Chamberlin, 1969; Gresser, personal communication, 1969) it appeared as though a threshold degree of thermostability was required for antiviral activity (De Clercq *et al.*, 1970a). This corresponded to approximately 60 °C as measured in 0.15 *M* NaCl. Most animal virus-cell culture systems are maintained at 38 °C and hence this threshold temperature of thermal stability is some 20 °C higher than the temperature in our biologic assay system. It was not clear from our initial studies at what level this requirement was necessary. Stability of the polyribonucleotide might be an advantage in penetrating the cell, in resisting cellular ribonuclease action, or in interacting with the specific cellular interferon triggering site (De Clercq and Merigan, 1969). In our assay system [human skin fibroblasts (HSF) with vesicular stomatitis virus (VSV) as challenge virus] we found polydeoxyribonucleotides significantly less active than their ribo-counterparts (DeClercq *et al.*, 1970a). Polydeoxyribonucleotides have repeatedly been reported to be unable to stimulate either interferon in rabbits (Lampson *et al.*, 1967), in rabbit kidney cell cultures (Vilcek *et al.*, 1968), or resistance to Sindbis virus multiplication in chick embryo fibroblasts (Colby and Chamberlin, 1969). More recently, Kleinschmidt *et al.* (1970) demonstrated interferon production with T4 coliphage in the mouse and attributed the interferon inducing capacity to the DNA content of the phage. DNA extracted from T4 coliphage was inactive; Kleinschmidt *et al.* (1970) considered susceptibility to nucleases as a possible explanation for the lack of activity of the free DNA.

In the (HSF-VSV) assay system antiviral activity was demonstrated with those single homopolymer species which tended to self-anneal or form multistranded complexes, such as polyriboinosinic acid, polyriboguanylic acid, or polyxanthylic acid (De Clercq and Merigan, 1969). By manipulating the conditions of our tissue culture incubation medium (acidification, excess of Mg^{++}) we could alter the antiviral activity and structural stability of certain homopolymer pairs in parallel and promote the activity of certain homopolymers that tended to self-aggregate under these conditions (polyribocytidylic acid and polyriboadenylic acid) (De Clercq and Merigan, 1969). However, these findings did not clarify the mechanism by which this multistrandedness operated to promote antiviral activity.

INTERACTION WITH THE CELL

More recently we have been interested in the sequence of interaction of interferon-stimulating RNA with cells using radiolabeled materials. Polymers labeled in the uridine moiety rather than in the sugar or phosphorus were used in order to decrease the likelihood of incorporation of the label into nonnucleotide species. The first step we noted in tissue culture was an adsorption of the polynucleotide to the cell surface which occurred rapidly at 37 °C and more slowly at 4 °C (Bausek and Merigan, 1969). This was followed by a recognition step which occurred at 37 °C only. The recognition step either depended on the bulk of RNA present on the membranes of the cell surface or on a very minute quantity of RNA that penetrated into the cell presumably in phagocytic vacuoles (Bausek and Merigan, 1969). The RNA associated with cells was rapidly degraded and its components utilized in host RNA synthesis. The recognition step was followed by the production of interferon, and the utilization in host RNA synthesis did not seem essential in the process of interferon production (Bausek and Merigan, 1969).

The effects of metabolic inhibitors on the interferon production by synthetic polyribonucleotides are quite complex; certain dosages of agents inhibiting RNA or protein synthesis actually increase interferon production by cells. Such findings have led to the postulation of complex mechanisms of control, including a two-stage synthesis and activation mechanism (Ho and Ke, 1970) or a mechanism involving an inhibitor of interferon formation which is synthesized after interferon production has begun (Vilcek *et al.*, 1969; Vilcek, 1970). Whether the interferon production occurs through a *de novo* synthesis or release from a preformed material, is a point of controversy requiring further studies (Finkelstein *et al.*, 1968; Ho and Ke, 1970).

Cells producing interferon are resistant to virus infection, and subsequent to the initiation of interferon production, there is a resistance to restimulation ("tolerance") that may be related to the mechanism that turns off the initial response (Bausek and Merigan, 1970). Both the protective effect and this tolerance phenomenon are transient in cells, either in tissue culture or *in vivo*, wearing off over several days.

In studies in our laboratory employing human fibroblasts (Bausek and Merigan, 1970) and in studies in Vilcek's laboratory employing rabbit kidney fibroblasts (Vilcek, personal communication, 1970), differences were observed between polyribonucleotide-produced tolerance and viral-induced tolerance to repeated interferon stimulation. The viral-stimulated tolerance could be mimicked by the application of interferon but this did not influence a second stimulation by RNA. In addition, the RNA-induced tolerance wore off in tissue culture whereas the viral-stimulated tolerance persisted. The time of appearance of the RNA-induced tolerance suggested a relationship to the mechanism turning off initial RNA stimulation. In connection with these studies we postulated that the tolerance to repeated interferon production with RNA was due to stimulation of an intracellular ribonuclease activity (Bausek and Merigan, 1970). Although an increased ribonuclease activity in extracts of cells exposed to synthetic polyribonucleotides could not be demonstrated (Bausek and Merigan, 1970), the findings that synthetic RNA are degraded in cells indicate that such activity must exist although below the level of our present direct assay systems.

THIOPHOSPHATE-SUBSTITUTED POLYRIBONUCLEOTIDES

We have recently completed studies that suggest the importance of ribonuclease resistance of synthetic RNA in its ability to stimulate interferon. We have reported that substituting sulphur for oxygen in the phosphate linkages of the double-stranded poly r(A-U) (an alternating copolymer of riboadenylic acid and ribouridylic acid) (Figure 1) causes a significant increase in its interferon-inducing capacity (De Clercq *et al.*, 1969). Introducing of the sulphur does not alter the thermal stability of the polyribonucleotide (Eckstein and Gindl, 1970), but does decrease its sensitivity to enzymatic degradation by several orders of magnitude (Eckstein and Gindl, 1970; De Clercq *et al.*, 1969, 1970c). We have also studied the partially and completely thiophosphate-substituted analogs of poly r(A-U) and poly r(I-C) (an alternating copolymer of riboinosinic and ribocytidylic acid) (De Clercq *et al.*, 1970c). In all cases, substituting thiophosphate for phosphate resulted in a significant increase in the ability to induce resistance to virus infection and interferon production *in vitro* as well as interferon production *in vivo*.

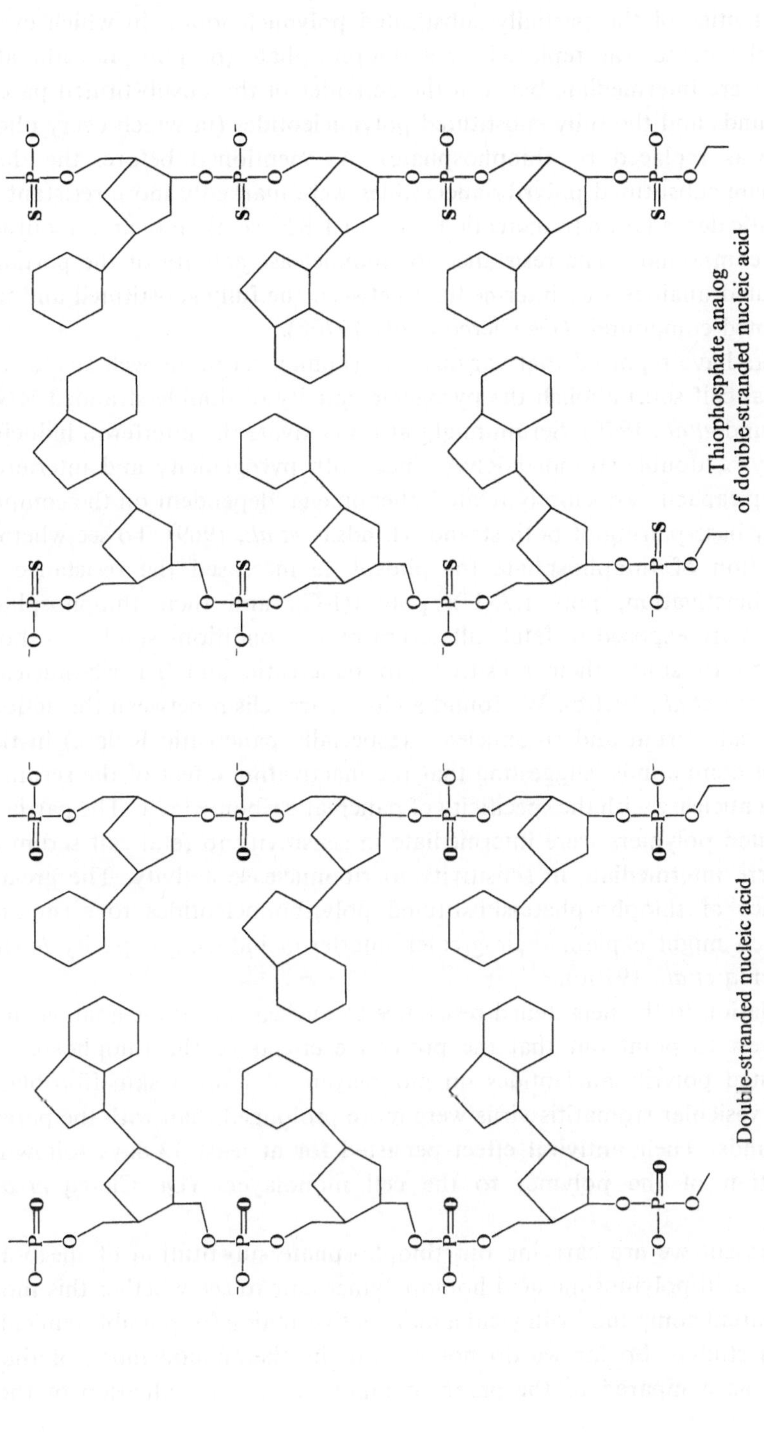

Fig. 1. Chemical structure of a polyribonucleotide with alternating base sequence [for example, poly r(A-U), poly r(I-C)], with and without thiophosphate groups substituted for phosphate groups.

The activities of the partially substituted polynucleotides, in which every other phosphate was replaced by a thiophosphate (or phosphorothioate) group, were intermediate between the activities of the unsubstituted parent compounds and the fully substituted polynucleotides (in which every phosphate was replaced by thiophosphate). As mentioned before, the thiophosphate-substituted polyribonucleotides were markedly more resistant to enzymatic degradation (pancreatic RNase, T1 RNase) than their unmodified parent compounds. The resistance to ribonuclease activity of the partially substituted analogs was intermediate between the fully substituted and unsubstituted compounds (De Clercq *et al.*, 1970c).

Others have reported that various sera (human serum as well as chicken and fetal calf sera) abolish the pyrogenic activity of double-stranded RNA (Nordlund *et al.*, 1970). Serum might also inactivate the interferon inducing capacity of double-stranded RNA since both pyrogenicity and interferon inducing capacity are closely related phenomena, dependent on the complementary base-pairing of both strands (Lindsay *et al.*, 1969). To see whether substitution of thiophosphate for phosphate increased the resistance to serum inactivation, poly r(A-U), poly r(I-C), and their thiophosphate analogs were exposed to fetal calf serum under conditions similar to those employed to study their sensitivity to pancreatic and T1 ribonuclease (De Clercq *et al.*, 1970c). We found a close parallelism between the actions of fetal calf serum and ribonuclease (especially pancreatic RNase) in this series of compounds, suggesting that the inactivating effect of the serum is due to a nuclease with the specificity of pancreatic ribonuclease. The partially substituted polymers were intermediate in sensitivity to fetal calf serum as they were intermediate in sensitivity to ribonuclease activity. The greater resistance of thiophosphate-substituted polyribonucleotides to serum inactivation might explain their greater interferon inducing capacity *in vivo* (De Clercq *et al.*, 1970c).

In relation to the heightened resistance to nucleases of these analogs, it is of interest to point out that the protective effects of the thiophosphate-substituted polyribonucleotides on monolayers of human skin fibroblasts against vesicular stomatitis virus were more prolonged than with the parent compounds. Their antiviral effect persisted for at least 13 days following application of the polymer to the cell monolayers (De Clercq *et al.*, 1970c).

At present we are carrying out thiophosphate substitution of the polycytidylic acid/polyinosinic acid homopolymer pair to see whether this more active parent compound will yield a more active analog for possible clinically directed studies. So far we do not known the therapeutic index of these analogs as compared to the parent compounds as an evaluation of their relative toxicities. We plan these studies in the near future because they will be critical in deciding whether these compounds offer any real advantages other than being more active on a weight basis.

THERMAL ACTIVATION

We have also developed another line of evidence as to the importance of the nuclease sensitivity in polynucleotide stimulation of interferon and antiviral resistance. This has come from a study of the mechanism underlying a chance observation made in our laboratory that preincubation at 37 °C rendered various alternating ribonucleotide copolymers and (to a lesser extent) homopolyribonucleotide pairs significantly more active in reducing virus plaque formation in our human skin fibroblast assay. Among the alternating copolymers studied, poly r(A-U) was most influenced by this thermal activation step, being potentiated by some 6 orders of magnitude in its ability to interfere with vesicular stomatitis virus plaque formation (De Clercq *et al.*, 1970b). Thermal activation of poly r(A-U) was regularly produced in MEM (minimal Eagle's medium) at a neutral or slightly alkaline *p*H after a short period of incubation (1 minute up to 2 hours) at 37 °C (as well as at 32 and 56 °C); the process clearly depended on the presence of divalent cations (Ca^{++}, Mg^{++}) (De Clercq *et al.*, 1970d). Thermal activation was also observed with the alternating polydeoxyribonucleotides poly d(A-T), poly d(I-C), and poly d(G-C); after heating at 37 °C in MEM, their activity was increased to values corresponding to those for the unheated (poly rA)·(poly rU) and (poly rI)·(poly rC) (De Clercq *et al.*, 1970b). Preincubation at 37 °C rendered poly r(A-U) and poly r(I-C) even more active in reducing VSV plaque formation in HSF than substituting thiophosphate for phosphate groups (Table 1). However, the completely substituted analogs, poly r($_s$A-$_s$U) and poly r($_s$I-$_s$C), were considerably more active than the preheated poly r(A-U) and poly r(I-C) when tested for interferon production either *in vitro* (HSF) or *in vivo*, following intravenous injection into rabbits (Table 1). The dependence of the thermal activation on the presence of divalent cations suggests that the phenomenon might be related to the binding of one part of the polynucleotide strand to another. Baldwin and his associates have described in alternating copolymers a phenomenon called "slippage" whereby one part of the strand would creep along on another part through shifting of one set of base-pairings at a time in such a way as to remove side chain loops and shift from a multilooped structure to a completely base-paired hairpin structure (Figure 2) (Inman and Baldwin, 1962; Spatz and Baldwin, 1965; Scheffer *et al.*, 1968). Careful measurements were made of viscosity, buoyant density in Cs_2SO_4, analytical sedimentation velocity, and optical density before and after heat activation of a preparation of poly r(A-U), but no significant changes were observed (De Clercq *et al.*, 1970d). When poly r(A-U) was studied for its sensitivity to endo- and exonuclease, its sensitivity to cleavage by pancreatic ribonuclease was markedly decreased after heating at 37 °C, whereas no difference could be

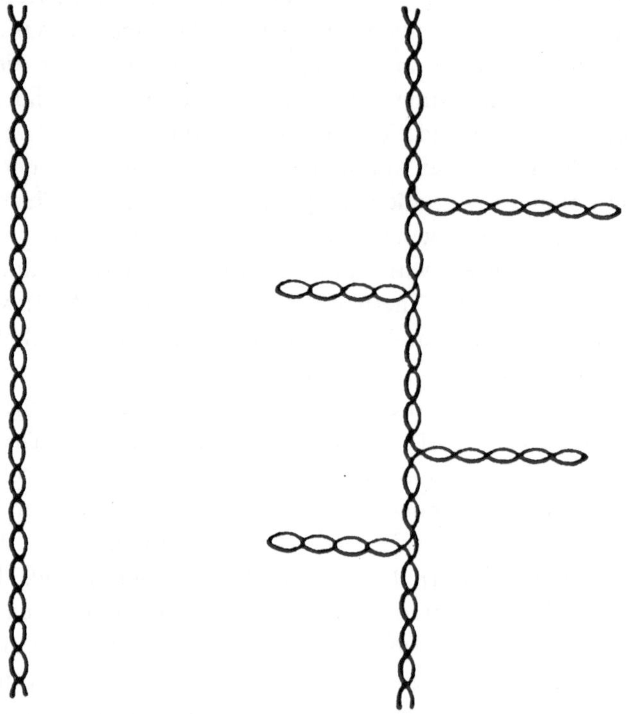

Fig. 2. Shift from a multilooped to a straight-chain (completely base-paired) helical structure in alternating copolymers (slippage: Scheffler *et al.*, 1968).

detected in the susceptibility to snake venom phosphodiesterase (De Clercq *et al.*, 1970d). Since our preparations were ribopolymers and not of extremely high molecular weight, it is not surprising that we failed to observe physical changes following incubation at 37 °C. On the other hand, the changes in endonuclease susceptibility indicate that unpaired regions, such as those associated with branching points or loops, were removed by the heat activation step.

We have been able to demonstrate with polymer cell binding experiments, conducted at either 4 or 37 °C, that preheated ^{3}H-poly r(A-U) was bound

Table 1. Comparative Study of the Effects of Thermal Activation and Substitution of Thiophosphate for Phosphate on the Antiviral Activity of the Alternating Polyribonucleotides Poly r(A-U) and Poly r(I-C)

Polymer	Thermal activation		Substitution of thiophosphate for phosphate (Preincubation at 25 °C)	
	Preincubation for 2 hours at 0 °C	Preincubation for 2 hours at 37 °C	Unsubstituted	Substituted
1. *Cellular resistance to vesicular stomatitis virus in human skin fibroblasts*[a]				
Poly r(A-U)	>1	0.000004	4	0.001
Poly r(I-C)	>10	0.00004	>4	0.0001
2. *Interferon production in human skin fibroblasts*[b]				
Poly r(A-U)	4	20	<1	20
Poly r(I-C)	2	18	<1	60
3. *Interferon production in the rabbit*[c]				
Poly r(A-U)	$\leqq 10$	60	$\leqq 15$	1000
Poly r(I-C)	$\leqq 10$	30	$\leqq 20$	16,000

[a] Minimal inhibitory concentration (µg/ml) or concentration of polymer required to reduce virus plaque formation by 50%.

[b] Interferon titer (units/4 ml tissue culture fluid) 8 hours after exposure of HSF monolayers (in 60 mm petri dishes) to 20 µg of poly r(A-U) or 30 µg of poly r(I-C) (for measurement of effect of thermal activation) and 8 µg of poly r(A-U) or poly r($_s$A-$_s$U) or 4 µg of poly r(I-C) or poly r($_s$I-$_s$C) (for measurement of effect of substitution of thiophosphate for phosphate). The polymers were removed after 3 hours incubation, the cells exhaustively washed with MEM and further incubated with 4 ml MEM per petri dish.

[c] Interferon titer (units/4 ml serum) 2 hours after intravenous injection of 20 µg of poly r(A-U) or poly r($_s$A-$_s$U) and 4 µg of poly r(I-C) or poly r($_s$I-$_s$C) in 4 ml MEM per rabbit.

more rapidly to human skin fibroblasts than a nonheated preparation of the same polymer (De Clercq *et al.*, 1970d). In addition, a significantly greater portion of the cell-associated (preheated) polymer in contrast to the unheated polymer was removed by ribonuclease treatment of cells, 1 hour after the initial incubation period (De Clercq *et al.*, 1970d). Therefore, preheated poly r(A-U) not only showed a greater affinity for the cell surface but also persisted in a polymeric state at the cell surface for a longer time. The presence of more intact polymer at the cell surface might be associated with a smaller portion of intact polymer within the cell. Hence, these studies do not allow us to decide whether the signal is a large amount of polymer on the cell surface or a very small fraction of polymer penetreated into the cell.

Table 2. Thermal Activation of Poly r(A-U)[a]

I. *Experimental conditions*	
Time:	1 min–5 hr
Temperature:	32–56 °C
*p*H:	7.5–9.0
Ionic requirements:	≧0.001 *M* Ca^{++}, Mg^{++}
Polymer concentration:	10^{-5}–1.6 μg/ml
II. *Physicochemical properties*	
Ultraviolet spectrum:	unchanged
Viscosity:	unchanged
Cs_2SO_4 buoyant density:	unchanged
Sedimentation velocity:	unchanged
Susceptibility to endonuclease:	decreased
Susceptibility to exonuclease:	unchanged
III. *Cell interaction*	
Rate of binding to cells:	increased
Persistence at the cell surface:	increased

[a] See De Clercq *et al.* (1970d).

But, on the other hand, they clearly indicate that heating at 37 °C produced a configurational change in the molecule which rendered it more resistant to endonucleolytic attack and increased its half-life in contact with the cell. The greater affinity of preheated poly r(A-U) for, and longer persistence at, the cell surface would seem to be related to its increased interferon stimulating ability.

The experimental conditions of thermal activation and the physicochemical properties and cell-interaction of heated and nonheated poly r(A-U) are summarized in Table 2.

In conclusion, our studies with the thiophosphate-substituted and heat-activated polyribonucleotides suggest that the requirement of a stable secondary structure for synthetic polyribonucleotides to stimulate interferon is related to a greater resistance of the polymers to nuclease degradation. Unfortunately it is still difficult to decide whether interferon stimulation relates to polymer acting at the cell surface or within the cell.

SUMMARY

Studies of the fate of synthetic RNA in contact with cells in tissue culture indicated that the synthetic RNA is degraded and its components incorporated into host RNA. However, at the time the commitment to interferon production is made, the bulk of the applied synthetic ribonucleotide is on the surface of the cell in a polymeric state. Two sets of observations lead to the conclusion that resistance to ribonuclease is important in the antiviral activity of polyribonucleotides. First, thiophosphate analogs of polyribonucleotides, which are more nuclease resistant, are markedly potentiated in their antiviral and interferon-stimulating abilities. Secondly, a preincubation at 37 °C increases the ability of polyribonucleotides to confer antiviral resistance in tissue culture; the configurational change induced by heating is associated with an increased resistance to ribonuclease degradation. When the kinetics of cell binding of (heated and unheated) ^{3}H-poly r(A-U) was followed after it was put in contact with cells, it was observed that the heated polymer was bound more rapidly and remained intact at the cell surface for a longer period of time than the unheated polymer. These findings suggest that the requirement for a stable secondary structure and a multistranded state for synthetic polyribonucleotide to stimulate interferon is related at least in part to the greater resistance of polymers to nuclease degradation and persistence at the cell surface.

References

Barnard, E. A. (1969). Ribonucleases. Ann. Rev. Biochem. **38:** 677–732.

Bausek, G. H. and Merigan, T. C. (1969). Cell interaction with a synthetic polynucleotide and interferon production *in vitro*. Virology **39:** 491–498.

——— (1970). Two mechanisms of cell resistance to repeated stimulation of interferon. Proc. Soc. Exp. Biol. Med. **124:** 672–616.

Colby, C. and Chamberlin, M. J. (1969). Specificity of interferon induction in chick embryo cells by helical RNA. Proc. Natl. Acad. Sci. U.S. **63:** 160–167.

Davies, D. R. (1967). X-ray diffraction studies of macromolecules. Ann. Rev. Biochem. **36:** 321–364.

De Clercq, E. and Merigan, T. C. (1969). Requirement for a stable secondary structure for the antiviral activity of polynucleotides. Nature **222:** 1148–1152.

——— (1970). Current concepts of interferon and interferon induction. Ann. Rev. Med. **21:** 17 46.

De Clercq, E., Eckstein, F., and Merigan, T. C. (1969). Interferon induction increased through chemical modification of a synthetic polyribonucleotide. Science **165:** 1137–1139.

——— (1970a). Structural requirements for synthetic polyanions to act as interferon inducers. Ann. N.Y. Acad. Sci. **113:** 444–461.

De Clercq, E., Wells, R. D., and Merigan, T. C. (1970b). Increase in antiviral activity of polynucleotides by thermal activation. Nature **226:** 364–366.

De Clercq, E., Eckstein, F., Sternbach, H., and Merigan, T. C. (1970c). The antiviral activity of thiophosphate substituted polyribonucleotides *in vitro* and *in vivo*. Virology **42:** 421–428.

De Clercq, E., Wells, R. D., Grant, R. C., and Merigan, T. C. (1970d). Thermal activation of the antiviral activity of synthetic double-stranded polyribonucleotides. J. Mol. Biol., in press.

Eckstein, F. and Gindl, H. (1970). Polyribonucleotides containing a phosphorothioate backbone. Eur. J. Biochem. **13:** 558–564.

Finkelstein, M. S., Bausek, G. H., and Merigan, T. C. (1968). Interferon inducers *in vitro*: Difference in sensitivity to inhibitors of RNA and protein synthesis. Science **161:** 465–468.

Ho, M. and Ke, Y. H. (1970). The mechanism of stimulation of interferon production by a complexed polyribonucleotide. Virology **40:** 693–702.

Inman, R. B. and Baldwin, R. L. (1962). Helix-random coil transitions in synthetic DNA's of alternating sequence. J. Mol. Biol. **5:** 172–184.

Kleinschmidt, W. J., Douthart, R. J., and Murphy, E. B. (1970). Interferon production by T4 coliphage. Fed. Proc. **29:** 635.

Lampson, G. P., Tytell, A. A., Field, A. K., Nemes, M. M., and Hilleman, M. R. (1967). Inducers of interferon and host resistance. I. Double-stranded RNA from extracts of *Penicillium funiculosum.* Proc. Natl. Acad. Sci. U.S. **58:** 782–789.

Lindsay, H. L., Trown, P. W., Brand, J., and Forbes, M. (1969). Pyrogenicity of poly I·poly C in rabbits. Nature **223:** 717–718.

Nordlund, J. J., Wolff, S. M., and Levy, H. B. (1970). Inhibition of biologic activity of poly I·poly C by human plasma. Proc. Soc. Exp. Biol. Med. **113:** 439–444.

Scheffler, I. E., Elson, E. L., and Baldwin, R. L. (1968). Helix formation by dAT oligomers. I. Hairpin and straight-chain helices. J. Mol. Biol. **36:** 291–304.

Spatz, H. Ch. and Baldwin, R.-L. (1965). Study of the folding of the dAT copolymer by kinetic measurements of melting. J. Mol. Biol. **11:** 213–222.

Vilcek, J. (1970). Metabolic determinants of the induction of interferon by a synthetic double-stranded polynucleotide in rabbit kidney cells. Ann. N.Y. Acad. Sci. **173:** 390–404.

——— Ng, M. H., Friedman-Kien, A. E., and Krawciw, T. (1968). Induction of interferon synthesis by synthetic double-stranded polynucleotides. J. Virol. **2:** 648–650.

——— Rossman, T. G., and Varacalli, F. (1968). Differential effects of actinomycin D and puromycin on the release of interferon induced by double-stranded RNA. Nature **222:** 682–683.

Watson, J. D. and Crick, F. H. C. (1953). A structure for deoxyribose nucleic acid. Nature **171:** 737–738.

Wilkins, M. H. F., Stokes, A. R., and Wilson, H. R. (1953). Molecular structure of deoxypentose nucleic acids. Nature **171:** 738–740.

THE SPECIFICITY OF INTERFERON INDUCTION

CLARENCE COLBY,* MICHAEL J. CHAMBERLIN, PETER H. DUESBERG, and MELVIN I. SIMON

Department of Biology
University of California, San Diego
La Jolla, California
and
Molecular Biology and Virus Laboratory
University of California, Berkeley
Berkeley, California

INTRODUCTION

The interferon response may be induced by a variety of nonreplicating polyanionic macromolecules (Ho *et al.*, 1967). Double-stranded RNA from a variety of sources is a potent inducer of interferon (Lampson *et al.*, 1967; Tytell *et al.*, 1967; Field *et al.*, 1967a). The demonstration that double-stranded synthetic polyribonucleotides are also effective interferon inducers (Field *et al.*, 1967b) stimulated the research presented in this paper.

The most basic question to which we initially addressed ourselves was with the nature of the interferon induction mechanism. That is, does induction occur by a nonspecific process requiring only that a large polyanion enter the cell, or are there specific physical and chemical properties that the inducer molecule must satisfy in order to be effective? If the latter proved to be the case, we hoped to probe the nature of the specificity of induction by studying a variety of synthetic polynucleotides, the chemical and physical properties of which could be modified by changing either the structure of the mononucleotide subunits or the sequence of the bases.

* Senior Dernham Postdoctoral Fellow of the American Cancer Society. These studies were supported by USPHS research grants CA 04774, CA 05619, CA 11426, and CA 10802 from the National Cancer Institute, by grant GM 12010 from the Institute of General Medical Studies and by grant 15655 from the National Science Foundation.

POLY I · POLY C-INDUCED INTERFERON

When aged primary cultures of chick embryo cells were treated with 10 µg/ml of poly I · poly C, the cultures became resistant to infection with Sindbis virus, vesicular stomatitis virus, and vaccinia virus (Colby and Chamberlin, 1969). However, no interferon could be detected in the medium of these cultures and lower concentrations of the polynucleotide were not effective. These results were quite discouraging since Field *et al.* (1967b) had reported antiviral resistance in rabbit kidney cells using as little as 0.001 µg/ml poly I · poly C. Following the example of Dianzani *et al.* (1968), we found an impressive enhancement of the activity of $I_n \cdot C_n$ in chick embryo cells using DEAE-dextran. The inclusion of 10 µg/ml of the polycation in the induction medium allowed detection of antiviral activity stimulated by 0.001 µg/ml poly I · poly C. Furthermore, the culture medium removed from cells treated with 10 µg/ml $I_n \cdot C_n$ in the presence of DEAE-dextran was found to contain an antiviral macromolecule that was sensitive to trypsin, resistant to nucleases, and the synthesis of which was eliminated by pretreatment of the cells with actinomycin D. Thus, it appears that poly I · poly C stimulates the *de novo* synthesis of interferon in chick embryo cells.

We found that the rate of uptake of P^{32}-labeled poly I · poly C into chick embryo cells is increased 20-fold in the presence of DEAE-dextran and that the extent of uptake of the labeled polynucleotide is increased 25-fold (Colby and Chamberlin, 1969; Colby *et al.*, 1969). Therefore, we concluded that a primary role of the polycation in increasing the activity of the polynucleotides is to facilitate their entry into the cells.

INDUCTION BY OTHER POLYNUCLEOTIDES

Using 10 µg/ml DEAE-dextran in the induction medium, we tested 20 different synthetic polynucleotides for their ability to induce resistance to Sindbis virus. Our first experiments were done at a polynucleotide concentration of 10 µg/ml, this being 10,000-fold higher than the minimum concentration of poly I · poly C necessary for antiviral activity. Whenever a polynucleotide was found to be active, it was then tested at lower concentrations.

Table 1 presents the results of these experiments. The polynucleotides are divided into two classes: active and inactive inducers. The minimum concentration required for antiviral activity is also given for each of the active polymers. The following points should be noted. Single-stranded polyribonucleotides are inactive inducers. Secondly, helical polynucleotides containing deoxyribose residues on one or both chains are also inactive. Interferon induction was found with all helical polyribonucleotides tested. Finally, helical polyribonucleotides containing I·C or G·C base pairs are more active than these containing A·U base pairs.

Our first conclusion drawn from these studies is that the interferon induction mechanism is a specific one imposing definite chemical and physical restrictions on the inducing molecule. Secondly, these results indicate that both the secondary structure of the polynucleotides and the chemical nature of the sugar residues are involved in the specificity of induction. Finally, the chemical nature of the bases of the mononucleotide subunits does not

Table 1. Polynucleotides Tested as Interferon Inducers

Inducing polynucleotides		Noninducing polynucleotides	
Poly rI·poly rC	0.001 μg/ml	poly rI	poly rI·poly dC
Poly r(I-C)	0.005 μg/ml	poly rC	poly dI·poly rC
Poly r(I-BrC)	0.01 μg/ml	poly rG	poly dI·poly dC
Poly rG·poly rC	0.001 μg/ml	poly rA	poly dG·poly dC
Poly r(A-U)	0.01 μg/ml	poly rU	poly d(A-T)
Poly r(A-BrU)	0.01 μg/ml	poly rA·2 poly rU	poly (rA-dU)
Poly rA·poly rU	2.0 μg/ml	2 poly dI·poly dC	

appear to be important, as evidenced by there being no effect of bromination of the bases or of sequence specificity.

Based on these conclusions, we postulated the existence of an intracellular receptor site that can recognize and interact with the interferon-inducing molecule (Colby and Chamberlin, 1969). We suggested that the receptor site is a protein that can recognize both secondary structure and 2′-hydroxyl groups.

THE IMPORTANCE OF SECONDARY STRUCTURE

In their initial report, Field *et al.* (1967b) emphasized that the single-stranded polyribonucleotides, I_n, C_n, A_n, and U_n, are inactive at concentrations more than 10,000 times greater than that which allowed detectable viral interference with the helical homopolymer pair, poly I·poly C. Our results confirm these observations (see Table 1). Since single-stranded polyribonucleotides are much more sensitive to nuclease degradation than helical polynucleotides, one might theorize that sensitivity to nucleases is the essential factor in determining the efficiency of polynucleotides to act as interferon inducers. If this is true, then there should be a direct correlation between nuclease sensitivity and interferon induction amongst inducing polynucleotides.

We investigated the kinetics of induction by the helical alternating polynucleotides, $r(I\text{-}C)_n$ and $r(A\text{-}U)_n$. Not only was $r(I\text{-}C)_n$ found to be active at lower concentrations, but the appearance of the antiviral activity was more rapid (Colby *et al.*, 1969). When we studied the kinetics of cellular uptake and breakdown of P^{32}-labeled poly (I-C) and poly (A-U), we found no differences in the rate of uptake of the two polynucleotides. However, poly (A-U) is broken down more rapidly inside the cell than poly (I-C), suggesting that nuclease sensitivity may explain the difference in their inducing efficiencies.

Table 2. Ribonuclease Sensitivity of Helical Polyribonucleotides

Polynucleotide	Minimum concentration (μg/ml)	Ribonuclease sensitivity (μmoles/hr/mg RNase)
Poly I · poly C	0.001	3.8
Poly (I-C)	0.005	17
Poly (I-BrC)	0.01	1.4
Poly G · poly C	0.001	<0.02
Poly (A-U)	0.01	110
Poly (A-BrU)	0.01	41
Poly A · poly U	2.0	13

Support for this idea also comes from experiments by De Clercq *et al.* (1969). The efficiencies of interferon induction and ribonuclease sensitivities of two helical alternating copolymers were compared. Both polymers have alternating riboadenosyl and ribouridyl units. They differ from each other in that one of the polymers, $r(A^s\text{-}U)_n$, has one of the phosphate oxygens replaced by sulfur. These workers reported a remarkable enhancement of the ability of the thio-substituted polymer to induce viral resistance along with a concomitant increase in RNase resistance. The authors were tempted to relate these two phenomena causally. However, they point out that poly (A^s-U) may have a higher affinity than poly (A-U) for the intracellular receptor site.

Results of experiments done in our laboratories support the second alternative. We determined the ribonuclease sensitivities of the seven helical polyribonucleotides which we had found to be active interferon inducers. These results are presented in Table 2. Note that although poly G · poly C is extremely resistant to ribonuclease, it is no better than poly I · poly C as an inducer. Bromination of the cytosine residues in $(I\text{-}BrC)_n$ increases the RNase resistance of the polynucleotide 10-fold as compared with $(I\text{-}C)_n$, yet poly (I-C) is the better inducer. This same relationship holds qualitatively for poly (A-BrU) and poly (A-U). Finally, poly A · poly U is degraded by RNase at a rate 10 times lower than its alternating analog, poly (A-U), yet the latter is by far the superior inducer.

It is quite clear from the work presented at this Symposium that the secondary structure of the polynucleotide is a very important element with respect to the induction of interferon. Obviously, the polynucleotide must remain intact long enough to reach the intracellular site. However, the data do not support the idea that the sole requirement for inducing activity is that the polynucleotide survive RNase digestion. We would suggest that, in addition to the survival of a potential inducing molecule, the appropriate secondary structure is essential for its recognition by the receptor protein. Thus, even though poly A · poly U is more resistant to nuclease degradation, the fact that it tends to form a triple-stranded structure at physiological temperatures, which may lead to a lack of recognition by the receptor protein, seems a reasonable explanation for the observation that poly A · poly U is a less potent inducer of interferon than poly (A-U).

THE IMPORTANCE OF 2′-HYDROXYL GROUPS

If a stable helical secondary structure is all that is required of a polynucleotide so that it may act as an inducer of interferon, then we would expect the hybrid homopolymer pair $rI_n \cdot dC_n$ to be as effective an inducer as $rI_n \cdot rC_n$. This is clearly not the case. Poly rI · poly dC at 10 µg/ml on chick cells is inactive as an antiviral agent (Colby and Chamberlin, 1969). Vilcek *et al.* (1968) found that poly rI · poly dC induced resistance to vesicular stomatitis virus in rabbit kidney cell cultures. However, the concentration required was more than 10^4 times greater than that required for poly rI · poly rC (Field *et al.*, 1967b). We tested six different helical polynucleotides which contained deoxyribose as the sugar moiety on one or both chains (see Table 1). One of these, poly (rA-dU) has deoxyribose alternating with ribose. All of these polynucleotides were inactive as interferon inducers at 10 µg/ml. Since x-ray diffraction studies (Chamberlin and Patterson, 1965) have revealed no differences in the three-dimensional helical structure of poly rI · poly rC and poly rI · poly dC, we assumed that the receptor site protein must be able to recognize 2′-hydroxyl groups.

Based on these results, we postulated that DNA viruses which induce interferon do so by some agent other than a DNA-RNA hybrid (Colby and Chamberlin, 1969). Subsequently, we looked for and found RNase-resistant RNA in chick embryo cells infected with vaccinia virus (Colby and Duesberg, 1969). The RNase-resistant RNA was thoroughly characterized as double-stranded RNA by a variety of physicochemical methods and by its ability at very low concentrations to induce antiviral activity (Colby and Duesberg, 1969; Duesberg and Colby, 1969). The double-stranded RNA is vaccinia virus specific (Colby and Duesberg, 1969) and appears to be made via a DNA-dependent mechanism (Duesberg and Colby, 1969; Colby *et al.*, 1971 a). The vaccinia virus double-stranded RNA may also be made in an *in vitro* reaction using the RNA polymerase in the virion particles (Colby *et al.*, 1971 a).

The presence of double-stranded RNA in cells infected with DNA-containing viruses is not limited to eucaryotic cells. We have found virus-specific double-stranded RNA in *E. coli* infected with phage T4 (Jurale *et al.*, 1970).

The issue concerning DNA-RNA hybrids has recently been complicated by the finding of Nemes *et al.* (1969) that a DNA-RNA hybrid preparation was an active interferon inducer in both rabbits and rabbit kidney cells in culture. The amount of hybrid required was intermediate between those for poly rI·poly rC and poly rI·poly dC. The hybrid sample used by Nemes *et al.* (1969) was prepared by one of us (Chamberlin) by transcribing f1 phage single-stranded DNA with *E. coli* RNA polymerase. Robertson and Zinder (unpublished data) found that the products of such a reaction mixture contain three nucleic acid components: DNA-RNA hybrid, single-stranded RNA, and double-stranded RNA. We (Colby *et al.*, 1971 b) have confirmed these results using antibodies (Schwartz and Stollar, 1969; Stollar and Stollar, 1970), specific for DNA-RNA hybrids and for double-stranded RNA.

A reaction mixture of f1 phage DNA, *E. coli* RNA polymerase, and nucleoside triphosphates was incubated at 37 °C for 20 minutes and the nucleic acids were purified by phenol extraction and ethanol precipitation. Such a preparation is capable of inducing antiviral activity in chick cells. Equilibrium centrifugation in Cs_2SO_4 gradients allowed us to separate these nucleic acids on the basis of their buoyant densities. Purified f1 DNA-RNA hybrid, banding at 1.57 g/cm³ and cross-reacting only with the antibody specific for DNA-RNA hybrid is inactive as an interferon inducer at a concentration of 10 µg/ml. The antiviral activity resides solely in that portion of the gradients corresponding to the density of double-stranded RNA (1.64 g/cm³). This material gives a positive reaction only with the antibody specific for double-stranded RNA using a micro-complement fixation assay (Colby *et al.*, 1971 b). Therefore, we conclude that a DNA-RNA hybrid containing all eight of the mononucleotide units is an impotent interferon inducer, just as the synthetic hybrid-like polynucleotides we tested previously.

De Clercq *et al.* (1970) have recently reported increasing antiviral activity of polynucleotides by thermal activation. The most striking activation was obtained with the alternating polyribonucleotides, poly r(A-U) and poly r(I-C). These investigators also reported antiviral activity with the alternating poly deoxyribonucleotides, poly d(A-T), poly d(I-C), and poly d(G-C) at concentrations between 0.01 and 0.1 μg/ml. Thus, it appears that under certain conditions, polynucleotides containing deoxyribose may be able to induce interferon.

We have just begun to study this phenomenon and have obtained the following preliminary results that indicate that there may be significant differences between the experimental systems used. The data of De Clercq *et al.* (1970) state that in the absence of thermal activation, greater than 1 μg/ml poly r(A-U) and greater than 10 μg/ml poly r(I-C) are the minimum inhibitory concentrations required in human fibroblasts. The minimum concentrations required in chick embryo cells (see Table 1) are 0.01 μg·ml poly r(A-U) and 0.005 μg/ml poly r(I-C). We have not been able to enhance the antiviral activity of any of the seven helical polyribonucleotides listed in Table 1 by incubation at 37 °C with the exception of poly r(I-C). The activation of $(I\text{-}C)_n$ amounted to less than a one-log reduction in virus yield as compared with $(I\text{-}C)_n$ which has not been thermally activated. We also attempted to thermally activate the synthetic hybrid pairs poly rI · poly dC and poly dI · poly rC as well as the deoxy polymers, poly dI · poly dC and poly d(A-T) at 10 μg/ml. We found that a 2 hour incubation at 37 °C had no effect on their ability to induce the antiviral state.

There are two obvious explanations for the discrepancy of results from these two laboratories. Either the polynucleotides used in our labs are quite different or there is a fundamental difference in the specificity of interferon induction between human skin fibroblasts and chick embryo cells. Fortunately, both possibilities may be experimentally attacked and we are confident that collaborative efforts between our two laboratories will give us the answers to these questions.

SUMMARY

At low polynucleotide concentrations we have presented evidence for a high degree of specificity for the processes involving interferon induction. Helical polyribonucleotides are very efficient inducers, whereas both natural and synthetic DNA-RNA hybrids are not. In connection with this, we have found double-stranded RNA in vaccinia virus-infected cells. We believe that the helical secondary structure of the interferon inducing molecule is important not only for intracellular stability, but also for recognition by the receptor site protein. The role of 2′-hydroxyl groups in contributing to the specificity of induction remains unknown. It is clear that the receptor protein can distinguish between ribose- and deoxyribose-containing polymers. However, we do not know whether the recognition involves the chemical nature of the sugar moeity itself or the physical differences in secondary structure of polynucleotides containing different sugars.

References

Chamberlin, M. J. and Patterson, D. L. (1965). Physical and chemical characterization of the ordered complexes formed between polyinosinic acid, polycytidylic acid and their deoxyribo-analogs. J. Mol. Biol. **12:** 410–428.

Colby, C. and Chamberlin, M. J. (1969). The specificity of interferon induction in chick embryo cells by helical RNA. Proc. Natl. Acad. Sci. **63:** 160–167.

Colby, C. and Duesberg, P. H. (1969). Double-stranded RNA in vaccinia virus infected cells. Nature **222:** 940–944.

Colby, C. Chamberlin, M. J., and Duesberg, P. H. (1969). The induction of interferon. Third International Symposium on Medical and Applied Virology, Ft. Lauderdale, Florida.

Colby, C., Jurale, C., and Kates, J. R. (1971a). Studies on the mechanism of synthesis of vaccinia virus double-stranded RNA *in vivo* and *in vitro*. J. Virol., in press.

Colby, C., Stollar, B. D., and Simon, M. I. (1971b). Interferon induction: DNA-RNA hybrid or double-stranded RNA? Nature, in press.

De Clercq, E., Eckstein, F., and Merigan, T. C. (1969). Interferon induction increased through chemical modification of a synthetic polyribonucleotide. Science **165:** 1137–1139.

De Clercq, E., Wells, R. D., and Merigan, T. C. (1970). Increase in antiviral activity of polynucleotides by thermal activation. Nature **226:** 364–366.

Dianzani, F., Cantagalli, P., Gagnoni, S., and Ritz, G. (1968). Effect of DEAE-dextran on production of interferon induced by synthetic double-stranded RNA in L cell cultures. Proc. Soc. Exptl. Biol. Med. **128:** 708–710.

Duesberg, P. H. and Colby, C. (1969). On the biosynthesis and structure of double-stranded RNA in vaccinia virus infected cells. Proc. Natl. Acad. Sci. **64:** 396–403.

Field, A. K., Lampson, G. P., Tytell, A. A., Nemus, M. M., and Hilleman, M. R. (1967a). Inducers of interferon and host resistance. IV. Double-stranded replicative form RNA from *E. coli* infected with MS2 coliphage. Proc. Natl. Acad. Sci. **58:** 2102–2108.

Field, A. K., Tytell, A. A., Lampson, G. P., and Hilleman, M. R. (1967b). Inducers of interferon and host resistance. II. Multistranded synthetic polynucleotide complexes. Proc. Natl. Acad. Sci. **58:** 1004–1010.

Jurale, C., Kates, J. R., and Colby, C. (1970). Isolation of double-stranded RNA from T4 phage infected cells. Nature **226:** 1021–1029.

Ho, M., Fantes, K. H., Burke, D. C., and Finter, N. B. (1967). *In* Interferons. Ed. by N. B. Finter, John Wiley, New York, 181–201.

Lampson, G. P., Tytell, A. A., Field, A. K., Nemes, M. M., and Hilleman, M. R. (1967). Inducers of interferon and host resistance. I. Double-stranded RNA from extracts of *Penicillium funiculosum*. Proc. Natl. Acad. Sci. **58:** 782–789.

Nemes, M. M., Tytell, A. A., Lampson, G. P., Field, A. K., and Hilleman, M. R. (1969). Inducers of interferon and host resistance. VII. Antiviral efficacy of double-stranded RNA of natural origin. Proc. Soc. Exptl. Biol. Med. **132:** 784–789.

Schwartz, E. F. and Stillar, B. D. (1969). Antibodies to polyadenylate-poly-uridylate copolymers as reagents for double-stranded RNA and DNA-RNA hybrid complexes. Biochem. Biophys. Res. Comm. **35:** 115–120.

Stollar, V. and Stollar, B. D. (1970). Immunochemical measurement of double-stranded RNA of uninfected and arbovirus-infected mammalian cells. Proc. Natl. Acad. Sci. **65:** 993–1000.

Tytell, A. A., Lampson, G. P., Field, A. K., and Hilleman, M. R. (1967). Inducers of interferon and host resistance. III. Double-stranded RNA from reovirus type 3 virions. Proc. Natl. Acad. Sci. **58:** 1719–1722.

Vilcek, J., Ng, M. H., Friedman-Kien, A. E., and Krawciw, T. (1968). Induction of interferon synthesis by synthetic double-stranded polynucleotides. J. Virol. **2:** 648–650.

STRUCTURAL REQUIREMENTS OF RIBOPOLYMERS FOR INDUCTION OF HUMAN INTERFERON: EVIDENCE FOR INTERFERON SUBUNITS*

William A. Carter and Paula M. Pitha

Departments of Medicine and Microbiology
The Johns Hopkins University
School of Medicine

Is the double helix necessary to trigger the signal for production of human interferon (IF)? How rapid is the triggering event? How do polybasic enhancers work? What is the nature of the end product of this reaction IF? Specifically, is the high MW product induced by artificial genomes structurally related to the lower MW, viral-induced, protein? Mechanisms of enhancement and structure of the IF protein will be emphasized in this report.

Human IF is released when a certain critical concentration of poly IC (about 5×10^{-5} *M*) is reached (Figure 1). The concentration (poly IC)-production curve is sigmoidal, suggesting a cooperative effect in binding or in a later step in production, once a threshold is reached. Intracellular protection is measurable at concentrations of ribopolymer 3 logs less than that necessary to trigger extracellular production.

To determine the requirement of double helix in triggering the production mechanism, the complex formed from high MW poly C and low MW oligo I was used. These have the same chemical composition as poly IC, but the stability of the double helix differs, so that one can manipulate secondary structure, leaving the primary structure intact. Polynucleotides are bound

* This work was supported by USPHS Research Career Development Award (to William A. Carter) A 142565 and research grant CA 06973; and grants from the Council for Tobacco Research, USA, No. 694, and American Cancer Society, Maryland Division, No. 69-01.

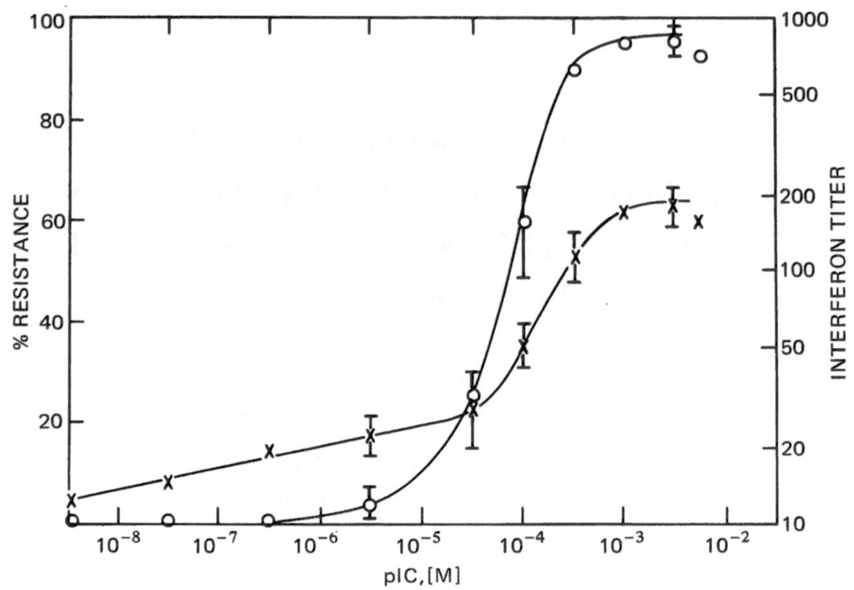

Fig. 1. IF production in monolayers of human skin cells as a function of the poly IC concentration. Monolayers were exposed to different poly IC concentrations for 1 hour at 37°, removed, and the medium replaced with fresh MEM. IF production and intracellular resistance were measured after 22 hours of incubation. *Legend:* ×, percent of cell resistance; ○, IF titer. The range of values for two independent experiments (made in duplicate) are given.

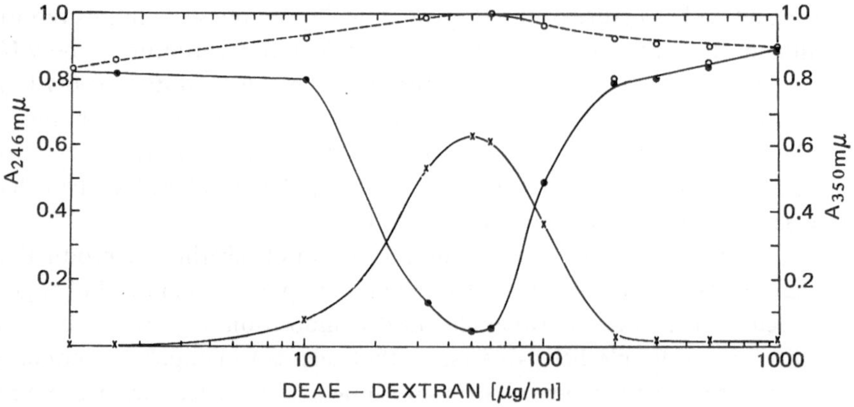

Fig. 2. The mixing curve obtained with a constant amount of poly IC (1.15×10^{-4} *M*) and a varying amount of DEAE-dextran (in 0.01 *M* sodium phosphate *p*H 7.4, 0.001 *M* $MgCl_2$, 0.15 *M* NaCl). *Legend:* ○, optical density, 246 mμ; ×, optical density, 350 mμ; •, optical density (at 246 mμ) in supernatant after centrifugation at 23,000 × g for 30 minutes.

well at low temperatures (that is, 2–10°), but further processing of these macromolecules is arrested at these low temperatures. Human fibroblasts were exposed to complimentary oligo:polynucleotide complexes above and below their melting points (Tm's) and then assayed for IF production at 37°. Poly C:I_6 (poly C complexed with hexainosinate) (Tm 5°) induces IF production only when cells are exposed to the complex below, but not above, its Tm. Thus, only the duplex component is active, and it is required for only a brief triggering event probably on the surface of the cell. It is probably not required for subsequent steps in production.

Further variation in secondary structure of ribopolymers can be achieved by polybasic enhancers, such as neomycin or DEAE-dextran. These changes in secondary structure augment IF yield. Neomycin increases the Tm of both poly IC and AU and protects the complexes against endonucleases. It potentiates human IF formation 10-fold. DEAE-dextran forms complexes with double-stranded ribopolymers, the solubility of which depends on the ratio of components (Figure 2). In this mixing curve the concentration of poly IC has been held constant (1.15×10^{-4} *M*). When the input ratio of DEAE-dextran nitrogen to phosphates of poly IC is approximately 1:1, a large increase in scattered light (350 mμ) is noted upon mixing, which indicates the formation of aggregates. The complexes are resolubilized by an excess of either reactant. Maximum aggregation occurs when equimolar amounts of positive and negative charges are present and coulombic repulsion is thus minimal.

Unlike neomycin, the addition of DEAE-dextran does not stabilize the secondary structure of poly IC (Figure 3). Poly IC alone displays a typical cooperative thermal transition (Tm 67°) and reannealing reaction. The Tm of poly IC–DEAE-dextran complex (formed with a phosphate-nitrogen ratio 1.0:0.2) is similar but at 70° an insoluble complex is formed. Below this temperature, the reaction is reversible. With the cationic component in excess (phosphate-nitrogen 1:4), the complex is actually destabilized. The Tm, although cooperative, is considerably lower (53°). This destabilization of secondary structure suggests DEAE-dextran may be different in its action from other polybasic enhancers which stabilize the poly IC complex.

Different poly IC–DEAE-dextran complexes vary systematically in their specific activity (Figure 4). The degree of intracellular protection and level of IF production rise coordinately with increasing complex formation. The net effect is to increase the "apparent" concentration of poly IC about 10-fold. The abrupt decrease in production with excess DEAE-dextran most likely indicates a masking of a portion of the polymer necessary to trigger the release mechanism. Specific poly IC-dextran complexes, isolated by density gradient sedimentation, are active.

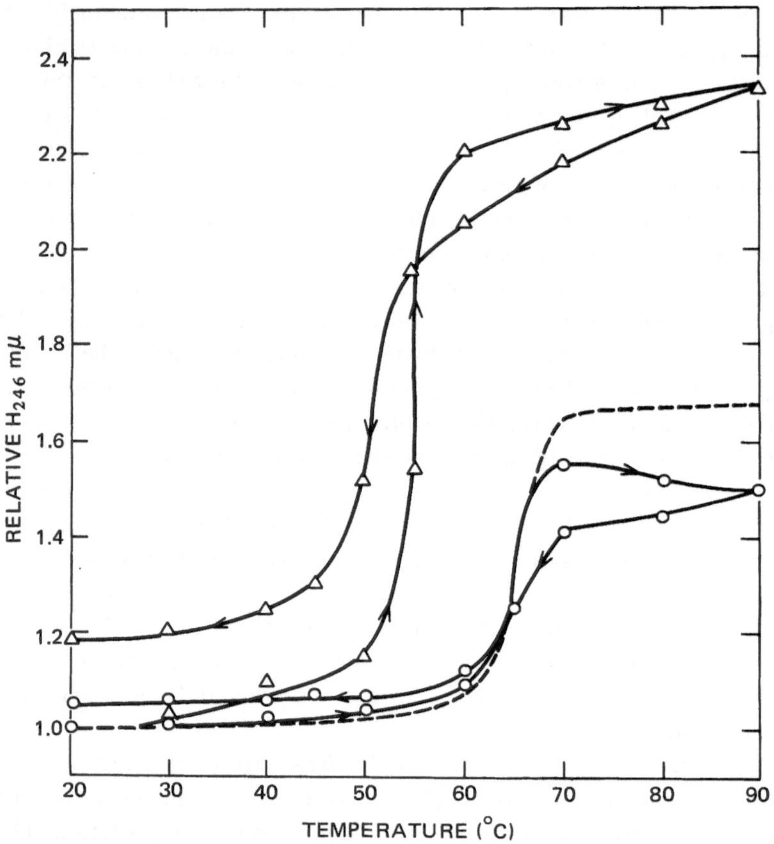

Fig. 3. Temperature-dependent heating and cooling curves of poly IC–DEAE-dextran complexes. $H_{246\ m\mu}$ is the relative hyperchromicity at 246 mμ. *Legend:* ---, poly IC alone; -○-, poly IC + DEAE-dextran (phosphate-nitrogen ratio 1:0.2); -△-, poly IC + DEAE-dextran (phosphate-nitrogen ratio 1:4).

Does DEAE-dextran protect against endonucleases or exonucleases? Poly IC–DEAE-dextran complex is protected against endonucleases such as pancreatic RNase (Figure 5, left panel). The initial rate of hydrolysis decreases as the amount of cationic component in the complex rises. In contrast, phosphorolysis (Figure 5, right panel), measured with the exonuclease, polynucleotide phosphorylase, is actually enhanced. The rate of phosphorolysis increases as more DEAE-dextran is present in the complex. Since polynucleotide phosphorylase prefers single-stranded structure, these results further support a destabilization in secondary structure. Direct comparison of the two reactions cannot be made since they are carried out

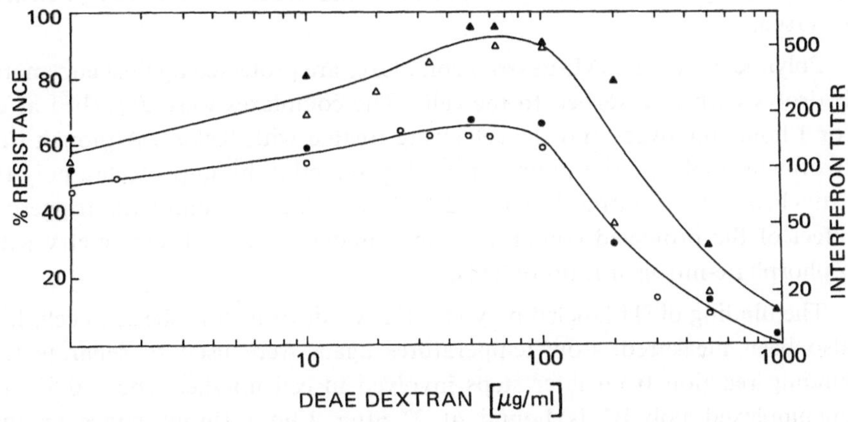

Fig. 4. IF production induced by different poly IC–DEAE-dextran complexes; the concentration of rI:rC is constant (1.15×10^{-4}). The open marks represent values obtained from cell-polymer interactions occurring during an incubation at 2°. The closed marks represent similar studies carried out by a 37° incubation. *Legend:* △ and ▲, IF titer; ●, percent cell resistance.

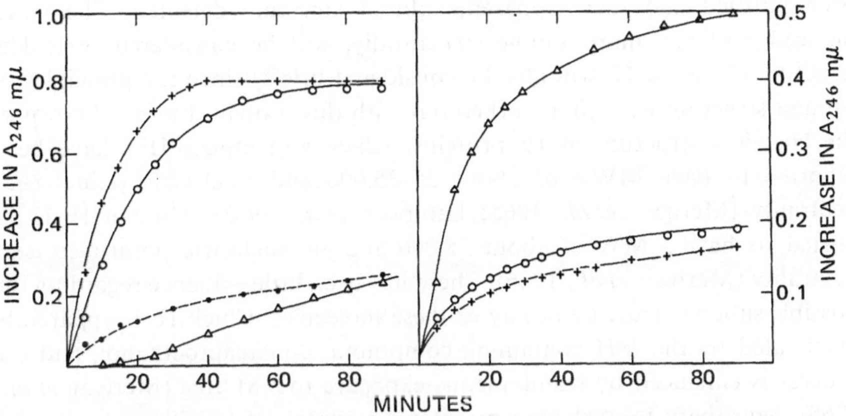

Fig. 5. The kinetics of enzymatic hydrolysis of poly IC–DEAE-dextran complexes. *Left panel:* The rate of endonucleolytic hydrolysis of poly IC-dextran complexes. Poly IC (10^{-4} *M*) was incubated with 0.8 μg pancreatic RNase (volume 3 ml, in 0.01 *M* phosphate *p*H 7.4, 0.001 *M* $MgCl_2$, 0.15 *M* NaCl, at 37°). Phosphate (poly IC)-nitrogen (dextran) ratios were varied: ×–×, ratio 1:0; ○–○, ratio 1:0.2; ●--●, ratio 1:1; △–△, ratio 1:4. *Right panel:* The rate of phosphorolysis of poly IC-dextran complexes using polynucleotide phosphorylase (*M. lysodeikticus*). A mixture (3 ml) containing 10 units enzyme, 10^{-4} *M* poly IC in 0.05 *M* Tris hydrochloride *p*H 8.2, 0.005 *M* $MgCl_2$, 0.017 *M* NaH_2PO_4 was incubated at 37°. The phosphate-nitrogen ratios are the same as those indicated in the left panel.

at different *p*H's and ionic strengths, but the order of reactivity is clearly reversed.

Polynucleotide–DEAE-dextran complexes are protected against enzymatic hydrolysis when adsorbed to the cells. The complexes were adsorbed at 2° for 1 hour, removed, and the cells were treated with RNase. Although this treatment abolishes the initiation of IF production by unprotected poly IC (attached but not intracellular at 2°), RNase does not blunt the triggering effect of the protected complex (formed under optimal stoichiometry with a phosphate-nitrogen ratio of 1:1).

The binding of ^3H labeled poly I:C-DEAE-dextran complexes to cells has also been measured. Low temperatures again were used to separate the binding reaction from later steps involved in cell uptake. About 0.5% of uncomplexed poly IC is bound at 2° after 1 hour (input concentration 6×10^{-4} *M*); 20 times more of the macromolecular aggregate is bound when DEAE-dextran is present. Thus, this enhancer potentiates IF release by delivering more of the active poly IC complex to the triggering site. It does so by increasing the uptake of the poly IC complex (probably by increasing its apparent mass) and rendering it less susceptible to endonucleases.

The IF proteins induced by both virus and synthetic ribopolymers have been purified and their apparent subunit structure detected. The viral-induced proteins, more simple structurally, will be considered first. The structure of mouse IF will also be considered briefly since the principles of subunit structure were first worked out with this protein. Little is known of the chemical structure of IF proteins. Chick and mouse IF's have been reported to have MW's of about 25–35,000 and isoelectric points near neutrality (Merigan *et al.*, 1965; Lampson *et al.,* 1963). Human IF is reported to have a MW of about 25,000 and an isoelectric point also near neutrality (Merigan *et al.*, 1966). There has been little evidence regarding the possible subunit structure of any of these molecules. Chick IF is apparently inactivated by the -SH containing compound, *β*-mercaptoethanol, and the process is enhanced by simultaneous exposure to 8 M urea (Merigan *et al.*, 1965). Sensitivity to -SH reagents may be explained by either a reduction of disulfide bonds or a dependence on metal ions for function.

As an initial step in the further characterization of IF's, the human IF's have been extensively purified by means of the new technique of electrofocusing in polyacrylamide gels. The viral-induced protein, although initially homogeneous on gel chromatography, displays two components on electrophoresis in *p*H gradients. MW measurements of the different forms are consistent with the hypothesis that native IF is a dimer of similar or identical subunits (Carter, 1970).

The proteins were generally recoverable in high yields (50–75%). It is difficult to measure specific activity accurately because of the low protein content (after isoelectric focusing) and thus to assign a minimal value. However, a lower limit of specific activity can be determined. By these estimates, the cumulative purification of human IF is 1500-fold (Form A)

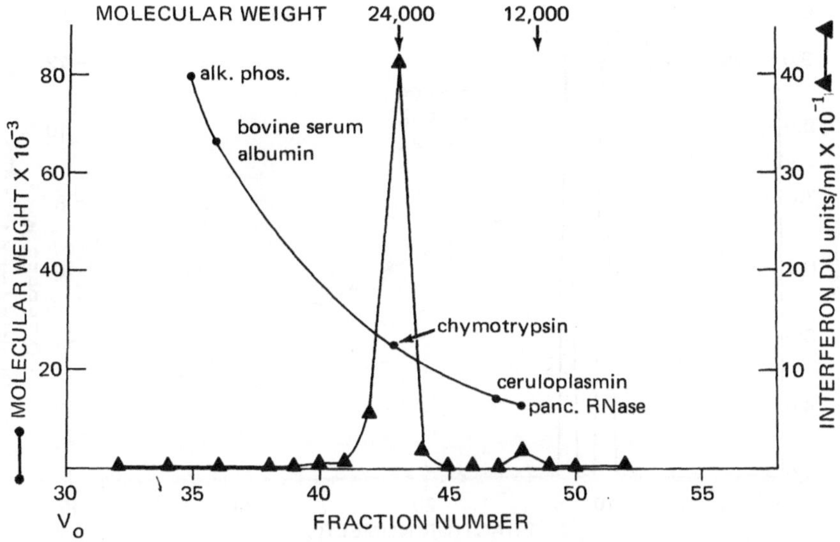

Fig. 6. Gel chromatography of human fetal IF. Unpurified human fetal IF (specific activity 2000 units/mg protein) was chromatographed on Sephadex®, G-200 at 4 °C. A solution of 0.05 *M* NaCl, 0.01 *M* Tris *p*H 7.4 was used as the eluent with a flow rate of about 2.5 ml/hr. Fractions containing 0.5 ml were collected and assayed in neonatal skin cells using the colorimetric method. About 85% of the 500 units applied was recovered. *Legend:* ●–●, MW standards; ▲–▲, IF (units/ml × 10^{-1}).

yielding a specific activity of 2.1 million units/mg protein, and 280-fold (Form B) with a specific activity of 386,000 units/mg.

In gel chromatography of unpurified IF's, mouse IF behaved as a single molecular species with a MW of 38,000. Human IF eluted predominantly in a position corresponding to a MW of 24,000 (Figure 6). Thus, both IF's exist largely as single molecular species at this stage of purification.

Isoelectric focusing of mouse and human IF was performed on the IF's obtained from gel chromatography. Mouse IF activity was found in two peaks (A and B) containing equal activity with isoelectric points of 7.15 and 7.35, respectively (Figure 7). Essentially 100% of the input activity was recoverable and the specific activity of the two peaks was the same. With increasing concentration of protein, peak A contained more activity than peak B. The observed *p*I's were constant over the range of protein used.

Human IF had more acidic *p*I's (Figure 8), three peaks were recovered with isoelectric points of 5.35, 5.60, and 5.70. These fractions, unlike those obtained with mouse, have different percentages of total activity. The major fractions, Forms A (*p*I 5.35) and B (*p*I 5.60), contained 99% of

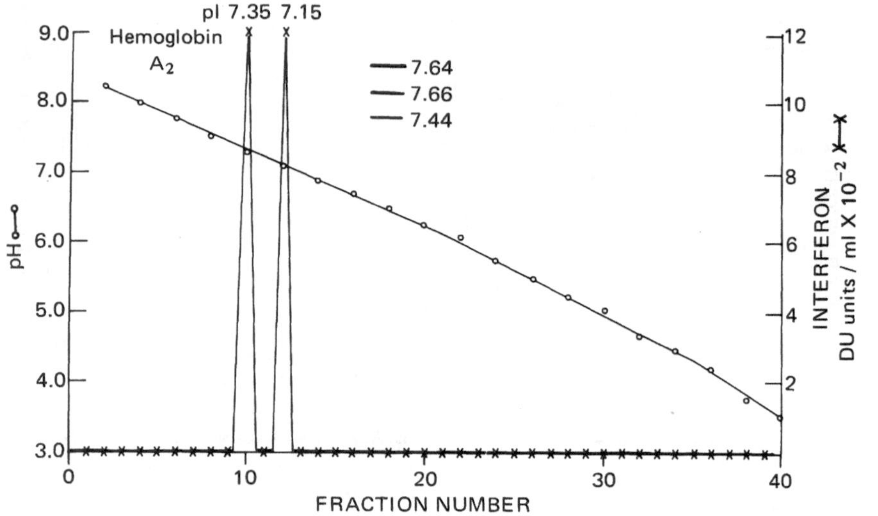

Fig. 7. Isoelectric focusing of mouse IF. Mouse IF (specific activity 100,000 units/mg) was focused on *p*H 3–10 ampholyte gradients. After electrophoresis, 7-mm gel segments were cut out for assay. Two thousand four hundred units (25 μg protein) were applied and all the activity was recovered in two equally active fractions. *Legend:* ○–○, *p*H gradient; ×–×, IF (units/ml $\times$ 10^{-2}).

the activity. A minor fraction, Form C (*p*I 5.70), contained 1% of the total activity. The relative activity in Forms A or B varied but Form A usually contained 50–85% of the total activity. Thus, Form A appears to be the more stable molecular species.

MW measurements were made on IF peaks obtained from isoelectric focusing (Figure 9). Mouse Form B yield 86% of the original 38,000 MW species and 14% of a previously undetected 19,000 MW species. Form A yielded 90% of the 19,000 MW species and 10% of the original 38,000 MW form. Under similar conditions, crude mouse IF was an exclusively 38,000 MW species. Thus, electrofocusing has separated two forms of the molecule.

Similar experiments were performed with human IF. Form A (Figure 10) yielded 90% of a 12,000 MW species (previously detected in trace amounts) and 10% of the original 24,000 MW form. It has not been possible to study accurately human Form B by the techniques previously applied to the two forms of mouse IF because of the small amount of activity. Indirect evidence was obtained indicating that Form B (*p*I 5.60) is the undissociated 24,000 species by devising conditions in which the native molecule

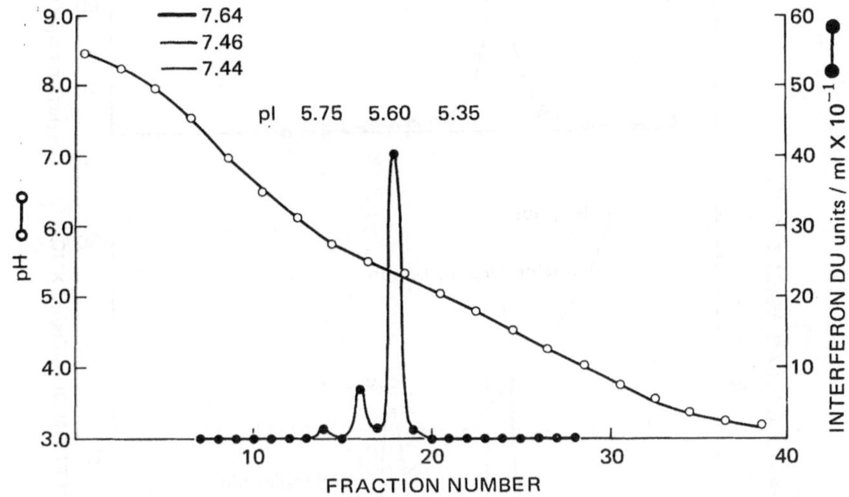

Fig. 8. Isoelectric focusing of human fetal IF. Human fetal IF (specific activity 20,000 units/mg protein) was focused on ampholyte gradients as described in Figure 7. Eight hundred units (40 μg) were applied with a recovery of 65%. In duplicate gel slices, protein content was estimated by peptide bond absorbance. Human fetal IF was usually 50–75% recoverable in the focusing step, but yields as low as 12% were obtained. *Legend:* o–o, *p*H gradient; •–•, human IF (units/ml × 10^{-1}).

(MW 24,000) could be electrophoresed without dissociation. Under these conditions only one species (*p*I 5.60) is formed, which has a MW of 24,000.

Both mouse and human IF's are separable into two active forms, one of which is twice the MW of the other. The simplest explanation is that the larger MW form is a dimer and the lower MW form the monomeric unit. Although it is possible that the monomeric units are different, it seems more likely that the dimers are made up of identical subunits. Only one monomeric species is seen, and it has exactly one-half the MW of the larger species. (The dimer to monomer interconversion may be symbolically written as $I_2 \rightleftarrows 2I$.)

To test the proposed transition, mouse IF Form B was isolated by isoelectric focusing, dialyzed versus low salt, and refocused at its isoelectric point. About one quarter of the activity was now present as Form A

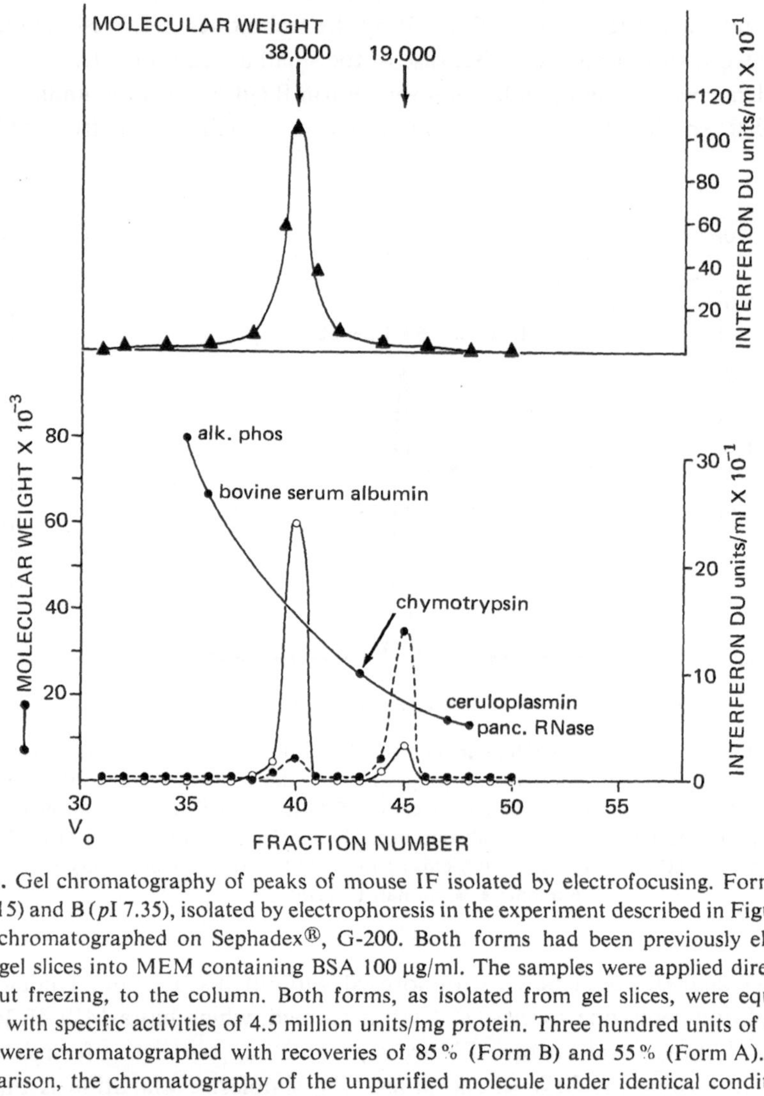

Fig. 9. Gel chromatography of peaks of mouse IF isolated by electrofocusing. Forms A (*p*I 7.15) and B (*p*I 7.35), isolated by electrophoresis in the experiment described in Figure 7 were chromatographed on Sephadex®, G-200. Both forms had been previously eluted from gel slices into MEM containing BSA 100 μg/ml. The samples were applied directly, without freezing, to the column. Both forms, as isolated from gel slices, were equally active with specific activities of 4.5 million units/mg protein. Three hundred units of each form were chromatographed with recoveries of 85% (Form B) and 55% (Form A). For comparison, the chromatography of the unpurified molecule under identical conditions is replotted in the top panel. *Legend: Top panel:* ▲–▲, unpurified IF (units/ml $\times$ 10^{-1}), *Bottom panel:* ●–●, MW standards; ●--●, IF (Form A) (units/ml $\times$ 10^{-1}); ○–○, IF (Form B) (units/ml $\times$ 10^{-1}).

(Figure 11). Form B can thus continue to dissociate into A, the presumptive monomeric species.

Something associated with the focusing procedure causes an apparent dissociation of IF. Could it be the low ionic strength which prevails through-

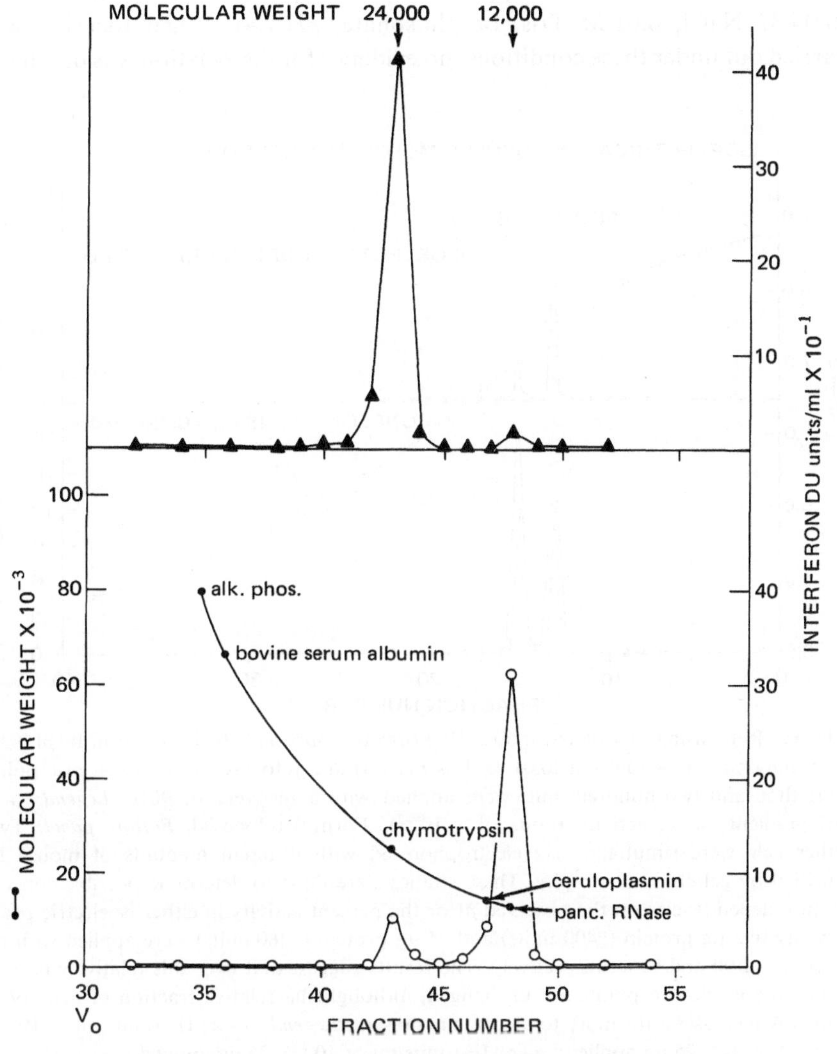

Fig. 10. Gel chromatography of human fetal IF (Form A) isolated by electrofocusing. Form A (*p*I 5.35), isolated by electrophoresis, was chromatographed on Sephadex®, G-200. Four hundred units of Form A (specific activity 2×10^6 units/mg protein) were applied and all of the activity was recovered. In the top panel, the chromatography of unpurified human fetal IF under identical conditions, is replotted for comparison. *Legend: Top panel:* ▲–▲, crude human IF (units/ml $\times 10^{-1}$), *Bottom panel:* ●–●, MW standards; ○–○, purified human IF Form A (units/ml $\times 10^{-1}$).

out the electrofocusing procedure? The idea of a salt dependency for stability of the dimeric form was supported by attempts to focus in higher salt (0.05 *M* NaCl, 0.01 *M* Tris, or phosphate, *p*H 7.4). When focusing was carried out under these conditions, no evidence for dissociation was obtained.

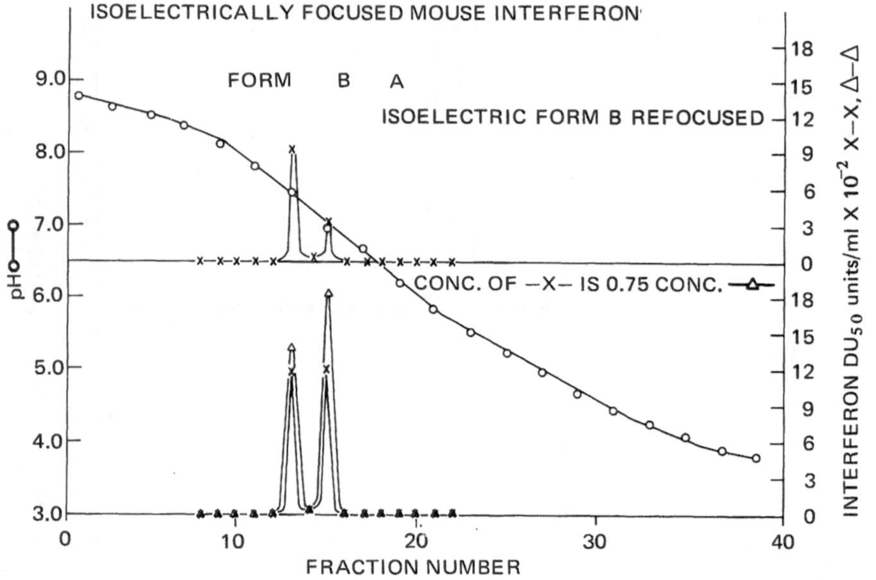

Fig. 11. Refocusing of isolated mouse IF Form B. *Top panel:* Isolated Form B (*p*I 7.35) was dialyzed versus low salt (0.01 *M* Tris *p*H 7.4) and refocused at its isoelectric point. One thousand two hundred units were applied with a recovery of 90%. *Legend:* o–o, *p*H gradient; ×–×, activity (units/ml × 10^{-2}), Form B refocused. *Bottom panel:* Two other gels were simultaneously electrophoresed with different amounts of mouse IF purified by gel chromatography. These studies were done to determine possible concentration dependencies on the observed *p*I or the percent activity in either isoelectric peak. Twenty-five μg protein (2400 units) and 35 μg protein (3360 units) were applied with recoveries of 90 and 95%, respectively. The results suggest that over this relatively narrow range the isoelectric points are unchanged, although the relative fraction of activity in Form A increases with more total protein present. *Legend:* ×–×, IF (units/ml × 10^{-2}), 25 μg applied; Δ–Δ, IF (units/ml × 10^{-2}), 35 μg applied.

Mouse IF yielded only Form B and human IF yielded Form B (*p*I 5.6) as the principal species.

These observations suggest that low salt conditions prevailing during focusing allow the detection of subunits. If this is so, dissociation should be demonstrable by gel chromatography. Mouse IF Form B was dialyzed versus low salt and chromatographed (Figure 12). The dialysis step alone

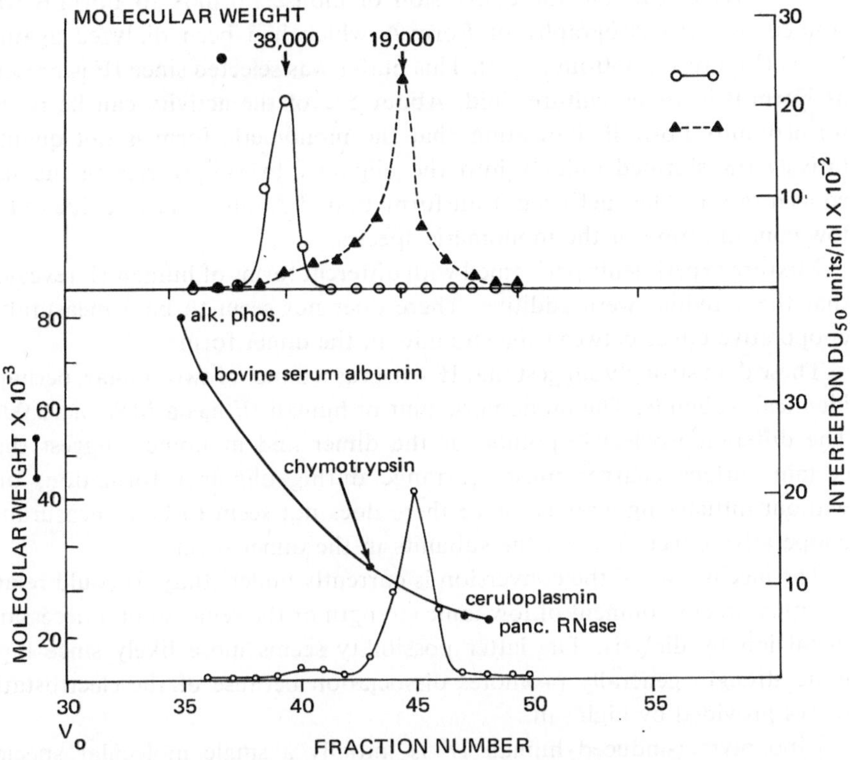

Fig. 12. Gel chromatography of mouse IF Form B dialyzed versus low salt. *Top panel:* Mouse IF was dialyzed versus high (0.05 *M* NaCl, 0.01 *M* Tris *p*H 7.4, 0.001 *M* $MgCl_2$) or low (0.01 *M* Tris *p*H 7.4) salt for 16 hours before chromatography on Sephadex®, G-200 (equilibrated with 0.05 *M* NaCl, 0.01 *M* Tris *p*H 7.4). In each instance, crude IF (specific activity 16,000 units/mg protein) was used, and over 95% of the 2800 units applied to the column were recovered. No detectable loss of activity occurred on dialysis versus low salt. *Legend:* ○–○, mouse IF (units/ml $\times$ 10^{-2}), dialyzed versus high salt; △–△, mouse IF (units/ml $\times$ 10^{-2}), dialyzed versus low salt. *Bottom panel:* To study the proposed reverse reaction, Form A→B, Form A was isolated in the experiment described in the top panel and dialyzed versus higher salt (0.05 *M* NaCl, 0.01 *M* phosphate *p*H 7.4 and 0.001 *M* $MgCl_2$) with and without fetal calf serum (6% v/v) for 20 hours. The proteins were rechromatographed on Sephadex ®, G-200. The results were unchanged by the presence of FCS. *Legend:* ●–●, MW standards; ○–○, mouse IF Form A (units/ml $\times$ 10^{-2}), after dialysis versus high salt containing 6% (v/v) FCS.

was sufficient to dissociate the molecule, as evidenced by the displacement of IF to a 19,000 MW peak. The 38,000 MW region contained little activity. There is no concomitant loss of activity, indicating that the smaller molecular form must be equally active. It provides additional support for subunits of similar structure.

The reverse reaction, the conversion of mouse Form A to Form B was studied by chromatography of Form A which had been dialyzed against MEM (Figure 12, bottom panel). This buffer was selected since IF is present as Form B in tissue culture fluid. About 5% of the activity can be transformed into Form B, indicating that the monomeric form is not quantitatively transformed quickly into the oligomer by simply raising the salt concentration. This inefficient transformation of A into B may be due to the low concentration of the monomeric species.

Mixture experiments performed with different forms of human IF revealed that the activities were additive. There does not seem to be a measurable cooperative effect between the subunits in the dimer form.

These data strongly suggest that IF exists as a dimer of two similar, perhaps identical, subunits. The monomeric unit of human IF has a MW of 12,000. The different isoelectric points for the dimer and monomer suggest that certain surface charges must rearrange during oligomer formation, but without influencing activity, since there does not seem to be a measurable cooperative effect between the subunits in the dimer form.

The mechanism of the conversion is currently under study. It could relate to either an environment of low ionic strength or the removal of a necessary metal ion by dialysis. The latter possibility seems more likely since high ionic strength generally promotes dissociation because of the electrostatic forces provided by high salt.

Ribopolymer-induced human IF is initially a single molecular species with a MW of 96,000 (Figure 13, top panel). This protein was induced by the complex of poly I:C-dextran formed under optimal complex conditions. Like the IF protein induced by virus, it was dissociable into smaller MW forms on dialysis versus low salt (Figure 13, bottom panel). On exposure to low salt, the predominant molecular form eluted in a position corresponding to a MW of 24,000. Small amounts of activity were noted corresponding to MW's of 96,000 and 12,000. (The protein is recoverable in yields of 20–65%.) These results suggest that polymer-induced human IF is a higher oligomeric form of the protein induced by virus. This explanation should be considered tentative until surface charges of these different molecular species are determined. As with viral induced IF, there is no measurable cooperative effect between subunits in the high oligomeric form.

The results indicate that native IF is not a group of heterogeneous proteins. Rather, native IF may exist in multiple oligomeric forms. Different inducers seem to trigger the release of different oligomeric species. On the basis of similarities in activities of homologous IFs induced by different signals, one could have deduced that there must be, at the very least, large regions of shared polypeptide sequence. Further studies will have to clarify the im-

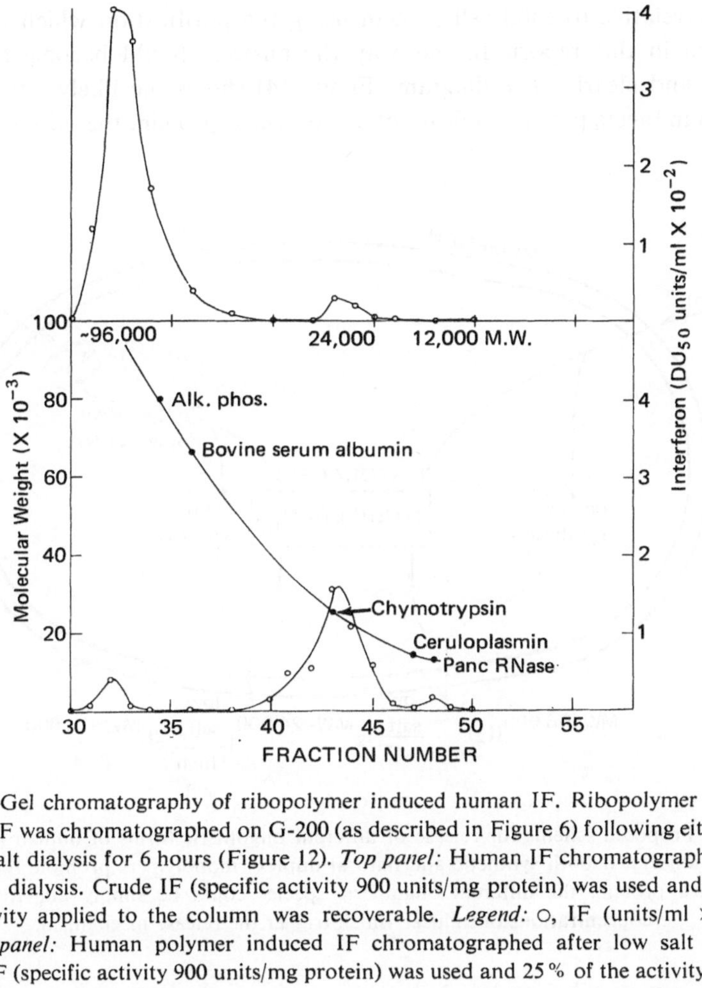

Fig. 13. Gel chromatography of ribopolymer induced human IF. Ribopolymer induced human IF was chromatographed on G-200 (as described in Figure 6) following either high or low salt dialysis for 6 hours (Figure 12). *Top panel:* Human IF chromatographed after high salt dialysis. Crude IF (specific activity 900 units/mg protein) was used and 90% of the activity applied to the column was recoverable. *Legend:* o, IF (units/ml $\times 10^{-2}$). *Bottom panel:* Human polymer induced IF chromatographed after low salt dialysis. Crude IF (specific activity 900 units/mg protein) was used and 25% of the activity applied to the column was recoverable. In other experiments 50–60% of the activity was recoverable with similar results. *Legend:* ●-●, MW standards; o-o, IF (units/ml $\times 10^{-2}$).

portant relationships between subunit structure and the biological properties of IF.

What is the mechanism by which different inducers trigger the release of different oligomeric forms? Studies with inhibitors of RNA and protein synthesis suggest that IF production may follow either (a) de novo synthesis or (b) release of "preformed" polypeptide (Vilcek, 1970). It is hard to distinguish between the alternatives without performing the critical experiment: grow cells in the presence of labeled amino acids added substantially before, or after, exposure to inducer, and measure specific activity of IF protein relative to total cell protein using the purification which has been outlined in this report. In this way the answer should become available simply and clearly. The diagram (Figure 14) shows the likely alternatives. If both inducers promote release of a preformed protein, the different oligo-

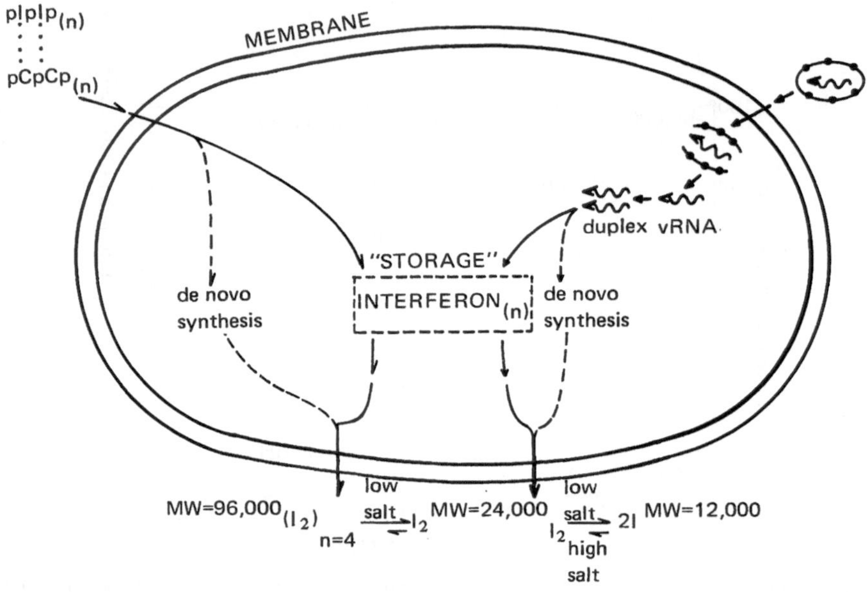

Fig. 14. Proposed scheme for release of different oligomeric forms of human IF by different inducers. If both synthetic and natural duplex ribopolymers promote release of a preformed protein, the different oligomeric species could be simply due to different concentrations of inducer which trigger the release mechanism.

meric species could be due to different concentrations of inducer which trigger the release mechanism. Are there repressors of IF? This seems likely but the evidence is indirect (see Vilcek, 1970) and not completely convincing. Nevertheless, this evidence constitutes the basis for an assay of repressor and with such an assay these activities can be characterized in simple, chemical terms.

ACKNOWLEDGEMENTS

The authors wish to thank Drs. Hamilton O. Smith, Bernard Weiss, and Thomas J. Kelly, Jr., for their helpful suggestions during the course of this work, and Mrs. Susan Doane for the preparation of this manuscript.

References

Carter, W. A. (1970). Interferon: evidence for subunit structure. Proc. Natl. Acad. Sci. (in press).

Lampson, G. P., Tytell, A. A., Nemes, M. M., and Hilleman, M. R. (1963). Purification and characterization of chick embryo interferon. Proc. Soc. Exp. Biol. Med. **112:** 468.

Merigan, T. C., Winget, C. A., and Dixon, C. B. (1965). Purification and characterization of vertebrate interferons. J. Mol. Biol. **13:** 679.

Merigan, T. C., Gregory, D. F., and Petralli, J. K. (1966). Physical properties of human interferon prepared *in vitro*. Virology **29:** 512.

Vilcek, J. (1970). Cellular mechanisms of interferon production. J. Gen. Physiol. (in press).

IMMUNOGENIC POLYNUCLEOTIDES*

L. D. Hamilton

Division of Microbiology, Medical Research Center, Brookhaven National Laboratory, Upton, New York

STRUCTURAL CONSIDERATIONS

X-ray diffraction of alkali-metal salts of DNA in the solid state has shown that the DNA molecule can exist in several well-defined conformations (Wilkins, 1963; Hamilton, 1968). Originally one thought that the highly crystalline *A* form of the Na salt of DNA was stable only at low humidities; at high humidities it transformed into the *B* form (Franklin and Gosling, 1953). It is now clear that excess salt in the fiber is necessary besides higher humidity for the $A \rightarrow B$ transition (Cooper and Hamilton, 1966). The *A*-form is an 11-fold double helix with pitch 28.15 Å (Fuller *et al.*, 1965). Double-helical RNA has a conformation similar to the *A* form of DNA. Unlike DNA, double-helical RNA retains the *A* conformation at high humidities. Synthetic polyribonucleotides—polyadenylic acid (poly A), polyuridylic acid (poly U), polyinosinic acid (poly I), polycytidylic acid (poly C), and polyguanylic acid (poly G)—can form fully ordered, two-stranded complementary helices very like those of native RNA's. These two-stranded complexes consist of an indeterminate number of complementary chains, so arranged that the numbers of complementary residues in each chain of the helix are equivalent, that is, there is virtually complete base-pairing. Since individual chains are unequally long, the nevertheless complete pairing in such a two-stranded helix, and the large increase in molecular weight after complexing of the homopolymers, must mean that each chain of the complex embodies several interruptions with the chains held together at these points by the complementary chain.

* Work at Brookhaven National Laboratory was supported by the U.S. Atomic Energy Commission.

Ionic strength-determined conformational transitions can occur in RNA as well as in DNA (Arnott *et al.*, 1968). Where salt is not in excess in the fiber, two-stranded poly (rI·rC) and poly (rA·rU) complexes are very similar to the conformation of *A* DNA. There are 11 nucleotides per turn of the helix with helix pitch 28 Å and diameter 22 Å. In the presence of excess salt, two-stranded poly (rI·rC) transforms from an 11- to 12-fold helix, that is, there are now 12 base-pairs per turn of the helix, the helix pitch is 36 Å, and the molecule's diameter is 23 Å. Poly (rA·rU) is complicated by its tendency to rearrange in media of high ionic strength into poly (rA·2rU) and poly rA. In this structure helix pitch is 35.5 Å and helices have a large diameter, ~28 Å. The helix is probably 12-fold.

Upon parenteral administration of poly (rI·rC) and poly (rA·rU), it is therefore probable that, given the salt concentration in physiological fluids and especially the presence of Mg^{++}, poly (rI·rC) presumably transforms into a 12-fold helix, and that poly (rA·rU) transforms into poly (rA·2rU) + poly rA. These reactions to ionic environment may decisively determine the effectiveness of two-stranded polynucleotides and account in part for the apparent lesser biological potency of poly (rA·rU) compared with poly (rI·rC) under physiological conditions. The high specificity of structural requirements for biological activity are underscored by the inactivity of poly (rI·dC) (Colby and Chamberlin, 1969); this has a structure isomorphous with 12-fold poly (rI·rC) (O'Brien and McEwan, 1970; Arnott, 1970), but one chain lacks the 2-OH groups on the sugar. Thus, it would appear that the 2-OH groups are necessary on both chains for activity.

BIOLOGICAL EFFECTS

Our interest in the ability of synthetic polynucleotides to form fully ordered, 2-stranded complementary helices very like those of native RNA's stemmed from their possible use as model compounds to study factors stabilizing the helix and drug interactions with nucleic acids. Moreover, the ability of polyribonucleotides to interact with polydeoxyribonucleotides to form 2- and 3-stranded complexes might serve as models for transcription of information from DNA by messenger RNA or from viral RNA to DNA. This interest was reinforced by the effectiveness of multistranded RNA (Lampson *et al.*, 1967; Tytell *et al.*, 1967; Field *et al.*, 1967a) and multistranded complexes of synthetic polyribonucleotides (Field *et al.*, 1967b) in inducing interferon production *in vivo* and *in vitro* and resistance to virus infections in animals, locally and systemically (Field *et al.*, 1967b; Park and Baron, 1968). Although it was not certain that the multistranded polynucleotides were effective because of their ability to induce interferon, since interferon was induced in one species and protection was largely observed in others, the results were intriguing enough for us to investigate the effects of 2-stranded complexes in several biological systems. Therefore, collaborative work was started with Dr. J. Richmond, U.S. Department of Agriculture Plum Island Animal Disease Laboratory studying foot-and-mouth disease virus (FMDV)—an RNA virus.

A. Antiviral. Foot-and-mouth disease virus causes highly contagious disease in cattle, sheep, hogs, and other animals; the effective control is slaughter. FMDV is endemic in many areas in South America; the last outbreak in 1967 caused losses exceeding three hundred million dollars. In the United States the last outbreak of the disease was in 1929. Strict quarantine on animals and animal products excludes the virus from the United States. It is estimated that an outbreak of FMDV in the United States could now lead to losses in billions of dollars.

Smaller animals, for example, mice, can be infected with FMDV; it produces a fulminating, rapidly fatal disease. Poly (rI·rC) induced host resistance in 9-day-old (suckling) and 23-day-old (weanling) mice when injected intraperitoneally in microgram amounts 18 hours before challenge with 100 LD_{50} of FMDV (strain Asia-1). With doses of 150 and 75 μg/mouse, ~100% survived; at 8 μg/mouse, the survival was ~50%. Neither poly (rI) nor poly (rC) individually protected the mice against similar virus inoculations. About 50% of mice survived when poly (rI·rC) (150 μg/mouse) was injected 48 hours before virus inoculation and 80% at 24 hours before or 18 hours after virus inoculation. If poly (rI·rC) was delayed until 24 hours after virus inoculation, the mice were already too ill for the infection to be arrested. Protection was diminished if the poly (rI·rC) was given about the time of virus inoculation. In 9-day-old mice intraperitoneal injection of 150 μg/mouse poly (rI·rC) protected 80% of mice against 500 LD_{50} FMDV, Asia-1; and even at a challenge dose of 600 LD_{50}, more than 50% survived. Poly (rI·rC) intravenously was more effective than intraperitoneally. As little as 1.5 μg/mouse, 90 minutes before virus challenge, protected 50% of 23-day-old mice inoculated with 100 LD_{50} FMDV Asia-1. A graded response, shown by titration of polynucleotide complex, directly related survival and serum interferon titer in these mice. Correlation of dose, survival at day 10, and interferon titer suggests that the efficacy of the poly (rI·rC) is due to its ability to induce interferon. Additional mechanisms cannot be excluded (Richmond and Hamilton, 1969).

Preliminary experiments extending these findings to guinea pigs and cattle indicate that protection against FMDV may be similarly induced with this synthetic 2-stranded polynucleotide complex. Further investigations in these and other animals are under way with this and other double-stranded complexes.

Protection by 2-stranded polynucleotide complexes extends to DNA viruses. Thus, in collaboration with Dr. C. Southam, Sloan-Kettering Institute, 2-stranded poly (rA·rU) as well as poly (rI·rC) protected against systemic herpes simplex infection in mice, herpes simplex-induced cytopathogenicity in human cell line HEp2 tissue cultures, and herpes simplex-induced keratoconjunctivitis in rabbits (Hamilton *et al.*, 1969). Some rabbits were pretreated ~2 hours before virus inoculation with 0.1 mg poly (rA·rU) or poly (rI·rC) solution (0.1 ml dropped onto the cornea and into the conjunctival sac). This dose was repeated 3–4 hours after the virus and 3 times a day thereafter to a total of 14 doses over 5 days. Other rabbits received their first dose of polynucleotide complex 24 hours after virus inoculation and 3 times a day thereafter to a total of 12 doses over 4 days. Some animals had both eyes treated and some one eye. Control eyes were untreated. All eyes appeared normal or slightly irritated after day 1 (24 hours after virus inoculation). By day 2 many untreated eyes were inflamed and had a viscous exudate which increased during the ensuing few days, usually reaching maximum severity in untreated eyes at days 3 or 4 with steady improvement thereafter to almost complete healing by day 10. Treatment with either poly (rA·rU) or poly (rI·rC) decreased severity and duration of the keratoconjunctivitis in treated eyes. Rabbits in which only one eye was treated showed the typical course of keratoconjunctivitis in the untreated eye. Thus, topical administration was indeed localized. Some rabbits whose eyes had responded to topical treatment with the polynucleotide died of cerebral involvement with the virus—again indicating localization of the therapeutic response.

In the HEp2 tissue culture experiments the polynucleotide complexes were added to a final concentration of 0.1 mg/ml ~2 hours before addition of virus. In some experiments additional doses of polynucleotide complexes were given 24 and 48 hours after virus. Poly (rA·rU) and poly (rI·rC) delayed cytopathogenicity (CPE) and protected against herpes simplex virus. The results were variable; repeated treatments with polynucleotide complexes appeared to give much greater protection than those treated only once, indicating that single doses did not saturate or exhaust the protective mechanism.

For study of systemic herpes simplex virus infection, mice were injected with $2^1/_2$ times the intraperitoneal LD_{50} dose as determined by prior and simultaneous titration in untreated mice (Table 1). The polynucleotide complexes poly (rI·rC) and poly (rA·rU) were usually given as 10 150-μg doses twice daily beginning 2–4 hours before administration of virus and

Table 1. Effect of Nucleotide Homopolymers on Systemic HSV Infection in Mice

Agent	Treatment				Mortality		Mean survival of mice that died (days)
	μg per dose	No. of doses	Days	Route	Fraction	%	
Diluent	...	10	0–4	i.p.	16/20	80	8.6
Poly (rI·rC)	150	9	0–4	i.p.	4/30	13	13.0
	100	10	0–4	i.p.	1/19	5	14.0
	50	10	0–4	i.p.	0/19	0	...
	20	10	0–4	i.p.	1/20	5	9.0
	5	10	0–4	i.p.	0/20	0	...
	150	10	0–4	s.c.	9/20	45	9.8
	150	6	0–2	i.p.	4/20	20	10.0
	150	6	2–4	i.p.	13/17	76	10.8
Poly (rA·rU)	150	10	0–4	i.p.	15/29	51	9.6
	100	10	0–4	i.p.	11/20	55	10.5
	50	10	0–4	i.p.	6/20	30	10.8
	20	10	0–4	i.p.	11/20	55	10.0
	5	10	0–4	i.p.	6/20	30	10.8
	150	10	0–4	s.c.	18/20	90	7.3
	150	6	0–2	i.p.	6/20	30	10.7
	150	6	2–4	i.p.	16/20	80	8.5
Poly (rA·2rU)	150	10	0–4	i.p.	7/40	18	10.1
Poly rI	150	10	0–4	i.p.	19/20	95	7.8
Poly rC	150	10	0–4	i.p.	16/20	80	8.8
Poly rA	150	10	0–4	i.p.	17/20	85	7.0
Poly rU	150	10	0–4	i.p.	19/20	95	8.0

continuing through day 4. A 6-dose schedule was used with 150 μg poly (rI·rC) or poly (rA·rU) i.p. starting either at 0, as a test of a shorter duration of treatment, or day 2 to test the efficacy of delayed treatment. Doses of poly (rI·rC) and poly (rA·rU) as low as 5 μg were used. Poly (rI·rC) given i.p. 150 μg/dose strikingly reduced virus mortality. Even 5 μg/i.p. afforded essentially complete protection; given subcutaneously the reduction in mortality was less impressive. With the shorter courses of treatment given on days 1 and 2 only, protection was nearly as good as with 5 days. When treatment was delayed until 48 hours after virus inoculation, mortality was not reduced, but mean survival time was 25% longer than in untreated controls. Poly (rA·rU) also significantly protected when 10 doses were given i.p. even at 5 µg; however, it was less protective than poly (rI·rC). Given s.c., poly (rA·rU) was ineffective, and where i.p. treatment was delayed until 48 hours after virus inoculation, there was no protection. The 3-stranded polymer poly (rA·2rU) gave essentially the same protection as the double-

Table 2. Precipitating Antibody in Guinea Pigs Sensitized and Skin-Tested[a] to *T. spiralis* and Treated with Poly (rI·rC)

Poly (rI·rC)		Sensitizing dose (total protein)	Ring test		
Amount (µg)	Time interval		Days after sensitization	No. positive / No. samples	Highest titer[b]
1500	4, 5, 6 days before sensitization	76 µg	8	3/3	1:200
3000	4, 2, 3 days after sensitization	76 µg	8	1/3	1:400
3000	at 4th sensitization (2-week intervals)	1.5 mg	1	3/3	1:3200
5000	at 4th sensitization (2-week intervals)	1.5 mg	8[c]	6/6	1:3200
None		76 µg	8	2/2	1:100

[a] Skin test dose = 76 µg total protein 7 days after sensitization.
[b] Titer = reciprocal of antigen dilution.
[c] First day tested.

stranded poly (rA·rU). None of the four single-stranded polymers given protected individually (Southam *et al.*, 1970).

B. Immunity. To study the effects of polynucleotide complexes on the delayed hypersensitivity reaction, poly (rI·rC) (1500–3000 µg) was injected intravenously (500–1000 µg/day) in guinea pigs either before (4, 5, 6 days) or after (1, 2, 3 days) a single sensitization with *Trichinella spiralis* antigen (76 µg total protein of crude saline extract) plus complete adjuvant. There was no apparent definitive effect with these doses at these time intervals on the skin test at 7 days after sensitization; the reaction, typically erythematous and indurated, peaked between 22–24 hours as in control animals. No acceleration of the appearance of antibody was detectable by passive cutaneous anaphylaxis (PCA) and precipitin ring test. Antibody was first detected by PCA 7–8 days after sensitization and by the ring test 8 days after sensitization. In control animals not given poly (rI·rC), antibody was first detected by PCA and the ring test at 8 days after sensitization. However,

Table 3. PCA Reactions in Guinea Pigs Sensitized and Skin-Tested[a] to *T. spiralis* and Treated with Poly (rI·rC)

Poly (rI·rC) Amount (µg)	Poly (rI·rC) Time interval	Sensitizing dose (total protein)	Days after sensitization	No. positive / No. samples	Average size (mm)
1500	4, 5, 6 days before sensitization	76 µg	8	3/3	17 × 40
3000	1, 2, 3 days after sensitization	76 µg	7	1/3	10 × 30
3000	at 4th sensitization (2-week intervals)	1.5 mg	7 1	2/3 3/3[b]	30 × 90 20 × 70[b]
None		76 µg	8	3/3	20 × 58

[a] Skin test dose = 76 µg total protein 7 days after sensitization.
[b] Serum diluted 1:80.

in the ring test, the titer was higher in those given poly (rI·rC) than in controls (Table 2).

When a single injection of poly (rI·rC) (3000 µg) was given to animals sensitized 4 times at 2-week intervals (1.5 mg total protein plus adjuvant), immediately after the last (fourth sensitization), PCA reactions were negative with day 1 serum samples; 2/3 of the animals died from anaphylactic shock within minutes with samples from day 2 and thereafter. When day 1 serum samples were diluted 1:80 and tested, PCA reactions were positive; there was no anaphylaxis and no inhibition of PCA reactions was noted (Table 3). Indeed, diluted samples from day 1 and thereafter evoked positive reactions. The anaphylactogenicity of undiluted serum began to subside by day 7 as evidenced by less deaths, although there was fatal anaphylaxis through day 21. Inhibition of PCA reactions also subsided by day 7 at which time reactions were first positive with undiluted serum (Table 3). Apparently poly (rI·rC) boosted the levels of 7Sγ2 to inhibit PCA reactions, and γ1 to provoke fatal anaphylaxis. The antibody level (1:3200) was sufficient to be detected by ring test even at day 1 after sensitization (Table 2).

C. Antitumor. The effectiveness of the 2-stranded polynucleotide complexes in protecting against certain viruses led to their trial (in collaboration with Dr. G. Tarnowski and Dr. C. Stock, Sloan-Kettering Institute) against experimental animal tumors. With some tumors, for example, RADA1 leukemia in A mice, inhibition by the polynucleotide complexes have been inconstant and minimal. With other tumors—especially solid tumors, for example, Ridgeway osteogenic sarcoma—the complexes have been consistently active (Tarnowski *et al.*, 1970). The Ridgeway osteogenic sarcoma arose spontaneously as an inguinal mass in a male AKR mouse (Karnofsky *et al.*, 1950). The original tumor contained solidly packed round cells with foci of bone formation. In the course of passages by the subcutaneous route in AKR mice it lost bone structure, but retained high alkaline phosphatase activity. This mouse osteogenic sarcoma grows rapidly—14 days after transplantation, tumors are $\sim 15 \times 20 \times 10$ mm of firm consistency with almost no areas of hemorrhage or necrosis (Sugiura, 1965).

The effects of treatment with poly (rI·rC) were scored by tumor diameter. Small fragments of tumors were transplanted by subcutaneous inoculation by trocar into the right axillary region of AKD2F1 male mice. Treatment was begun 5 days after transplantation and consisted of 6 daily intraperitoneal injections of poly (rI·rC). Two perpendicular tumor diameters were measured on the 13th and 20th day of tumor growth and the average diameters of each tumor computed. Average tumor diameters were pooled for each group and the diameters of treated groups (T) were compared with those of corresponding control groups (C). This calculated T/C ratio was then compared with the minimum acceptance T/C value at 99.5% confidence limits obtained from an analysis of $\sim$100 untreated tumors. Toxicity of poly (rI·rC) to the tumor host was scored from observed mortality and from loss of body weight of tumor-bearing mice (Schmid *et al.*, 1966). The average weight change (AWC) in grams was determined for each observation period starting with the beginning of the experiment. Table 4 shows a typical titration experiment with one of the most potent poly (rI·rC) preparations. The data are given for the average diameter T/C at the end of treatment, that is, day 13, and 1 week after treatment, day 20. At 10 mg/kg, tumor inhibition was considerable with minimal weight loss at day 13. The weight changes are given in absolute amounts and in relative weight. These data also clearly show the relationship between the average tumor diameter T/C against relative weight. This relationship between tumor diameter, relative weight, and varying dose is shown more clearly in Figure 1 with the linear

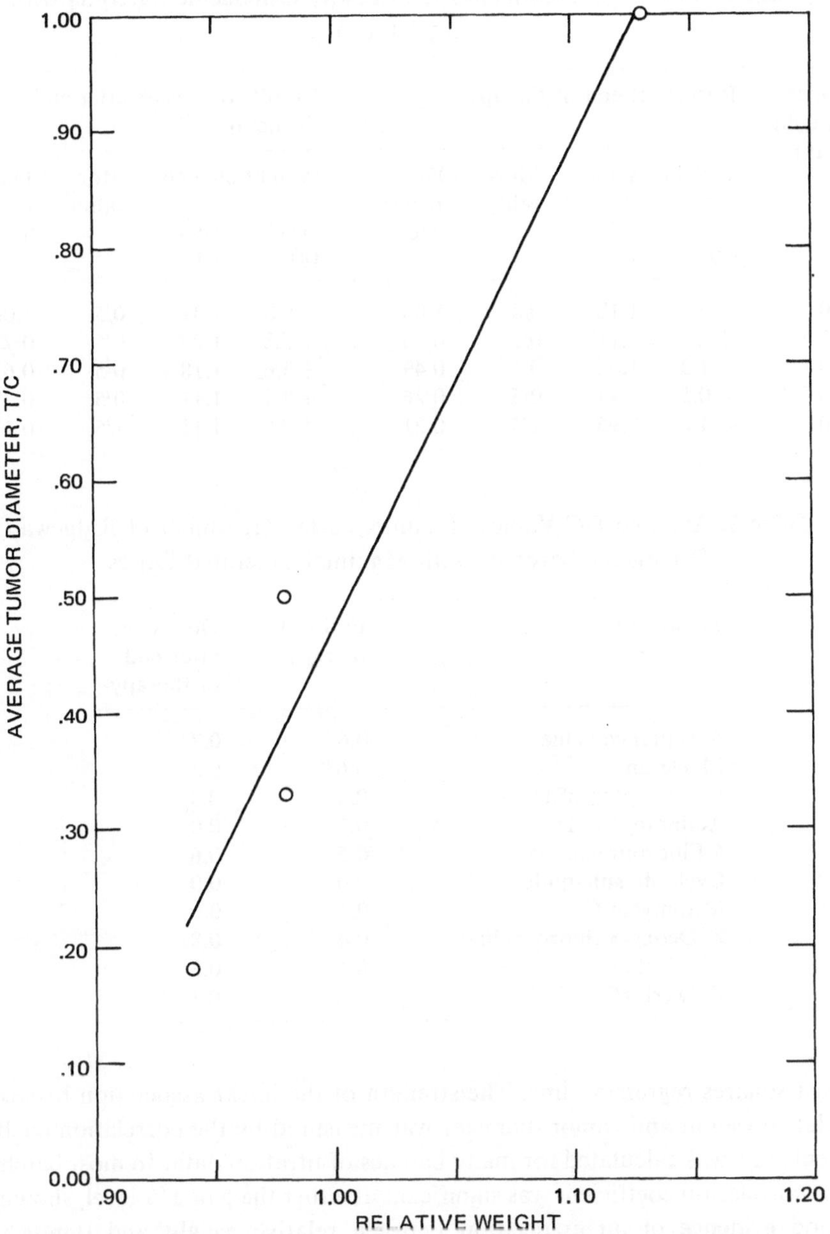

Fig. 1. The least-squares linear regression line where X = relative weight of mice (sum of the weights of mice at the end of therapy divided by the sum of weights of mice at the time of inoculation of tumors), Y = ROS average tumor diameter, treated tumors/control tumors (T/C). The points from the bottom up represent poly (rI·rC) doses of 20 mg, 10 mg, 5 mg, and 0 mg/kg/day.

Table 4. Effects of Treatment of Ridgeway Osteogenic Sarcoma with Poly (rI·rC)

Dose (mg/kg/day)	Results at end of therapy				Results one week after end of therapy			
	Weight change		Mortality	Diameter T/C	Weight change		Mortality	Diameter T/C
	AWC (g)	Rel. wt			AWC (g)	Rel. wt		
0	+2.7	1.12	0/5	1.00	+6.8	1.31	0/5	1.00
2.5	+2.2	1.11	0/5	0.64	+5.8	1.29	0/5	0.77
5	+1.3	1.06	0/5	0.49	+3.6	1.18	0/5	0.62
10	—0.5	0.97	0/5	0.26	+2.1	1.11	0/5	0.27
20	—1.0	0.95	1/5	0.21	+2.0	1.11	1/5	0.14

Table 5. Average T/C Values of Tumors after Treatment of Ridgeway Osteogenic Sarcoma with Maximal Tolerated Doses

Treatment	End of therapy	One week after end of therapy
Acceptance value	0.6	0.7
Myeleran	0.6	0.3
6-Mercaptopurine	0.4	0.3
Actinomycin, D	0.1	0.0
5-Fluorouracil	0.5	0.6
Cyclophosphamide	0.0	0.0
Mitomycin C	0.3	0.3
2′-Deoxy-5-fluorouridine	0.4	0.8
Vinblastine	0.4	0.5
Poly (rI·rC)	0.2	0.1

least-squares regression line. The strength of the linear association between relative weight and tumor diameter was measured by the correlation coefficient; this was calculated for many batches of titration data. In most batches the correlation coefficient was significant at either the 5 or 1% level, showing good evidence of an association between relative weight and tumor inhibition. As the weight loss was within 5%, this correlation was not too detrimental.

Table 5 compares the effect of other agents on the average T/C values of tumors after treatment with maximum tolerated doses (Tarnowski *et al.*,

1966). This includes many agents currently used in palliation on cancer. It can be seen from this list that only actinomycin D and cyclophosphamide were more effective than poly (rI·rC) at the end and 1 week after end of treatment. Because of these results in experimental animal tumors and the selective toxicity of poly (rI·rC), one might expect poly (rI·rC) to have some useful palliative role in the treatment of cancer in man.

D. Toxicity and adrenal cortical function. The antiviral and antitumor effects of poly (rI·rC) led to the study of its toxicity in animals, prerequisite for clinical trial in man. These studies were carried out in collaboration with Drs. F. S. Philips, M. S. Zedeck, H. Marquardt, S. S. Sternberg, and M. Fleisher at the Sloan-Kettering Institute, New York (Philips *et al.*, this Symposium). In rats poly (rI·rC) sharply inhibited liver DNA synthesis; nuclear RNA was unaffected. Since adrenal glucocorticoids, as known, inhibit DNA synthesis in liver, this suggested similar studies in adrenalectomized animals (adx). We observed that adrenalectomized animals were much more sensitive than intact animals to the lethality of poly (rI·rC) (Zedeck *et al.*, 1970). Because of possible use of poly (rI·rC) in cancer chemotherapy and the fact that doses of the complex needed to inhibit tumors in mice generally exceeded those needed to induce resistance to virus infection and interferon in mice and rabbits, we measured the influence of the adrenals on its toxicity. Adrenalectomized rats were ~1000 times more sensitive to the lethality of poly (rI·rC) than were intact animals. Death occured in the adx rats between 3–4 to 24 hours after injection. Poly (rI·rC), 1.5 mg/kg, in adx rats, 2–3 hours later, inhibited thymidine incorporation into liver DNA 60%, small intestine 55%, and thymus 30%. In intact animals, 3–5 hours after poly (rI·rC), 1.5 mg/kg, DNA synthesis was unaffected; however, after 100 mg/kg, liver DNA synthesis was inhibited 70%; incorporation into nuclear RNA was unaffected. Corticosterone (as acetate), the natural corticoid for the rat (Dorfman, 1959), prevented death in adx animals given poly (rI·rC), in doses as large as 100 mg/kg. Epinephrine 50 μg/kg s.c., or desoxycorticosterone, up to 50 mg/kg s.c., did not protect. Since adx animals were more sensitive to poly (rI·rC), and adrenalcortical hormones have antitumor activity (Stock and Sugiura, 1958), increased secretion of the adrenalcorticoids might be involved in tumor inhibition by poly (rI·rC)—an additional mechanism to be kept in mind in considering how poly (rI·rC) protects against tumors.

E. Effect of homopolymers. The susceptibility of the adx animals permitted comparison of the toxicity of the homopolymers with that of the complex (Zedeck *et al.*, 1970). The two homopolymers were given alone or separately but immediately after each other. Experiments with individual homopolymers were important to rule out endotoxin contamination, perhaps retained during preparation and purification of poly (rI·rC), as responsible for the sensitivity of adx rats to the poly (rI·rC). Adx rats given poly rI or poly rC, 10 mg/kg, were not killed; 10 mg/kg of the complex would have killed all adx animals. In contrast, injection of poly rI, 5 mg/kg, followed immediately but separately by poly rC 5, 0.5, or 0.05 mg/kg was lethal; likewise, the reverse combinations. Administration of poly rI 5 mg/kg, then

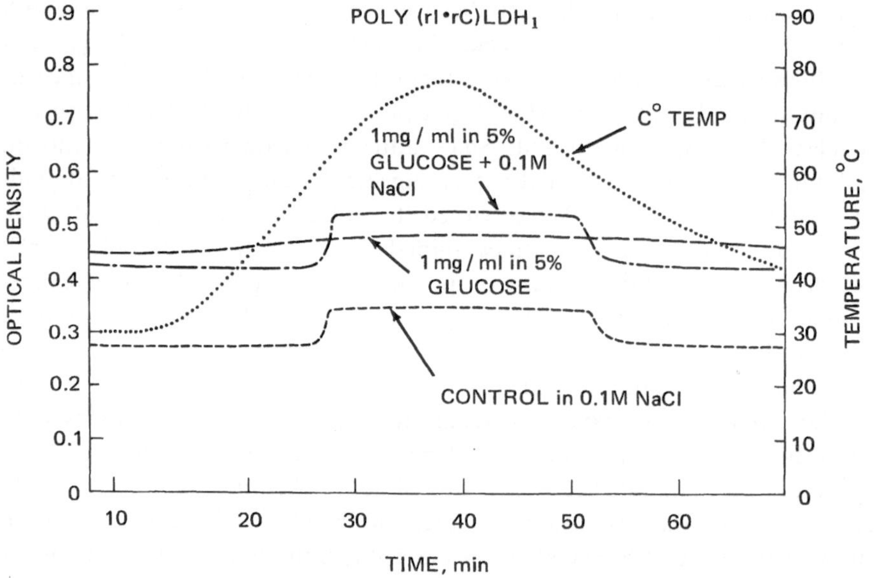

Fig. 2. Melting and reannealing of poly (rI·rC) H1 in 0.1 *M* NaCl and in 5% glucose + 0.1 *M* NaCl. Note the flat, nonmelting curve given by poly (rI·rC) H1, 1 mg/ml in 5% glucose only, indicating strand separation. This strand separation was completely reversible on addition of counter-ion.

1 hour later, poly rC 5 mg/kg or vice versa, was not lethal. Poly (rI·rC), 0.1 mg/kg, was not lethal; however, poly (rI·rC), 0.1 mg/kg, given separately but immediately after poly rI or poly rC, 5 mg/kg, was lethal. Therefore, the two homopolymers when given separately but immediately after each other are probably reversibly associated *in vivo* in equilibrium with poly (rI·rC), that is, the 2-stranded complex determines activity. Since poly (rI·rC), 0.1 mg/kg, was not lethal in adx rats, the fact that poly rC, 0.05 mg/kg, + poly rI, 5 mg/kg, was lethal probably means that the amount of poly (rI·rC) formed from the association of the individual homopolymers may have enhanced activity in the presence of a large amount of one homopolymer.

Additional evidence that the two homopolymers given separately come together under physiological conditions when injected simultaneously, parenterally, is unexpectedly afforded by experiments reported by Dr. Philips at this Symposium (see Philips *et al.*, Figure 3). When poly (rI·rC) was dis-

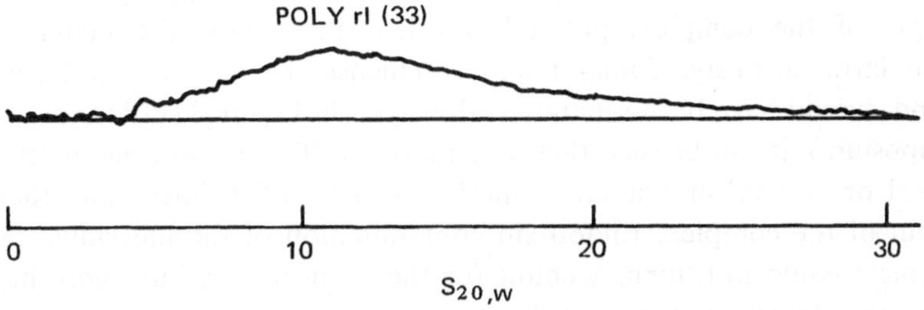

$S_{20,w}$	I (33) PERCENT	I (35) PERCENT
0		
0 – 5	4.5	23.3
5 – 10	25.5	61.7
10 – 15	41.5	13.4
15 – 20	21.9	1.7
20 – 25	6.5	0.0

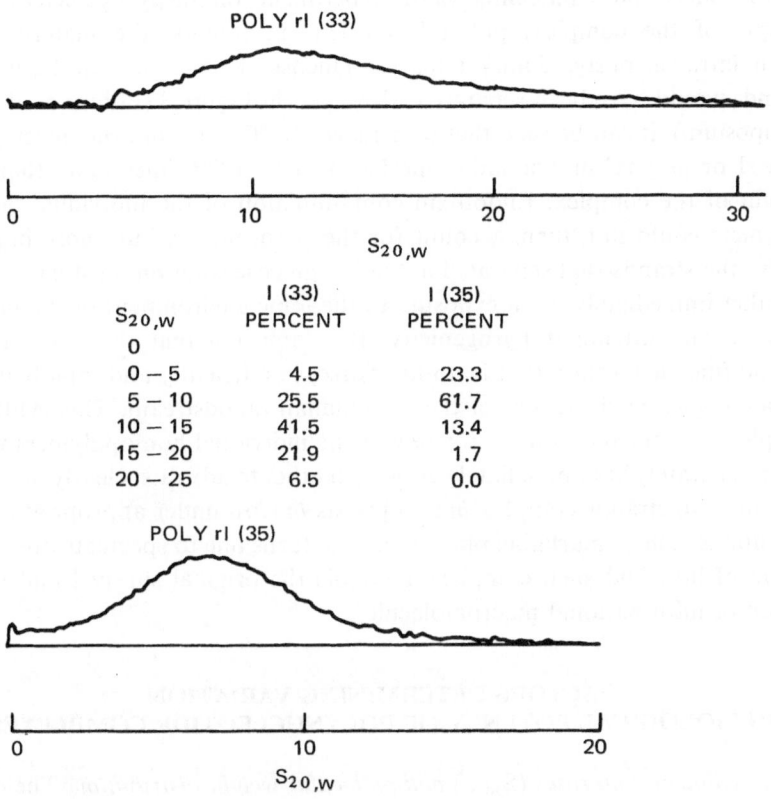

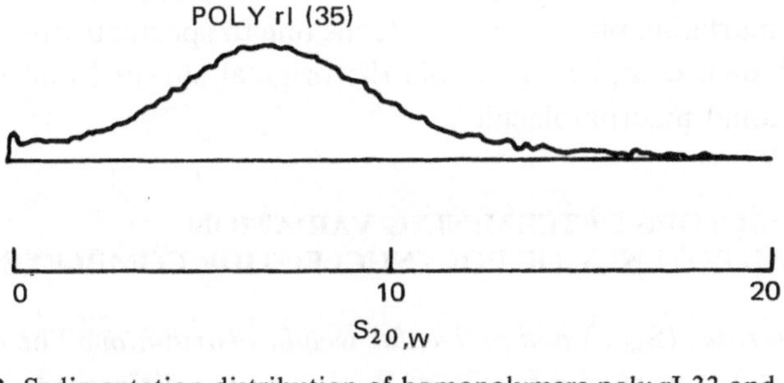

Fig. 3. Sedimentation distribution of homopolymers poly rI 33 and 35. The percentage of total homopolymer sedimenting within the indicated interval of $S_{20,\,w}$ is given for each batch of poly rI.

solved in 5% glucose, the two strands separated because the negative charges on the phosphates were not neutralized by any counter-ion. This was not at once apparent because dilutions for spectral reading of the dissolved poly (rI·rC) in 5% glucose were made by the addition of an ion-containing buffer. Upon addition of this ion the two strands immediately rejoined giving the original spectrum of the complex. In the absence of a counter-ion, poly (rI·rC) in 5% glucose gave the spectrum of the separated poly rI combined with poly rC. Figure 2 shows that poly (rI·rC) H 1, the sample whose toxicity Dr. Philips describes at this Symposium, when dissolved in 5% glucose gives a completely flat, that is, nonmelting curve. In contrast, the complex in 0.1 *M* NaCl gives a regular melting curve with reannealing of the separated strands on cooling. Similarly, the complex dissolved in 5% glucose to which 0.1 *M* NaCl has been added gives the same melting curve and reannealing. In the experiments on the pyrogenicity of this sample of the complex, poly (rI·rC) H1, in rabbits, the material was given intraveneously, 5 mg/ml in 5% glucose. In this situation the two strands would have been separated. However, in Figure 3 (Philips *et al.*, this Symposium), it can be seen that to achieve a 1 °C rise from the injection of poly rI or poly rC individually, one had to give 1000 times more than the weight of the complex. Endotoxin contamination of the individual homopolymers could not, then, account for the pyrogenicity, but, more importantly, the strands that separated in the 5% glucose solution must have come together immediately upon exposure to the ionic environment of the bloodstream, with attendant pyrogenicity. It is amazing that the two strands should find each other, that is, to hybridize, so efficiently and rapidly in the ultra-complex environment of the mammalian bloodstream. This evidence, coupled with the data on the toxicity of the individual homopolymers when given separately but immediately after each other to adx rats, clearly indicates that the two strands complex *in vivo* just as *in vitro* under appropriate ionic conditions. This remarkable phenomenon returns one to speculate about the origin of life: Did such complexing enable the original survival and replication of informational macromolecules?

FACTORS DETERMINING VARIATION IN BIOLOGICAL POTENCY OF POLYNUCLEOTIDE COMPLEXES

A. Sedimentation rates ($S_{20,w}$) *and molecular weight distribution.* The complexes of poly (rI·rC) consist of a family of duplexes with different lengths. This can readily be seen from the sedimentation $S_{20,w}$ profile of preparations of poly (rI·rC) (Figure 4). These banding-velocity sedimentations were made under nearly physiological conditions, 5×10^{-4} *M* EDTA, 0.1 *M* Na phosphate *p*H 7.8 in 50% D_2O (Commerford and Hamilton, 1970). The sedimentation patterns of the polynucleotide complexes were widely spread and asymmetrical; this is not due to diffusion but to a heterogeneous distribution of molecular weights, that is, duplexes with different lengths. Any one preparation is composed of a skewed distribution of polynucleotide chains with corresponding variation in $S_{20,w}$ values. There is no empirical relationship between $S_{20,w}$ values and molecular weights for duplex polynucleotides that have single strand breaks; they would tend to be more flexible in

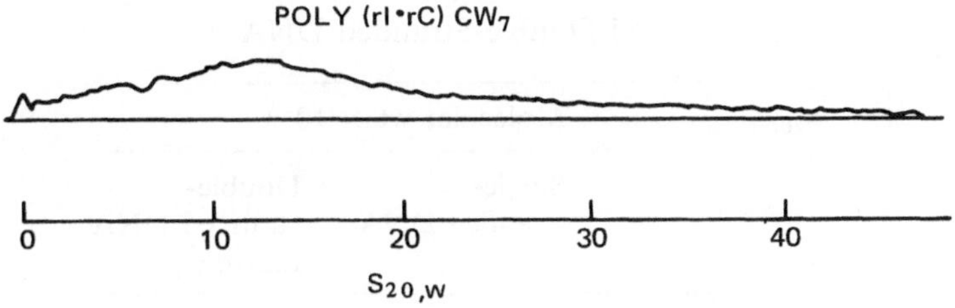

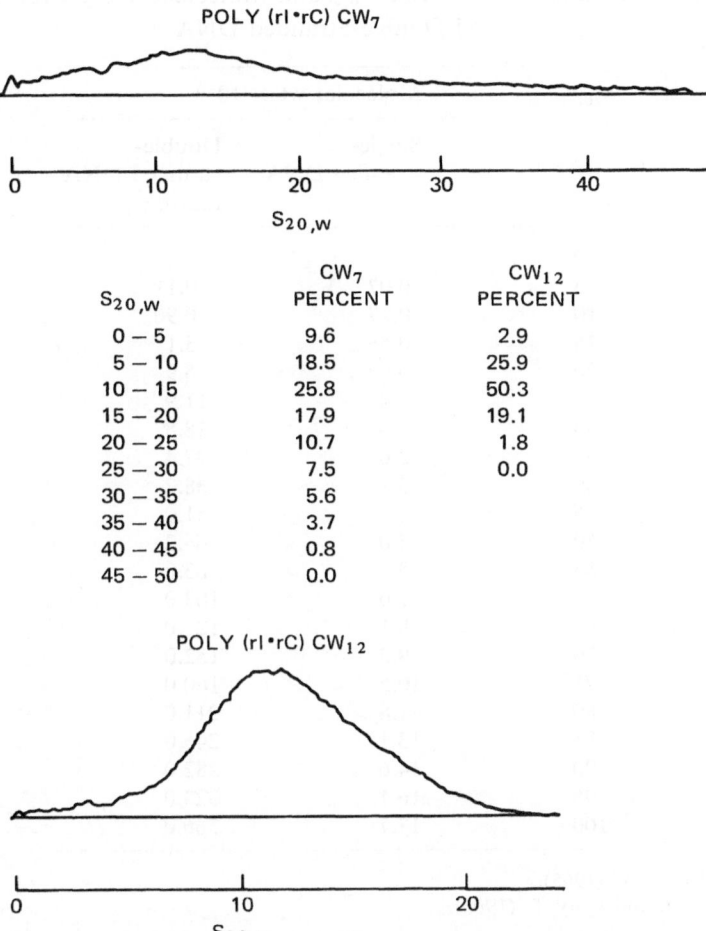

$S_{20,w}$	CW_7 PERCENT	CW_{12} PERCENT
0 – 5	9.6	2.9
5 – 10	18.5	25.9
10 – 15	25.8	50.3
15 – 20	17.9	19.1
20 – 25	10.7	1.8
25 – 30	7.5	0.0
30 – 35	5.6	
35 – 40	3.7	
40 – 45	0.8	
45 – 50	0.0	

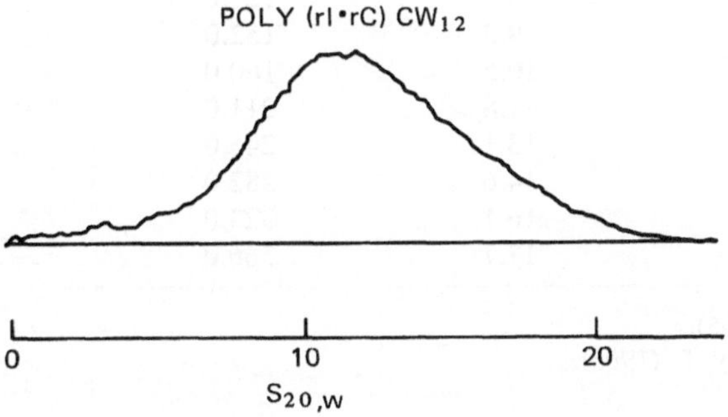

Fig. 4. Sedimentation distribution of poly (rI·rC) CW7 and CW12; poly rI 33 was used for CW7 and poly rI 35 for CW12. The percentage of total complex sedimenting within the indicated interval of $S_{20,\,w}$ is given for CW7 and CW12.

solution than double-stranded DNA. Table 6 gives the molecular weights of single-stranded and double-stranded DNA for corresponding $S_{20,\,w}$ values. The values for double-stranded DNA represent an upper limit of molecular weight for the corresponding $S_{20,\,w}$ values of duplex polynucleotides. The percent distribution of $S_{20,\,w}$ values can be characterized for each batch. Factors that determine the distribution of $S_{20,\,w}$ values of the complexes include the distribution of $S_{20,\,w}$ values in the original homopolymers composing the complexes. Thus (Figure 3), from the sedimentation distributions

Table 6. Relationship between $S_{20,w}$ and Molecular Weight for Single- and Double-Stranded DNA

$S_{20,w}$	Molecular wt × 10^{-6}	
	Single-stranded DNA ss[a]	Double-stranded DNA native[b]
0	0	
5	0.07	0.11
10	0.27	0.90
15	0.56	3.1
20	0.94	6.9
25	1.4	11.9
30	2.0	18.8
35	2.6	27.4
40	3.3	38.2
45	4.1	51.0
50	5.0	66.2
55	5.9	83.8
60	7.0	103.0
65	8.1	126.0
70	9.2	152.0
75	10.5	180.0
80	11.8	211.0
85	13.1	245.0
90	14.6	282.0
95	16.1	323.0
100	17.7	366.0

[a] Studier, F. W. (1965).
[b] Eigner, J. and Doty, P. (1965).

on two batches of poly rI 33 and 35, the major portion of I 33 has an $S_{20,w}$ of between 10–15, while in I 35 the major component has $S_{20,w}$ value 5–10; there is little over 20. These homopolymers were used to form the complexes CW 7 and CW 12; and, as seen, I 33, which had the higher $S_{20,w}$ distribution, gave rise to a complex with higher $S_{20,w}$ values than did the I 35 used for CW 12 (Figure 4). Other factors affecting the distribution of $S_{20,w}$ values in the complex are how the homopolymers were complexed. One method for complexing the homopolymers included heating the homopolymers to 80 °C for 15 minutes. The effect of this heat treatment of $S_{20,w}$ values is shown in Figures 5 and 6. Figure 5 shows the sedimentation distribution of poly (rI·rC) TY 19 and TY 18; TY 18 is identical with TY 19 except that it has been heated for 15 minutes. Similarly, in Figure 6, LDH 1 heated is the

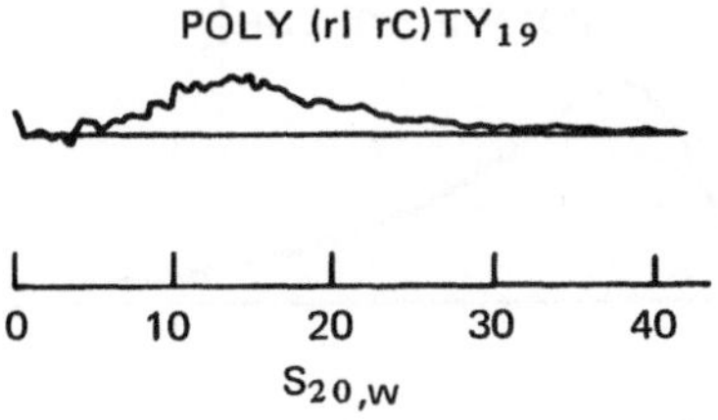

$S_{20,w}$	TY_{19} PERCENT	TY_{18} PERCENT
0 – 5	.03	7.4
5 – 10	1.8	18.1
10 – 15	7.9	43.7
15 – 20	16.3	24.1
20 – 25	21.3	6.2
25 – 30	20.4	.5
30 – 35	13.4	0.0
35 – 40	9.0	
40 – 45	6.0	
45 – 50	3.8	
50 – 55	0.0	

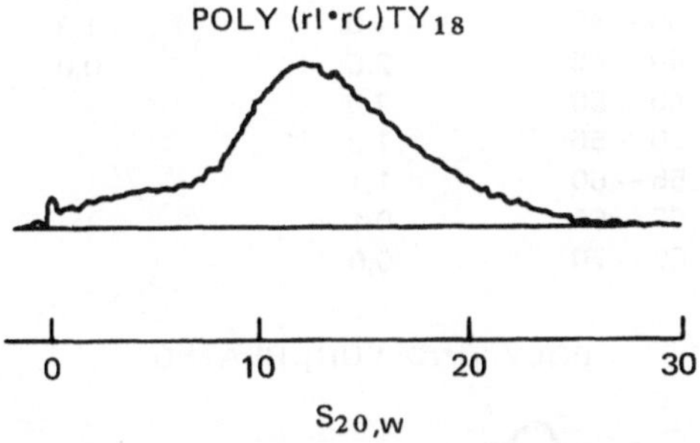

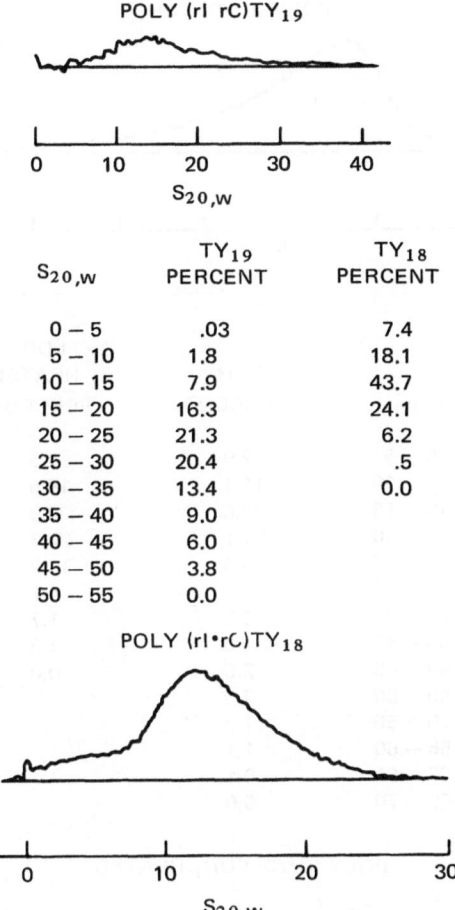

Fig. 5. Sedimentation distribution of poly (rI·rC) TY19 and TY18. TY19 and TY18 were prepared from identical homopolymers in the same way except that TY18 was heated to 80 °C for 15 minutes (0.15 *M* NaCl, 0.01 *M* Tris *p*H 7). The percentage of total complex sedimenting within the indicated interval of $S_{20,w}$ is given for TY19 and TY18.

same as LDH1 unheated except for the heat treatment for 15 minutes at 80 °C. In both figures, the heat treatment sharply lowered the $S_{20,w}$ values of the complex. Thus, in TY18 there was essentially no material with an $S_{20,w} > 25$ in contrast to TY19. Similarly, the distribution of H1 was shifted sharply to smaller size. Thus, heat shifts the molecular weight distribution of the polynucleotide complexes to lower values. H1 in Figure 6 is the complex used by Dr. Philips for toxicity studies.

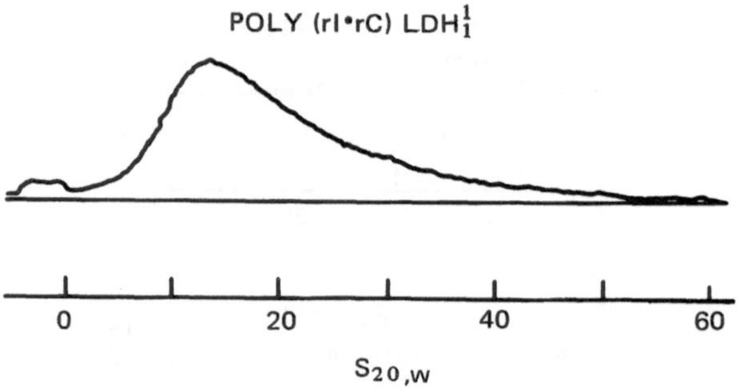

$S_{20,w}$	LDH_1 PERCENT	LDH_1 HEATED PERCENT
0 – 5	2.9	7.5
5 – 10	11.1	18.5
10 – 15	23.6	32.9
15 – 20	21.1	22.7
20 – 25	14.4	11.2
25 – 30	9.4	4.1
30 – 35	6.5	1.7
35 – 40	4.0	1.3
40 – 45	2.5	0.0
45 – 50	2.0	
50 – 55	1.3	
55 – 60	1.1	
60 – 65	0.1	
65 – 70	0.0	

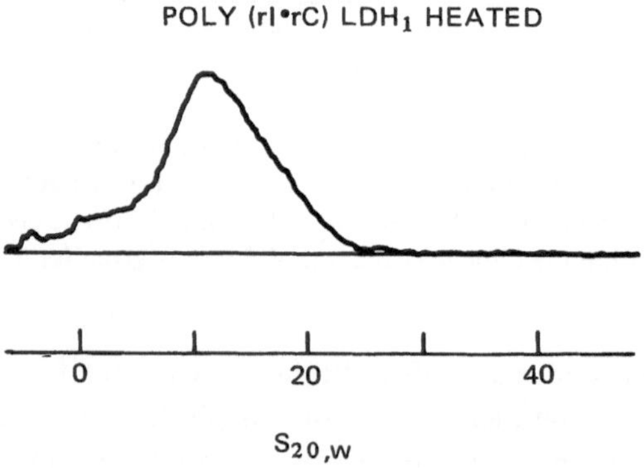

Fig. 6. Sedimentation distribution of poly (rI·rC) H1, heated and unheated. H1 heated was heated to 80 °C for 15 minutes (0.15 *M* NaCl, 0.01 *M* Tris *p*H 7). The percentage of total complex sedimenting within the indicated interval of $S_{20,\,w}$ is given for H1 and H1 heated.

B. Correlation of biological potency with $S_{20,w}$ distribution of complexes. Variations in the biological potency of different preparations of poly (rI·rC) were first noted in the LD_{50} doses in mammals. Similar variations were then noted in antitumor and antiviral activity. LD_{50} titration data were available for 10 batches of poly (rI·rC) in the male Charles River "pathogen-free" mouse. The probit transformation of percent survival was plotted against log dose. The maximum likelihood estimate of the LD_{50} with its standard deviations was calculated from the probit analysis. Similar data were available on 9 batches of poly (rI·rC) given at different dose levels to protect against FMDV virus. The percent survivors were plotted against log dose; the linear regression coefficients were significant at either the 5 or 1% level. To compare the biological potency of different batches of poly (rI·rC), the number of survivors at a common dose log 10 μg was computed from the regression lines. The data were also examined for significance of variation among the slopes of the lines within groups made from similar homopolymers and between groups of complexes made from different homopolymers by an analysis of covariance. Similar data were available for many batches of poly (rI·rC) against tumors. The average tumor diameter T/C at 10 mg/kg dose level was used to rank the different preparations; sometimes this value was the average of three experiments. There was not much difference in the relative weight of mice. Analysis of these data on the LD_{50}, FMDV protection at 10 μg, and T/C at 10 mg/kg are given in Table 7 for those samples in which all three systems were examined on the same batches of poly (rI·rC) and for which sedimentation data were available. Table 7 shows a good correlation between the increasing $S_{20,w}$ and the LD_{50}, that is, the complex with the highest molecular weight distribution required the least mg/kg for the LD_{50}; complexes with $S_{20,w} < 20$–25 seemed the least toxic. The antiviral and antitumor activity seemed to parallel each other; but here, although the complexes with the lowest molecular weight distribution were the least effective, the complexes that were more toxic, possibly because of the high sedimentation distribution, were less effective against the Ridgeway osteogenic sarcoma and the FMDV. The data are too few to be conclusive, but they suggest that chain lengths above a certain point may be more toxic and less biologically efficient with respect to antitumor and antiviral activity.

ACKNOWLEDGEMENTS

Apart from collaborators mentioned in the paper, I thank my colleagues, Drs. S. Arnott, S. L. Commerford, C. W. Kim, and K. H. Thompson for

Table 7. Correlation of Toxicity, Antitumor, and Antiviral Activity with $S_{20,w}$ Distribution

Male mice LD_{50} mg/kg S.D.	Sedimentation distribution $S_{20,w}$ in percent																		ROS 10 mg/kg T/C	Percent survivors FMDV 10 µg dose
	0	5	10	15	20	25	30	35	40	45	50	55	60	65	70	75	80	85		
	5	10	15	20	25	30	35	40	45	50	55	60	65	70	75	80	85	90		
26 ± 4	1	6	19	19	14	10	7	5	4	3	3	2	2	2	1	1	1	0.1	0.47 (4)[a]	45 (4)[a]
26 ± 10	1	4	16	23	20	15	10	6	3	2	0.5								0.40 (2)	60 (2)
34 ± 5	0	2	8	16	21	20	13	9	6	4									0.45 (3)	54 (3)
50 ± 6	10	19	26	18	11	7	6	4	1										0.25 (1)	62 (1)
55 ± 8	7	18	44	24	6	1													0.58 (6)	
86 ± 14	3	26	50	19	2														0.53 (5)	42 (5)

[a] ( ) Rank in order of biological potency.

helpful discussion, and Dr. E. P. Cronkite for encouragement. I also thank Mrs. C. Kendrick, Miss R. Klimaski, and Mrs. T. Young for their technical assistance in preparing the polynucleotide complexes.

References

Arnott, S. (1970). Geometry of nucleic acids, *In* Progress in Biophysics and Molecular Biology, **21:** 265–319. Pergamon Press, New York.

——— Fuller, W., Hodgson, A., and Prutton, I. (1968). Molecular conformations and structure transitions of RNA complementary helices and their possible biological significance. Nature **220:** 561–564.

Colby, C. and Chamberlin, M. J. (1969). The specificity of interferon induction in chick embryo cells by helical RNA. Proc. Natl. Acad. Sci. **63:** 160–167.

Commerford, S. L. and Hamilton, L. D. (1970). Unpublished work.

Cooper, P. J. and Hamilton, L. D. (1966). The *A–B* conformational change in the sodium salt of DNA. J. Mol. Biol. **16:** 562–563.

Dorfman, R. I. (1959). Comparative biochemistry of adrenocortical hormones, *In* Comparative Endocrinology. 613–623. John Wiley, New York.

Eigner, J. and Doty, P. (1965). The native, denatured and renatured states of deoxyribonucleic acid. J. Mol. Biol. **12:** 549–580.

Field, A. K., Lampson, G. P., Tytell, A. A., Nemes, M. M., and Hilleman, M. R. (1967a). Inducers of interferon and host resistance. IV. Double-stranded replicative form RNA (MS2-RF-RNA) from *E. coli* infected with MS2 coliphage. Proc. Natl. Acad. Sci. **58:** 2102–2108.

——— Tytell, A. A., Lampson, G. P., and Hillman, M. R. (1967b). Inducers of interferon and host resistance. II. Multistranded synthetic polynucleotide complexes. Proc. Natl. Acad. Sci. **58:** 1004–1010.

Franklin, R. E. and Gosling, R. G. (1953). The structure of sodium thymonucleate fibres. I. The influence of water content. Acta Cryst. **6:** 673–677.

Fuller, W., Wilkins, M. H. F., Wilson, H. R., and Hamilton, L. D. (1965). The molecular configuration of deoxyribonucleic acid. IV. X-ray diffraction study of the *A* form. J. Mol. Biol. **12:** 60–80.

Hamilton, L. D. (1968). DNA: Models and reality. Nature **218:** 633–637.

Hamilton, L. D., Babcock, V. I., and Southam, C. M. (1969). Inhibition of herpes simplex virus by synthetic double-stranded RNA (polyriboadenylic and polyribouridylic acids and polyriboinosinic and polyribocytidylic acids). Proc. Natl. Acad. Sci. **64:** 878–883.

Karnofsky, D. A., Patterson, P. A., and Ridgway, L. P. (1950). The growth and histology of a variety of mouse tumors explanted to the chorioallantoic membrane of the chick embryo. Cancer Res. **10:** 228–229.

Lampson, G. P., Tytell, A. A., Field, A. K., Nemes, M. M., and Hilleman, M. R. (1967). Inducers of interferon and host resistance. I. Double-stranded RNA from extracts of *Penicillium funiculosum*. Proc. Natl. Acad. Sci. **58:** 782–789.

O'Brien, E. J. and McEwan, A. W. (1970). Molecular and crystal structure of the polynucleotide complex: polyinosinic acid plus polydeoxycytidylic acid. J. Mol. Biol. **48**: 243–261.

Park, J. H. and Baron, S. (1968). Herpetic keratoconjunctivitis: therapy with synthetic double-stranded RNA. Science **162**: 811–813.

Philips, F. S., Fleisher, M., Hamilton, L. D., Schwartz, M., and Sternberg, S. (1970). Polyinosinic-polycytidylic acid toxicity. This Symposium.

Richmond, J. Y. and Hamilton, L. D. (1969). Foot-and-mouth disease virus inhibition induced in mice by synthetic double-stranded RNA (polyriboinosinic and polyribocytidylic acids). Proc. Natl. Acad. Sci. **64**: 81–86.

Schmid, F. A., Cappuccino, J. G., Merker, P. C., Tarnowski, G. S., and Stock, C. C. (1966). Chemotherapy studies in an animal tumor spectrum. I. Biologic characteristics of the tumors. Cancer Res. **26**: 173–180.

Southam, C. M., Babcock, V. I., and Hamilton, L. D. (1970). Inhibition of Herpes simplex virus in mice by complexes of homopolyribonucleotides. To be published.

Sugiura, K. (1965). Tumor transplantation, *In* Methods of Animal Experimentation, **2**: 171–222. Academic Press, New York.

Stock, C. C. and Sugiura, K. (1958). Screening steroids against a spectrum of tumors. Ann. N.Y. Acad. Sci. **76**: 720–728.

Studier, F. W. (1965). Sedimentation studies of the size and shape of DNA. J. Mol. Biol. **11**: 373–390.

Tarnowski, G. S., Schmid, F. A., Cappuccino, J. G., and Stock, C. C. (1966). Chemotherapy studies in an animal tumor spectrum. II. Sensitivity of tumors to fourteen antitumor chemicals. Cancer Res. **26**: 181–206.

Tarnowski, G. S., Stock, C. C., and Hamilton, L. D. (1970). Unpublished work.

Tytell, A. A., Lampson, G. P., Field, A. K., and Hilleman, M. R. (1967). Inducers of interferon and host resistance. III. Double-stranded RNA from Reovirus type 3 virions (Reo 3-RNA). Proc. Natl. Acad. Sci. **58**: 1719–1722.

Wilkins, M. H. F. (1963). Molecular configuration of nucleic acids. Science **140**: 941–950.

Zedeck, M. S., Marquardt, H., Sternberg, S. S., Fleisher, M., and Hamilton, L. D. (1970). Polyriboinosinic-polyribocytidylic acid and component homopolymers: toxicity and adrenal cortical function. Proc. Natl. Acad. Sci. **67**: 180–184.

DISCUSSION

DR. W. BRAUN: Dr. Merigan and Dr. Colby, have you had an opportunity to measure the cytotoxicity of your modified polynucleotides in relation to their capacity to stimulate interferon?

DR. T. C. MERIGAN: No, I have not studied that.

DR. W. BRAUN: Has anybody? It seems so apparent that the increased stability may also mean increased damage to membranes and there is increasing evidence that the stimulatory effects may be mediated by membrane events. This brings me to the second question. Dr. Colby talked about intracellular recognition. Does it have to be intracellular? Isn't it sufficient that you may have an interaction between a stable damaging polynucleotide and the membrane?

DR. C. COLBY: Even though we could not show any uptake of the labeled polynucleotide in the absence of DEAE-dextran, it is quite possible that the receptor site is at the surface. If the whole thing occurs on the surface of the cell, then the DEAE-dextran may simply expose a previously covered site that was not available to the polynucleotides. There are absolutely no data whatsoever to say that it is or is not the case.

DR. W. BRAUN: I am bringing this up because data to be presented later indicate that interferon stimulation represents a response to damage of the membrane, whereas for the stimulation of conventional immune responses a "tickling" of the membrane rather than damage to the membrane may suffice. Therefore, it seems to me that it is conceivable that whatever you do to stablize the polynucleotides will result in more membrane damage and, therefore, in more interferon stimulation. Is there anybody who has data that would contradict such a view?

DR. H. B. LEVY: I have some data comparable to those reported by Dr. Shugar of Poland. We introduced polymethoxy instead of the hydroxyl in the carbon 2 of the sugar. Dr. Shugar tells us that this compound is quite resistant to ribonuclease action but it appears to be quite ineffective as an interferon inducer when coupled with poly I. In other words, poly I-poly I methoxine, while stable to the enzymatic degradation, appears to be a very poor interferon inducer.

DR. T. C. MERIGAN: The point I would like to make is that there are many types of ribonucleases, and I think the 2′ *o*-methyl modification inhibition of ribonuclease susceptibility depends on a cleavage mechanism based on a 2, 3-intermediate which is postulated for pancreatic ribonuclease, but it certainly is not true with all ribonucleases, or even for endonucleases. I think the important thing we found out about our thiophosphate modification is that it produces resistance to a variety of different nucleases. Hence, I think the critical thing is the kind of ribonuclease that the cell has within it or on its surface.

DR. C. COLBY: The stability of the molecule is very important but there has to be something else. I don't think that any one of us would argue that the molecule must in fact be stable. But that is not the only consideration. I think there are two things going on.

DR. S. BARON: I would like to make one comment regarding the measurement of degradation of polynucleotides. The question always comes up as to whether you are measuring the critical fraction of the polynucleotide within the cell. It may be that only a small fraction of the material taken up is doing the job, and that the degradation rate is being measured on the wrong fraction.

DR. T. C. MERIGAN: I think that is a good point and I would like to extend it to say that if you could relate degradation to biologic activity, it would be useful in understanding mechanisms.

DR. R. M. MORRELL: My question is directed to Dr. Hilleman. I am interested in the pyrogenicity of poly IC in humans and I wonder if there is any information on the role of endogenous pyrogens.

DR. M. R. HILLEMAN: Dr. Charles Young of the Sloan-Kettering Institute carried out the clinical work in the studies I reported so I should like to ask him to respond to this question.

DR. C. W. YOUNG: I am a firm believer in the concept that most of the pyrogenicity of various biological molecules is mediated through endogenous pyrogen. Where is the endogeneous pyrogen coming from? Is this simply identical to an endotoxin reaction in man? There are certain similarities but there are also certain differences. In man, endotoxin gives $1^1/_2$ hours latent period between the administration of the endotoxin and the onset of the chill phase of the fever. Endotoxin fever is dose-related, but generally it lasts about 8 hours. With poly IC there has been a 6-hour latent period in the majority of our patients, with a maximal fever at 12–24 hours; and it has taken roughly 48 hours to disappear. I assume that the majority of administered poly IC is taken up in the reticulo-endothelial system, primarily in the liver and the spleen. This is, however, more a matter of belief than scientific conclusion based on experimental data.

DR. W. REGELSON: Cyclohexamate incubation with poly IC increases the yield of interferon and, therefore, the question arises as to whether this is related to an effect on preformed assembly and release of interferon or whether this involves protease activation. The question I ask in view of Dr. Braun's concept of greater degrees of damage is this: Has anybody ever looked at whether or not some of the differences in regard to interferon inducibility is a function of protease destruction of interferon?

DR. W. A. CARTER: The alternative hypotheses—*de novo* synthesis of interferon versus release of preformed polypeptide—has been studied by Dr. Jan Vilcek, who might like to comment along these lines.

DR. J. VILCEK: We have quite a few data relating to the effect of inhibitors of protein synthesis and RNA synthesis on the production of interferon, but much more work is needed to interpret them. Regarding the possible role of proteases, we have found that the inhibition of protease activity results in a decrease in interferon.

DR. J. P. EBEL: We have studied viral multiplication in the presence of RNA's that were extracted from the host cells. As long as these RNA's are not chemically modified, they have no action in viral multiplication, but when they are modified by various chemical reagents, such as methylation, halogenation, acetylation, or action of nitrous acid, they induce an inhibition of viral multiplication that is similar to that of poly IC. All of these modifications induce an alteration of the secondary structure of the RNA's and also produce a limited increase of resistance to nucleases, but we don't think that this increase is sufficient to explain the inhibitory effects on viral multiplication. I think there must be something else to explain these effects.

DR. C. COLBY: What is the source of your RNA and how much is required for the antiviral activity?

DR. J. P. EBEL: The RNA was extracted from chick embryo filbroblasts, the virus was sendaï, and the concentration of RNA was 25 µg.

DR. C. COLBY: This is a high concentration of RNA compared with the amounts of double-stranded natural or synthetic RNA's required to elicit antiviral activity in chick cells. Until you characterize the species of RNA that is active, I don't think that we can comment on the specificity of induction in your system.

DR. H. B. LEVY: Just to add a point of historical perspective, Isaacs, quite a number of years ago, found that RNA extracted from chick embryo filbroblast cells was ineffective in inducing interferon, but when treated with nitrous acid it became effective in inducing interferon in chick embryo cells.

DR. A. D. STEINBERG: Dr. Hilleman has suggested that poly I-poly C in peanut oil might be useful in conjunction with another antigen such as the viral antigen antibody production. We have immunized a variety of strains of mice with poly I-poly C in adjuvant and obtained antibodies to the polynucleotides. Of course, immunogenicity is not an absolute contraindication to the use of a drug but some caution, perhaps, is indicated in the use of poly IC along with the adjuvant in this regard.

DR. W. BRAUN: I think you are referring to an immunogenicity of poly IC in certain strains of mice. Poly IC alone or with oily adjuvant cannot evoke the formation of antibody in most strains of animals, I believe.

DR. A. D. STEINBERG: Poly I-poly C in adjuvant will induce antibodies directed to poly IC in all strains of mice tested.

DR. W. BRAUN: In all strains of mice tested?

DR. A. D. STEINBERG: Yes.

DR. C. W. YOUNG: I would like to add one comment on the question of antigenicity of poly IC, relating specifically to the experience with patients. There is little question that poly IC is a potent reticulo-endothelial stimulant, but in comparison, let's say, with protein antigens, it is not a particularly effective antigen. We have had one allergic reaction in about 80 patients.

I think that the National Cancer Institute has seen one also. If we were to take a protein antigen, we would expect allergic reactions in all of the patients based on the way we are giving the drug. Thus, I think that there is no question that it does have antigenic properties, but on a scale of 1 to 10. I would take it down around the level of 2.

DR. R. DOUTHART: Is interferon induction a one-hit phenomenon? If it is, how can you compare one inducer with another one if you compare them on a milligram basis without any data with respect to molecular weight?

DR. C. COLBY: The answer to the first question is, we don't know. As to the second question, probably the best we can do is to recognize that our polymers are heterogeneous with respect to molecular weight, but that there is probably an upper limit and a lower limit. There is probably a threshold size that must be met for a variety of reasons, for uptake, for interaction, and so on. There is probably an upper size limit above which the cell simply cannot handle the material, and my intuition would suggest that the polymers in between these upper and lower limits would be reasonably similar in their effect.

DR. R. DOUTHART: The point I was trying to make is this: If you would take a polymer, say of the molecular weight of 5000, and another polymer of a molecular weight of 50,000, and compare these on a weight basis, you could be misled. Do I make my point clear?

DR. C. COLBY: Your point is quite clear, and the simple fact is that none of us who has worked with these polymers has gone to the trouble to select an exact molecular size for our experiments, and therefore your criticism is perfectly valid.

DR. L. D. HAMILTON: We have begun to test if we can get a statistically significant correlation between biological effects, antiviral effects, antitumor effects, and molecular size, and it does look at the moment as though the molecular weight is important; in some potent preparations you seem to get up to, we will say, about 30,000,000; if you go above 30,000,000 you are running into a sort of negative situation and you are doing more harm than good as far as interferon and the human response is concerned.

DR. A. FORLANO: Dr. Hamilton, I was impressed with your data concerning the reduction in tumor size by some of the synthetic polynucleotides. However, one point strikes my attention. If we assume that the neoplastic processes are associated with a change in the genetic material of DNA molecules, then it becomes difficult for me to see how interferon will reduce the tumor size simply by interference with viral multiplication. Is it possible that interferon is in fact inhibiting the multiplication of tumor cells because of a similarity between the virus and the tumor cell? In other words, the implication here is that in the transformation process the tumor cell may have become somewhat virus-like. How about the effect of interferon on a tumor cell that has been produced by a carcinogenic chemical?

DR. L. D. HAMILTON: The antitumor effects we have observed were not necessarily due to the fact that we were stimulating interferon. We just don't know. We might just as well be stimulating, inducing, or promoting the release of other proteins, of which interferon is one of the more obviously measurable materials; for example, it was measurable in the hoof-and-mouth diseases experiment. But there could be many other mechanisms that are responsible in the regression of the tumor. I didn't mean to imply that it was an interferon mechanism.

DR. H. B. LEVY: With regard to the question of whether or not interferon is responsible for all of these effects, I think that the question is still very much up in the air, but I would point out that Dr. Griffith, using interferon in tissue culture, has found that interferon inhibits the growth of certain tumor cells fairly strongly in tissue culture, while not affecting the growth of normal cells. So at least in one system, interferon itself is able to affect the growth of a tumor cell in tissue culture.

DR. F. S. PHILIPS: I would like to point out that the kinds of doses that Dr. Hamilton and others have used in mice to demonstrate antitumor effect are doses that can quickly produce extreme cytotoxic effects in the mouse, for example, in the epithelium of the intestines, in which lesions can be seen as early as 1 hour after polynucleotide administration. It is possible that these are due to interferon, but I think not, and I rather suspect that there is just as much likelihood that there is something about the cytotoxicity of the polynucleotides *in vivo* that we simply don't understand, and this may have nothing to do with interferon release. The actual breakdown of the cell may be based on something else that we don't yet understand, and I suspect that if we would understand the antitumor responses, we might also understand the breakdown of certain normal cells that occurs at comparable doses.

DR. J. NIBLACK: Dr. Hilleman, in your discussion you suggested that it might be possible to avoid a hyporeactive state to poly IC by giving a subsaturating dose so that not all cells would be triggered to induction. Yet Dr. Levy presented data in which he dosed humans intravenously with as little as 0.07 mg poly IC per kilogram, and he seemed to get a hyporeactive state. Would you comment on the effects of his observations in your speculation?

DR. M. R. HILLEMAN: The concept is that one would administer repeated doses, an amount of poly IC that would suffice to induce general resistance to viral infections but would not trigger off all interferon-producing cells. By so doing, one might be able to maintain resistance with repeated small doses without running into the problem of hyporeactivity. We have evidence that this regime might work in man in that repeat doses, even fairly large ones, permitted retention of circulating interferon in man for several days. The data on repeated doses in animals presented by others support this finding. More data are required to test this but early data give a basis of hope that the approach will work in a practical sense.

DR. J. G. GALLAGHER: I would like to direct this question to Drs. Hilleman and Baron. What are the implications and the possible consequences in man of the hyporesponsive state, regardless of whether it is total or relative hyporeactivity? It seems to me that there might be a possibility in man of either making him hypersusceptible to viral agents or reactivating latent viruses by the strategy of using poly I-poly C for relatively innocuous viral infections. As a corollary to this, using poly I-poly C against rhinovirus, or influenza virus infections, at a dose effective for them, may well be totally inadequate for preventing, and may even predispose to, more serious viral infections; for example, lower respiratory tract infections, such as virus pneumonia frequently seen in children with chronic leukemia and also respiratory syncytial virus infection, and perhaps a host of other naturally occurring viral infections in man.

DR. M. R. HILLEMAN: This is an important question and is one that will need to be answered, in the clinical sense, by very careful studies. There is some evidence to indicate, however, that loss of interferon-induced resistance may correspond roughly in time with loss of the hyporeactive state. Additional stimulation of less than all the potential interferon-producing cells may provide a margin of safety.

DR. J. G. GALLAGHER: The reason I raised the question of possible undesirable consequences of the hyporeactive state in man is that from *in vitro* experiments it appears there are tremendous differences in the relative sensitivities of human viruses to human interferon. There may be a thousand-fold difference between the sensitivity of influenza virus versus human adenovirus or cytomegalovirus and this may pose more than a theoretical problem *in vivo*.

DR. M. R. HILLEMAN: This is true. However, my guess is that we all are repeatedly stimulated to induce interferon—probably quite extensively. Also, there is a good possibility that the course of natural events has already tested the question many times over. All we can do at this time is to conduct meaningful extensive investigations to seek answers to these questions.

DR. S. BARON: I would like to add a comment in relation to Dr. Gallagher's question. Hyporesponsiveness may pose another problem for effective use of interferon inducers for therapy of certain established viral infections. During virus infection the virus generates interferon and a hyporesponsive period. Use of an interferon-stimulating drug during the period of viral-induced hyporesponsiveness may give a very poor interferon response which could impede therapy.

Part III

Stimulation of Antibody Formation and Cell-Mediated Immunity

SPECTRUM AND MODE OF ACTION OF POLY A:U IN THE STIMULATION OF IMMUNE RESPONSES

WERNER BRAUN, M. ISHIZUKA, Y. YAJIMA,*
D. WEBB, and R. WINCHURCH

Institute of Microbiology
Rutgers University, The State University of New Jersey
*New Brunswick, New Jersey***

It is now almost 20 years since we recognized that nucleic acids, apart from their specific informative properties based on nucleotide sequences, also can exert important regulating effects that are basically independent of nucleotide sequences. First in studies with bacteria, and later in studies with mammalian cells (compare Braun and Firshein, 1967), we as well as a few others (for example, Johnson, 1968) observed that oligo- and polynucleotides, in contrast to ineffective mononucleotides, can alter the rate of a number of biosynthetic events including rates of cell multiplication. When synthetic polynucleotides became available we tested their effects in a system that had proved susceptible to stimulation by natural oligo- and polynucleotides, namely antibody formation. At first, using single-stranded polynucleotides, such as poly A, poly U, poly C, and poly G, we found no effects. We were about to discard these homopolymers as inactive, when we made a final trial to determine the possible effectiveness of a mixture of these homopolymers and discovered, in tests on antibody formation to sheep red blood cells (sRBC) in mice, a pronounced stimulation (Braun and Nakano, 1967). The rest is now history, documented by much of what is being reviewed at this Symposium: Complementary homopolymers, such as double-stranded

* Senior Research Scientist, Miles Laboratories, 1969/70.
** These studies have been supported by NIH grants AM-08742 and AI-09343, NSF grant B9-0301 R and ACS grant 501. This review was prepared during the senior author's tenure of an NIH Special Fellowship.

poly A:U, poly C:G, or poly I:C, influence the behavior of cells involved in immune responses. The requirement for double- or multiple-stranded molecules turned out to be due to the relative resistance of such molecules to rapid enzymatic degradation, whereas single-stranded homopolymers, which were inactive *in vivo*, but are active *in vitro*, are too rapidly degraded in the animal (Braun and Nakano, 1967).

Much of the work on the capacity of polynucleotides to stimulate antibody formation was carried out with mice or with murine spleen cultures, and one of the most frequently employed tests has been to assay the number of spleen cells that form antibodies to sRBC. It is relatively simple to perform such assays with the aid of the now practically ubiquitous technique of localized hemolysis in gel (Jerne *et al.*, 1963). When one performs such tests on spleen cell populations of mice immunized with sRBC, one finds a gradually increasing number of antibody-forming cells (AFC), which becomes easily detectable 24–48 hours after immunization and, as far as 19S antibody-forming cells are concerned, reaches a maximum 4–5 days after immunization. As illustrated in Table 1, the simultaneous administration of

Table 1. Stimulatory Effects of Different Polynucleotide Combinations on the Number of Spleen Cells Forming Anti-sRBC in CF1 Mice Immunized 48 Hours Earlier with 10^8 sRBC

Treatment of spleen donors	Average number (± S.E.) of hemolysin-forming cells per 10^8 nucleated spleen cells (48 hours after treatment)
None	23.3 ± 11.4
sRBC	347.3 ± 88.9
sRBC + A:U[a] (150 γ of each polynucleotide)	1168.9 ± 345.5
sRBC + A:U (50 γ of each polynucleotide)	531.1 ± 190.6
sRBC + A:U (10 γ of each polynucleotide)	352.2 ± 46.1
sRBC + G:C[a] (150 γ of each polynucleotide)	1258.3 ± 308.6
sRBC + G:C (50 γ of each polynucleotide)	634.2 ± 294.9
sRBC + G:C (10 γ of each polynucleotide)	284.0 ± 38.3
sRBC + I:C[a] (300 γ of the complex)	1755.1 ± 433.5
sRBC + I:C (100 γ of the complex)	874.9 ± 160.0
sRBC + I:C (20 γ of the complex)	900.1 ± 151.0

5 mice per group.

[a] A:U = poly A + poly U.
G:C = poly G + poly C.
I:C = poly I + poly C.

double-stranded synthetic polynucleotides enhances the early rate of such increases, the most pronounced difference being usually detectable 48 hours after immunization. The stimulatory polynucleotides do not substantially affect the terminal number of AFC but they seem to stimulate their performance, as indicated by higher titers of circulating antibodies. Table 2 gives an example of such stimulation, showing circulating precipitating antibodies to a synthetic polypeptide, (T, G)-A—L, in rabbits immunized either with the antigen in complete Freund's adjuvant (C.F.A.), as alum precipitate, or as alum precipitate plus poly A:U (data from unpublished studies by R. Maron, F. Fuchs, M. Sela, and W. Braun). It is obvious that the average initial response is elevated as effectively by nontoxic poly A:U as by the relatively toxic C.F.A. However, the prolonged maintenance of high titer is accomplished more effectively by C.F.A., which, by virtue of its oily vehicle, releases antigen and stimulators gradually over a prolonged period of time; poly A:U, as currently administered, exerts more short-lived, transient stimulatory effects. Even more impressive data on stimulatory effects of poly A:U on circulating titers in mice have been reported by Johnson and associates (Johnson, 1970) in studies with BGG as antigen and have been confirmed in some of our own studies on responses of mice to heterologous serum proteins. Other antigens that have been employed by us and others in successful tests on stimulatory effects of synthetic polynucleotides include various types of red blood cells, synthetic polypeptides, key hole limpet hemocyanin, viruses, tumor cell antigens, and bacterial toxins. Species in which stimulatory effects of synthetic polynucleotides have been observed include mice, rats, chickens, rabbits, and guinea pigs.

Polynucleotides can stimulate primary as well as secondary responses, 19S antibodies as well as 7S antibodies (Braun *et al.*, 1968). In addition, we have demonstrated (Braun *et al.*, 1970a) that poly A:U, given *without specific antigen* but in conjunction with a modifier of lymphocyte permeability, such as diluted antilymphocyte serum, can evoke significant booster-type responses to previously experienced antigenic stimuli. In this respect, polynucleotides, when combined with modifiers of lymphocyte permeability, can mimic some of the apparently nonspecific stimulatory effects of bacterial endotoxins (Braun *et al.*, 1969).

Polynucleotides not only enhance antibody formation in normally responding animals, but can also stimulate responses under conditions in which they are ordinarily weak or impaired. Thus, in newborn mice antibody formation to sRBC ordinarily does not occur in detectable amounts until the animals are 8–12 days old, the actual time depending on the strains employed (Hechtel *et al.*, 1965). However, if the antigen is administered in conjunction with poly A:U or poly I:C (Table 3), a premature initiation of antibody formation

Table 2. Effect of Poly AU on the Immune Response of Rabbits to a Synthetic Antigen, (T, G)-A—L. Results of Immune Responses are Expressed as Optical Density at 280 mμ of Immune Precipitates Obtained at Equivalence with 0.5 ml of Antiserum

	i	i	i	b	i	i	i	b	b	b	i	i	b	b
Day	0	3	3	8	10	11	13	19	25	34	50	51	58	65
Group 1 *C.F. Adj.*	1 mg in C.F.A. i.d.			— — — —	1 mg in C.F.A. i.d.			0.400 0.375 0.165 0.095	0.540 0.420 0.120 0.065	0.900 0.550 0.225 0.130	1 mg in C.F.A. i.d.		0.830 0.700 0.210 0.090	n.d.
Mean max. precip.								0.258	0.286	0.451			0.457	
Group 2 *Alum*	1 mg in Alum i.d.			— — — —	1 mg in Alum i.d.			0.320 0.070 0 0	0.320 0.030 0 0	0.075 0 0 0	1 mg in Alum i.d.		0.100 0.100 0 0	0.055 0.020 0 0
Mean max. precip.								0.100	0.057	0.018			0.050	0.018
Group 3 *Alum + pAU*	1 mg in Alum + 1 mg pAU i.d.	1 mg pAU i.v.	1 mg pAU i.v.	— — — —	1 mg in Alum + 1 mg pAU i.d.	1 mg pAU i.v.	1 mg pAU i.v.	0.330 0.290 0.200 0.080	0.230 0.200 0.090 0	0.085 0.050 0 0	1 mg in Alum + 1 mg pAU i.d.	1 mg pAU i.v.	0.150 0.130 0.055 0.012	0.070 0.060 0 0
Mean max. precip.								0.225	0.130	0.034			0.087	0.032

i: immunization.
b: bleeding.

Table 3. Effect of Poly I:C, Poly A:U, and *Serretia marcescens* Endotoxin (LPS) on the Initiation of Antibody Formation to sRBC in 4-day-old C57 Bl Mice (From Winchurch and Braun, 1969.)

Treatment of spleen donors	Average number (± S.E.) of antibody-forming cells per 10^8 nucleated spleen cells 4 days after immunization
10^9 sRBC	6.2 ± 6.0
10^9 sRBC + 30 γ poly I-C	364 ± 45.4
10^9 sRBC + 30 γ poly A:U	264.5 ± 72.5
10^9 sRBC + 15 γ LPS	755.7 ± 101.4

Number of animals per group: 4–6.

is produced (Winchurch and Braun, 1969). The same effect can be obtained after transfer of peritoneal macrophage populations from adult animals into the newborn (Braun and Lasky, 1967; Argyris, 1968) or after the administration of bacterial endotoxin (Winchurch and Braun, 1969), a known stimulator of macrophage activity. Thus, these experiments, together with others to be mentioned subsequently, indicate that members of the macrophage population are one target of stimulation by polynucleotides.

As illustrated in Table 4, polynucleotides also restore impaired antibody responses in aging C57 B1 mice (Braun, 1970), and our earlier finding (Braun *et al.*, 1963) that oligonucleotides derived from natural sources can postpone the onset of spontaneous mammary carcinoma in aging C3H mice

Table 4. Influence of Poly A:U, Poly G:C, or Poly I:C on the Response to sRBC in Over 13-Month Old Retired C57 Bl Breeders (From Braun, 1970.)

Treatment of spleen donors	AFC/10^8 spleen cells	
	Day 2	Day 3
	41.2 ± 16.5	
sRBC (10^8)	39.8 ± 12.5	63.3 ± 6.9
sRBC + poly A:U[a]	328.1 ± 81.5	706.3 ± 30.7
sRBC + poly G:C[a]	289.7 ± 57.7	741.5 ± 256.8
sRBC + poly I:C[a]	251.3 ± 36.3	1206.5 ± 281.8

[a] 150 γ of each of the homopolymers i.v./mouse.

may represent another example of such stimulatory effects in animals that have lost their capacity to muster normal immune responses. It is interesting to note that antibodies to nucleic acids are known to occur spontaneously in aging C57 B1 animals (Friou and Teague, 1964) at about the time that we observed impaired antibody responses to sRBC. Since it has been demonstrated that stimulatory nucleic acids are released by spleen cells in the course of normal antibody responses (Nakano and Braun, 1967), it is possible that such stimulators are prerequisite to any "normal" response and that their inactivation by antibody in the aging C57 B1 animals might be the cause of their impaired responses. Similarly, the functions of the humoral thymus factor in normal immune responses (Small and Trainin, 1967; Rosemoer *et al.*, 1970) may be similar to those of stimulatory polynucleotides, particularly since synthetic polynucleotides have been shown to repair at least some of the immunosuppresive effects of thymectomy (Johnson, 1970). Polynucleotides also can repair the damaging effects of X-radiation on antibody formation (Jaroslow, 1968) and can assist in the production of normal responses in genetically low responder strains (Mozes *et al.*, 1970). The capacity of multiple-stranded synthetic polynucleotides to produce respectable responses to normally weak immunogens is also reflected in their ability to depress the rate of growth of syngeneic tumor cells (Figure 1), but it remains to be established whether this effect, which in our studies has so far been limited to the early period of tumor development, is dependent on a stimulation of cytotoxic antibodies, cell-mediated immune responses, or other antitumor factors.

Synthetic polynucleotides can stimulate both humoral antibody formation as well as cell-mediated immune responses. The latter effect is illustrated by the finding that guinea pigs, sensitized to tuberculin, will show positive skin reactions 1 week later if they receive one injection of either poly A:U (Webb and Braun, unpublished data) or complete Freund's adjuvant (Maguire, 1970) 3 days after sensitization with plain tuberculin. Without the administration of such stimulators, comparable skin reactions will not develop for 14 days. Others (H. Cantor, personal communication; Turner *et al.*, 1970) have observed stimulatory effects of synthetic polynucleotides on graft versus host responses and on isograft rejection.

In most of our studies we have preferred to employ poly A:U rather than poly I:C because the former, in contrast to the usually employed concentrations of the latter, is nontoxic and nonpyrogenic. This difference appears to be associated with the slow rate of depolymerization of poly I:C and attendant cytotoxic effects, whereas poly A:U is depolymerized more rapidly and thus is changed more rapidly from a cytotoxic molecule to noncytotoxic derivatives (Braun, 1969). It also appears that the cytotoxicity of poly I:C

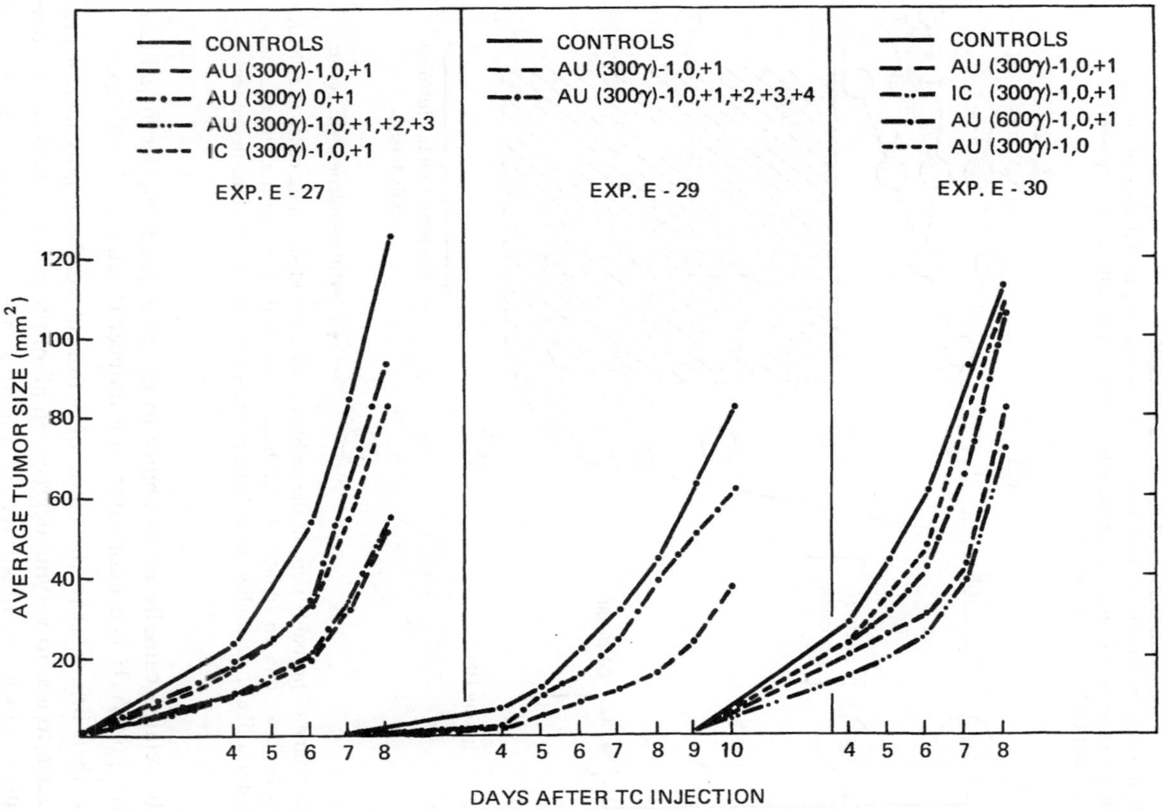

Fig. 1. Effect of poly A:U and poly I:C on the growth of MCDV-12 ascites tumor cells in syngeneic Balb/c mice. Ascites cells (2×10^6) were inoculated intradermally and the polynucleotides were injected intravenously on the days and in the amount indicated.

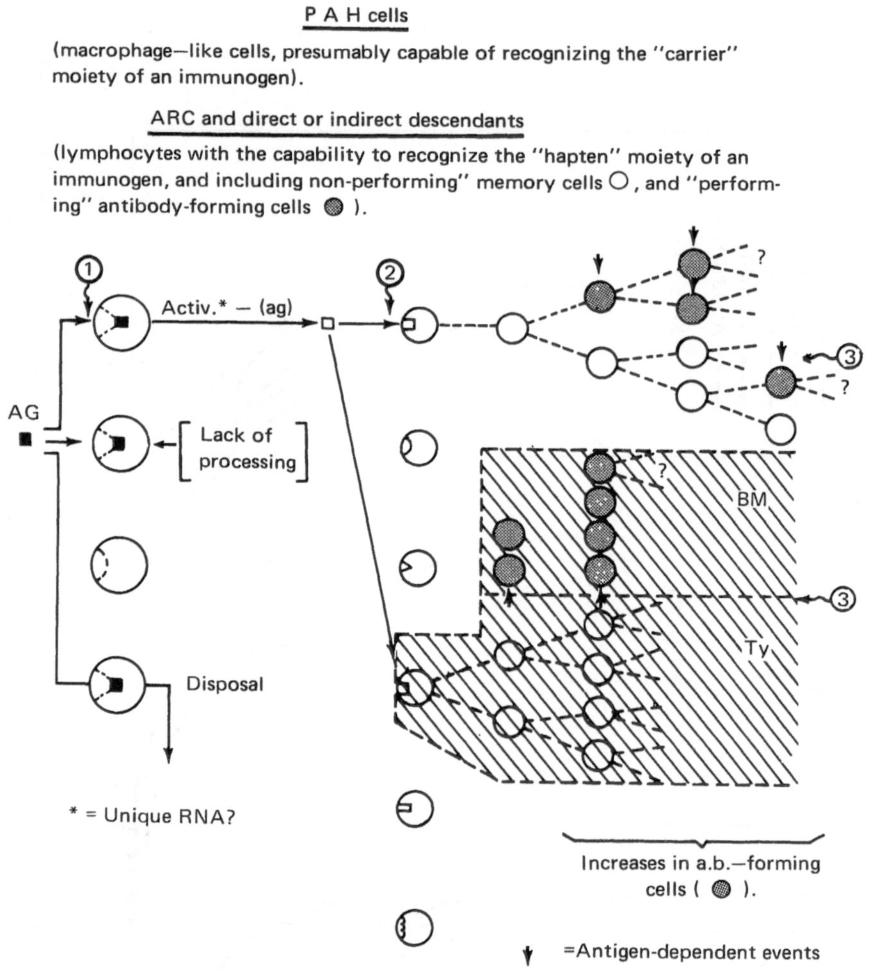
P A H cells
(macrophage—like cells, presumably capable of recognizing the "carrier" moiety of an immunogen).
ARC and direct or indirect descendants
(lymphocytes with the capability to recognize the "hapten" moiety of an immunogen, and including non-performing" memory cells ○, and "performing" antibody-forming cells ●).
1
2
3
Activ.* – (ag)
AG
Lack of processing
Disposal
* = Unique RNA?
BM
Ty
?
Increases in a.b.—forming cells (●).
=Antigen-dependent events

Fig. 2. Outline of probable steps in the activation of antibody-forming cells. The shaded portion indicates one possibility, the unshaded section above it indicates an alternative, but less likely, possibility. BM: bone-marrow derived; Ty: thymus-dependent.

may be related causally to its ability to act as a good interferon inducer, whereas poly A:U is a poor interferon inducer (Braun *et al.*, 1970a).

This then brings us to the mode of action of synthetic polynucleotides. They seem to act, to varying degrees, on all of the cells involved in immune responses, which, as illustrated in Figure 2, to the best of our present knowledge, include (1) primary antigen handling cells (PAH cells) that are part of the macrophage population, (2) thymus-dependent precursor cells of antibody-forming cell populations (which perhaps are also the cells

Table 5. Influence of Poly A on the Number of Antibody-Forming Cells (AFC) in 3-Day-Old Spleen Cell Cultures Initiated with Different Numbers of Cells. The Antigen Used is sRBC, 1% v/v. All Values are Based on Duplicate Determinations on Duplicate Cultures

Cultures initiated with			AFC/10^6	
Cells	Ag	Poly A	Exp. 59	Exp. 63
2×10^7	—	—	16.0	16.1
1.5×10^7	—	—	7.3	9.1
1×10^7	—	—	9.5	7.1
2×10^7	+	—	118.3	146.1
1.5×10^7	+	—	131.6	106.1
1×10^7	+	—	81.2	111.9
2×10^7	+	10 γ	269.1 (2.3)[a]	233.5 (1.6)
1.5×10^7	+	10 γ	158.9 (1.2)	199.0 (1.9)
1×10^7	+	10 γ	245.7 (3.0)	270.7 (2.4)
2×10^7	+	1 γ	205.9 (2.3)	228.3 (1.6)
1.5×10^7	+	1 γ	284.6 (2.2)	135.3 (1.3)
1×10^7	+	1 γ	158.4 (2.0)	141.1 (1.3)

[a] Values in parentheses represent the ratio of AFC in poly A-supplemented cultures to AFC in poly A-free cultures.

participating in cell-mediated immune responses), and (3) bone-marrow-derived antibody-forming cells.

Polynucleotides (1) stimulate functions of PAH cells, (2) increase the rate of multiplication of precursor cells of antibody-forming cells and possibly of the antibody-forming cells themselves, and (3) affect the step leading from precursor cell to antibody-forming cell (③ in Figure 2). Some effects of the polynucleotides, especially the early effects on PAH cells, appear to be due to the polyanionic nature of the initial form of the stimulatory material; other effects appear to be due to oligonucleotides and possibly also to other derivatives that are produced as the result of enzymatic alterations of polynucleotides *in vivo* and *in vitro*. The polymerized anionic molecule, like many other polyanions (Cohn and Parks, 1967; Braun *et al.*, 1970b), act on membranes, whereas the derivatives may act intracellularly, perhaps in part on transfer RNA molecules (Braun *et al.*, 1968).

These conclusions are based both on *in vivo* studies and on *in vitro* tests, particularly on tests in which it has been possible to analyze separately the effects of polynucleotides on macrophages and lymphocyte-rich spleen cell populations. There is one important difference between *in vivo* tests and

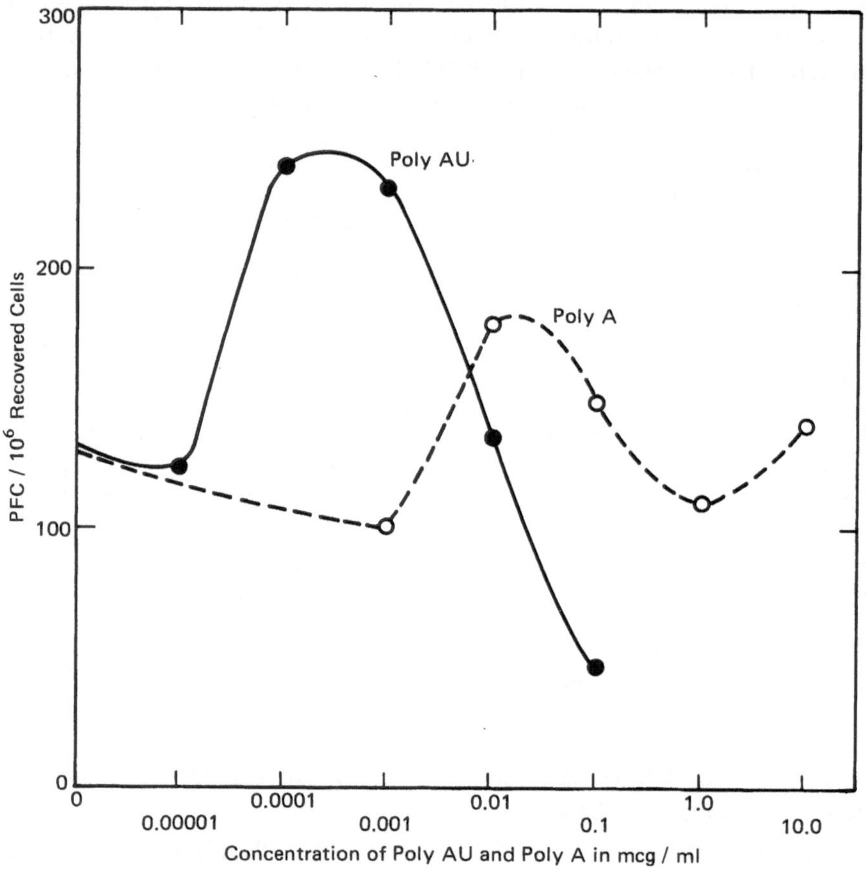

Fig. 3. Effects of poly A and poly A:U on antibody formation to sRBC *in vitro*. The results represent averages of triplicate cultures assayed on day 3.

in vitro tests: *In vivo* only double- and multiple-stranded polynucleotides are active; *in vitro*, presumably due to quantitative and qualitative differences in enzymes, single-stranded synthetic homopolymers, such as poly A, are active (Table 5), and so are poly I, poly C, poly G, and poly U; they produce enhancing effects over a wide range of concentrations, but stimulatory effects of double-stranded molecules, such as poly A:U, also have been obtained at concentrations that are much lower than those required for single-stranded polynucleotides (Figure 3). Others, specifically Stout and Johnson (1970), have reported stimulatory effects of poly A:U *in vitro* on rosette-forming lymphocytes from primed animals.

The following *in vivo* and *in vitro* observations shed some additional light on the mode of action of polynucleotides. To begin with, the ability of

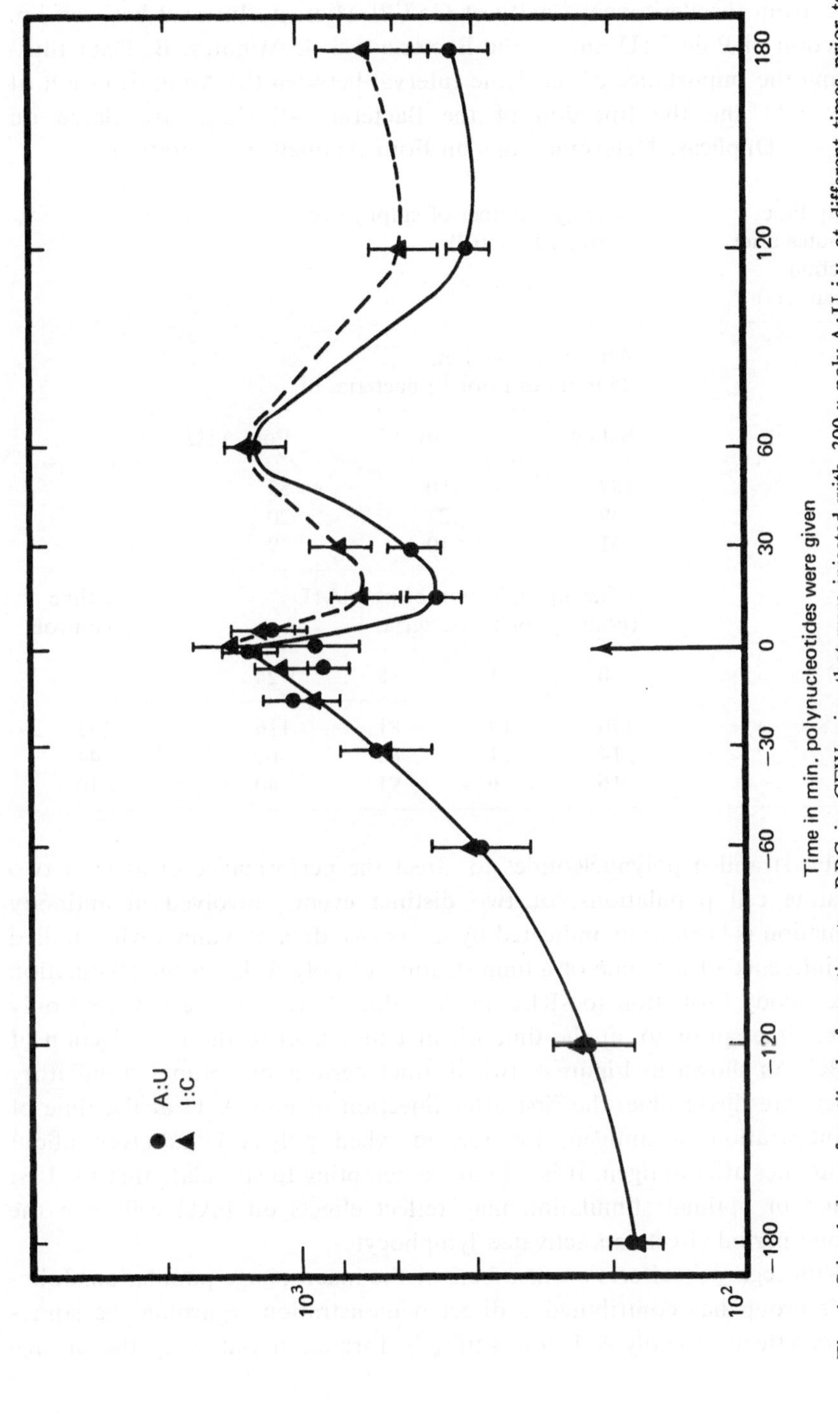

Fig. 4. Stimulation of antibody formation to sRBC in CFW mice that were injected with 300 γ poly A: U i.v. at different times prior to or subsequent to i.v. administration of sRBC.

Table 6. Influence of Poly A:U (300 γ/Mouse) on the Clearance of Staphylococci from the Peritoneal Cavity of C57/Bl Mice. **A.** Interval between i.p. Injection of Poly A:U and of the Bacteria was 45 Minutes. **B.** Data Illustrating the Importance of the Time Interval between the Administration of Poly A:U and the Injection of the Bacteria. All Data Are Based on Duplicate Determinations on Four Animals per Notation

Assay time (minutes after injection of bacteria)	Average number of staphylococci recovered ($\times 10^3$)				
A.	After i.p. injection, 45 minutes prior to bacteria, of				
	Saline	Poly A	Poly A:U		
21	187	150	53		
43	39	27	20		
63	31	10	9		
B.	After i.p. injection of poly A:U (hours prior to bacteria)				Saline controls
	0	1	8	24	
25	120	19	81	116	163
50	14	11	80	62	44
75	16	6	91	40	10

double-stranded polynucleotides to affect the performance of at least two separate cell populations, or two distinct events, involved in antibody formation is beautifully indicated by the *in vivo* data of Yajima who studied the influence of the time of administration of poly A:U on the stimulation of antibody formation to sRBC *in vivo*. Poly A:U was injected (i.v.) only once, either prior to, at the time of, or subsequent to the i.v. injection of sRBC. As shown in Figure 4, two distinct periods of optimal stimulatory effects are discernible, the first after injection of poly A:U at the time of administration of antigen, the second when poly A:U is given about 90 minutes after antigen. It is, of course, tempting to speculate that the first period of optimal stimulation may reflect effects on PAH cells and the second period effects on activated lymphocytes.

With regard to effects on members of the macrophage population, Johnson's group has contributed a direct demonstration regarding the stimulatory effects of poly A:U on antibody formation following the *in vitro* exposure and subsequent *in vivo* injection of peritoneal macrophage preparations (Johnson and Johnson, 1968), and we have observed that *in vivo* clearance of bacteria by peritoneal cells is greatly stimulated, provided 45–60 minutes elapse between i.p. poly A:U injection and the i.p. injection of bacteria (Table 6).

The effect of polynucleotides on lymphocytes is indicated by the *in vivo* data on enhanced rate of increase in antibody-forming cells in animals receiving antigen and poly A:U, and by the effects of poly A:U plus ALS

Table 7. Results from Two Separate Experiments Illustrating Stimulating Effects of Delayed Addition of Poly A (1 γ/ml) to sRBC-containing Spleen Cell Cultures

Culture conditions	AFC/10^6		
	Exp. 64	Exp. 75	
	(1)	(1)	(2)
No antigen	7.1	15.6	7.1
Antigen	111.9	256.4	298.2
Antigen + poly A at 0 hrs	255.5	338.9	307.4
Antigen + poly A at 24 hrs	219.4	249.8	440.5
Antigen + poly A at 48 hrs	218.1	410.9	497.2

(1) Data obtained from 3-day-old cultures.
(2) Data obtained from 4-day-old cultures.

on memory cells to which we referred earlier. Also, the *in vitro* data of Stout and Johnson (1970) regarding effects of poly A:U on rosette-forming cells belong into this category.

The effects of poly A on lymphocytes *in vitro* is supported by the already mentioned preliminary observations on the effects of separate treatment of macrophage- and lymphocyte-rich populations, and also by the ability of poly A to stimulate AFC after delayed addition of the polynucleotide to cultures (Table 7). Recent tests have revealed that an exposure of spleen cells *in vitro* for as little as 2 hours, followed by washing of the cells to remove any non-cell-associated polynucleotides, suffices to produce a significant enhancement of subsequent increases in AFC. Such effects can be produced when poly A is added to the culture either 4 or 24 hours after initiation of the culture. Furthermore, spleen cell populations of low density that are rich in lymphocytes yield unusually high numbers of AFC *in vitro*, and cultures initiated with such subpopulations can be stimulated by poly A. We, as well as others (Shortman *et al.*, 1970) have observed that glass-adhering cells can interfere *in vitro* with the performance of lymphocytes. Poly A, at least at some concentrations that stimulate antibody-formation *in vitro*, has a selective inhibitory effect on the adhering cells and thus its overall stimulatory effects *in vitro* may well be a combination of direct stimulatory effects on lymphocytes and inhibitory effects on cells that normally interfere with optimal performance of the lymphocytes.

We have also explored the effects of poly A on phytohemagglutinin (PHA)-induced lymphocyte transformation (Bach and Hirschhorn, 1964) *in vitro*. The effects were assessed by measuring the uptake and incorporation of tritiated thymidine. Addition of poly A with PHA at the beginning of the 3-day culture period yielded only occasional stimulatory effects, delayed addition of poly A to the culture always produced inhibitory effects. Such inhibiting effects were not detectable when tritiated cytidine, instead of thymidine, was used. Oligomers consisting of 2 or 4 adenylic groups stimulated PHA-induced effects on lymphocytes and proved to be more effective than oligonucleotides of larger size. In a related study, Friedman *et al.* (1969) have reported that poly A:U can enhance tuberculin-initiated increases in DNA synthesis of leukocyte cultures from tuberculin-positive human donors

Table 8. The Effects of cAMP (200 γ/Mouse i.p.) and of Theophylline (200 γ/mouse i.p.) on Antibody Formation to sRBC in CFW Mice, Tested in the Absence and Presence of Concommitant Administration of Poly A:U (3–300 γ/Mouse i.v.) or of ALS (0.1 ml/Mouse i.p.). Five Animals per group (After Braun *et al.*, 1970c.)

Treatment of spleen donors	Average number (± S.E.) of AFC per 10^8 nucleated spleen cells after 48 hours
A. Unimmunized controls	61.1 ± 10.4
sRBC (10^8)	642.3 ± 75.0
sRBC (10^8) + cAMP	1303.2 ± 284.9
sRBC (10^8) + cAMP + ALS	3027.5 ± 238.6
sRBC (10^8) + ALS	487.4 ± 124.6
sRBC (10^8) + poly A:U	5904.7 ± 475.8
sRBC (10^8) + poly A:U + cAMP	4040.6 ± 705.7
B. Unimmunized controls	30.3 ± 4.6
sRBC (10^8)	493.1 ± 46.7
sRBC (10^8) + poly A:U (3 γ)	346.8 ± 38.9
sRBC (10^8) + poly A:U (3 γ) + theophylline	707.0 ± 55.7
sRBC (10^8) + poly A:U (300 γ)	1757.0 ± 267.5
sRBC (10^8) + poly A:U (200 γ) + theophylline	3230.7 ± 208.7

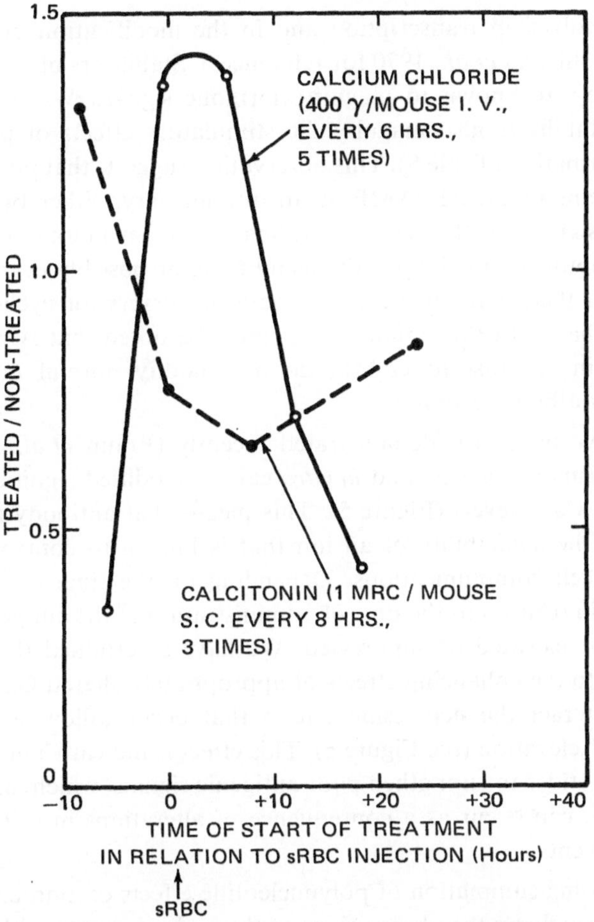

Fig. 5. The influence of calcium chloride and calcitonin, administered at different times prior to and after immunization with sRBC, on the number of antibody-forming spleen cells (AFC) in mice 48 hours after immunization. The results are given in terms of the ratio of average number of AFC in treated groups to the average number of AFC in the control group receiving sRBC only. Each group contained 5 CFW females, the figures adjacent to some of the points indicate average AFC/10^8 spleen cells $\pm$ S.E.

and also in mixed leukocyte interactions; in contrast, these double-stranded polynucleotides depressed PHA effects in such leukocyte cultures.

We have recently shown (Ishizuka *et al.*, 1970) that cells involved in antibody formation, just as hormone-dependent cells, are stimulated *in vivo* as well as *in vitro* by cyclic AMP. This finding suggests that as in the case of hormones, a membrane reaction between stimulus (antigen) and receptor site on the membrane may activate membrane-associated adenyl cyclase, which converts ATP into cyclic AMP, which in turn activates intracellular enzymes involved in transcription and in the modification of various enzymes (see Ishizuka *et al.*, 1970 for references). Stabilizers of cAMP, such as theophylline, are known to magnify hormone signals. We have observed that such stabilizers also magnify the stimulatory effects of poly A:U on antibody formation (Table 8). This observation suggests that polynucleotides act on systems involving cAMP as an intermediary, either by magnifying membrane signals, in the sense of adding further membrane effects to the antigen-receptor interaction on the membrane, or possibly even by furnishing, through their derivatives, appropriate precursors for cyclic nucleoside monophosphates. In this connection it must be noted that AMP, ATP, or adenosine, in contrast to cAMP, do not modify normal or poly A:U-stimulated antibody responses.

Finally, we have also demonstrated recently (Braun *et al.*, 1970c) that antibody formation *in vivo* and *in vitro* can be modified significantly by an elevation of Ca^{++} levels (Figure 5). This means that antibody formation is affected by the availability of an ion that is known to control membrane events and cell communications. Depending on the time of alteration of Ca^{++} levels in relation to the time of administration of the antigen, responses can either be elevated or suppressed. We have ascertained that poly A:U cannot add to the enhancing effects of appropriately altered Ca^{++} levels but it can counteract the depressing effects that occur following "early" or "late" Ca^{++} elevation (see Figure 5). This effect is indicative of the capacity of polynucleotides to strengthen poor antigenic signals, which, as these Ca^{++} data suggest, can occur as a consequence of alterations in cell membrane-associated events.

The foregoing summation of polynucleotide effects on immune responses has leaned heavily on the observations of the author's group without wishing to slight the important contributions of others reviewed elsewhere in this volume. However, we hope that this abbreviated compilation suffices to indicate that the exploration of the effects of polynucleotides on immune responses continues to have two agreeable features: It permits a study of basic mechanisms at the same time as it promises to provide us with better means for the control of immune responses in human and veterinary medicine.

References

Argyris, B. F. (1968). Role of macrophages in immunological maturation. J. Exp. Med. **128:** 459–467.

Bach, F. and Hirschhorn, K. (1964). Lymphocyte interaction: A potential histocompatibility test *in vitro*. Science **143:** 813–814.

Braun, W. (1969). Relationships between the effects of poly I, poly C and endotoxin. Nature Lond. **224:** 1024–1025.

——— (1970). Some causes and repair of altered antibody formation in aged animals. *In* Aging and Autoimmunity. Ed. by M. Sigel. C. C. Thomas, Springfield, Ill.

Braun, W. and Firshein, W. (1967). Biodynamic effects of oligonucleotides. Bact. Rev. **31:** 83–94.

——— Ishizuka, M., and Seeman, P. (1970c). Suppression and enhancement of antibody formation by alteration of Ca^{++} levels. Nature Lond. **226:** 945–946.

——— and Lasky, L. J. (1967). Antibody formation in newborn mice initiated through adult macrophages. Fed. Proc. **26:** 642.

——— and Nakano, M. (1967). Antibody formation: Stimulation by polyadenylic and polycytidylic acids. Science **157:** 819–821.

——— Regelson, W., Yayima, Y., and Ishizuka, M. (1970b). Stimulation of antibody formation by pyran copolymer. Proc. Soc. Exp. Biol. Med. **133:** 171–175.

——— (1968). Stimulation of antibody-forming cells by oligonucleotides of known composition. *In* Nucleic Acids in Immunology. 347–363. Ed. by O. J. Plescia and W. Braun. Springer-Verlag, New York.

——— Yajima, Y., Jimenez, L., and Winchurch, R. (1970a). Activation, stimulation and the occasional non-specificity of antibody formation. *In* Developmental Aspects of Antibody Formation and Structure. Ed. by J. Sterzl. Academic Press, New York.

Friedman, H. M., Johnson, A. G., and Pan, P. (1969). Stimulatory effect of polynucleotides on short term leukocyte cultures. Proc. Soc. Exp. Biol. Med. **132:** 916–918.

Friou, G. J. and Teague, P. O. (1964). Spontaneous autoimmunity in mice. Antibodies to nucleoprotein in strain A/J. Science **143:** 1333–1334.

Hechtel, M., Dishon, T., and Braun, W. (1965). Hemolysin formation in newborn mice of different strains. Proc. Soc. Exp. Biol. Med. **120:** 728–732.

Ishizuka, M., Gafni, M., and Braun, W. (1970). Cyclic AMP effects on antibody formation and their similarities to hormone-mediated events. Proc. Soc. Exp. Biol. Med. **134:** 963–967.

Jaroslow, B. N. (1968). Nucleic acids and induction of antibody synthesis in inhibited systems. *In* Nucleic Acids in Immunology. 404–413. Ed. by O. J. Plescia and W. Braun. Springer-Verlag, New York.

Jerne, N. K., Nordin, A. A., and Henry, C. (1963). The agar plaque technique for recognizing antibody-producing cells. *In* Conference on Cell-Bound Antibodies. Ed. by B. Amos and H. Koprowski. Wistar Institute Press, Philadelphia.

Johnson, A. G., Schmidtke, J., Merritt, K., and Han, I. (1968). Enhancement of antibody formation by nucleic acids and their derivatives. *In* Nucleic Acids in Immunology. 379–385. Ed. by O. J. Plescia and W. Braun. Springer-Verlag, New York.

Johnson, A. G. (1970). *In* Developmental Aspects of Antibody Formation and Structure. Ed. by J. Sterzl. Academic Press, New York.

Johnson, H. G. and Johnson, A. G. (1968). Enhancement of antibody synthesis in mice by macrophages stimulated *in vitro* with antigen and poly A-U. Bacteriol. Proc. **68**: 75.

Maguire, H. C. (1970). A unique rapid enhancement by adjuvant of delayed hypersensitivity to protein antigens in the guinea pig. Int. Arch. Allergy N.Y. **38**: 427–441.

Mozes, E., Shearer, G. M., Sela, M., and Braun, W. (1971). This Volume.

Nakano, M. and Braun, W. (1967). Cell-released nonspecific stimulators of antibody-forming spleen cell populations. J. Immunol. **99**: 570–575.

Rosenoer, V. M., Biano, G., and Brown, B. L. (1970). The thymus and immunocompetence: The target cell of the thymic factor. Proc. Soc. Exp. Biol. Med. **133**: 394–397.

Shortman, K., Diener, E., Russell, P., and Armstrong, W. D. (1970). The role of nonlymphoid accessory cells in the immune response to different antigens. J. Exp. Med. **131**: 461–482.

Small, M. and Trainin, N. (1967). Increase in antibody-forming cells of neonatally thymectomized mice receiving calf thymus extract. Nature Lond. **216**: 377–379.

Turner, W., Chan, S. P., and Chirigos, M. A. (1970). Stimulation of humoral and cellular antibody formation in mice by poly Ir: Cr. Proc. Soc. Exp. Biol. Med. **133**: 334–338.

Winchurch, R. and Braun, W. (1969). Antibody formation: Premature initiation by endotoxin or synthetic polynucleotides in newborn mice. Nature Lond. **223**: 843–844.

STIMULATION OF THE IMMUNE SYSTEM BY HOMOPOLYRIBONUCLEOTIDES*

A. G. Johnson, R. E. Cone, H. M. Friedman, I. H. Han, H. G. Johnson, J. R. Schmidtke, and R. D. Stout

The Department of Microbiology, The University of Michigan Ann Arbor, Michigan

Stimulation of the immune response by adjuvants, such as endotoxins, nucleic acids, and smaller molecular weight oligonucleotides, has been under investigation in our laboratory (Johnson *et al.*, 1968), with the hope that heretofore subliminal events leading to primary antibody synthesis would be magnified. The mechanism of action of such stimulation would likely be clarified more easily through the use of defined oligonucleotides, particularly since it was time-consuming to purify the products, active as adjuvants, following breakdown of nucleic acids with nucleases (Johnson and Hoekstra, 1967). Accordingly, when homopolyribonucleotides (polyadenylic acid complexed *in vitro* with polyuridylic acid, termed poly A:U) were shown to be effective adjuvants (Braun and Nakano, 1967), we initiated an in-depth study of their general properties and the cell type(s) affected. Our results to date are as follows.

GENERAL PROPERTIES

The adjuvant effect on antibody synthesis in the inbred mouse, BALB/aj strain, is seen in Table 1. Shortening of the induction period by injection of poly A:U with a single injection of bovine gamma globulin (BGG) as antigen is evidenced by the early appearance on days 3–5 of antibody in stimulated mice. In addition, the peak antibody response was elevated and both 19S

* These studies were supported by United States Public Health Service Grants AI 1524 and AM 14273.

Table 1. Adjuvant Activity of Polyadenylic-Polyuridylic Acid Complexes

Products injected[a] BGG (3 mg) +	Antibody titer on day					
	3	5	7	9	11	15
...	0	10	20	80	80	160
Poly A:U, 150 µg each	20	40	160	640	1280	1280
Endotoxin, 10 µg	20	40	160	1280	1280	1280

[a] Given intravenously to mice, BALB strain.

and 7S antibody titers were found to be raised. Similar enhancing action was observed with other antigens, for example, ferritin, bovine serum albumin, sheep red blood cells, and the Vi antigen of *Escherichia coli*; and poly A:U was shown to be effective also in the rabbit, rat, and guinea pig.

Complexes of polyinosinic acid (poly I) and polycytidylic acid (poly C), formed *in vitro*, (poly I:C) were also effective adjuvants. However, although poly I:C was far more potent than poly A:U at inducing interferon, it was no more effective than poly A:U at increasing antibody levels.

As was the case with other adjuvants (references in Schmidtke, 1969), the time of injection of poly A:U in relation to that of antigen was important in demonstrating an enhancing effect. When this was tested, poly A:U proved to be effective over only a very short period of time (Table 2).

Table 2. Effect of Time of Injection of Poly A:U in Relation to Antigen

Products injected BGG (1.0 mg) + Poly A:U, 150 µg each	Antibody titer on day 10
—3 days[a]	640
—2 days	320
—1 day	10
—18 hours	0
—12 hours	0
— 6 hours	320
— 2 hours	5120
with	2560
+12 hours	640
+24 hours	320
BGG (1.0 mg) alone	640

[a] Time of antigen injection = day zero.

Paradoxically, *suppression* of antibody synthesis occurred when poly A:U was given 1/2 to 1 day before BGG. Whether this feature is applicable with other antigens and in other species is under investigation. Evidence that poly A:U action was limited to a short time interval after injection also was gained from experiments in which the homopolyribonucleotides were incubated in normal mouse serum, either singly or in complex form, for 10- and 20-minute intervals (Table 3). In all cases, the adjuvant action was destroyed, indicating that its effectiveness on cells *in vivo* was limited to an interval less than 10 minutes.

Table 3. Effect of Incubation of Poly A:U in Normal Mouse Serum on Adjuvant Action

Products injected BGG (1.0 mg) +	Antibody titer on day	
	6	9
...	40	80
Poly A:U, 150 μg each	160	1280
Poly A:U, NMS[a], 10 min	10	0
Poly A + poly U, NMS, 10 min[b]	0	40
Poly A:U, NMS, 20 min	0	40
Poly A + poly U, NMS, 20 min[b]	0	10

[a] NMS = normal mouse serum.

[b] Poly A + poly U, NMS = poly A and poly U incubated separately in normal mouse serum and then combined.

There was a dramatic effect of poly A:U in rendering minute amounts of antigen immunogenic. Thus, 1 ng of BGG elicited readily reproducible titers of antibody under the stimulus of poly A:U (Table 4). This finding suggests the potential practicality of utilizing these polynucleotides as adjuvants when employing antigens that are necessarily low in quantity (for example, viruses), or those that are weakly antigenic (for example, autochthonous tumors or autoantigens).

It was found that poly A:U could be administered by a different route than antigen without impairing the adjuvant action (Table 5). Thus, although there was an increase in viscosity on mixture of the 2 homopolyribonucleotides, admixture was not a requirement for increased antibody synthesis.

Several studies of the physical properties of poly A:U necessary for adjuvanticity were carried out. Neither freeze-thawing, repeated 5 times, nor incubation for 14 days at 24 °C had an adverse effect on the adjuvant action.

Table 4. Adjuvant Action of Poly A:U on 10^{-9} g Antigen

Products injected i.v.	Antibody titer on day	
	9	11
BGG, 10 nanograms (ng)	0	0
BGG, 10 ng + poly A:U, 150 µg each	160	80
BGG, 10 ng + endotoxin, 10 µg	10	0
BGG, 1 ng	0	0
BGG, 1 ng + poly A:U, 150 µg each	160	160
BGG, 1 ng + endotoxin, 10 µg	0	0

Table 5. Effect of Routes of Injection for Poly A:U and Antigen

Products injected[a]	Antibody titer on day 10
BGG, i.p.[b]	40
BGG, i.p. + poly A:U, i.p.	640
BGG, i.v.	80
BGG, i.v. + poly A:U, i.v.	2560
BGG, i.p. + poly A:U, i.v.	640
BGG, i.v. + poly A:U, i.p.	1280

[a] BGG dose = 0.1 mg/mouse; poly A:U = 150 µg. Each polymer complexed *in vitro*.
[b] i.p. = intraperitoneally; i.v. = intravenously.

However, when poly A:U was denatured by heating at 100 °C for 10 minutes, cooled rapidly, and kept cold until injection with antigen, adjuvant activity was destroyed. Incubation of either the poly A:U complex or the individual polynucleotides with pancreatic ribonuclease at an enzyme-polynucleotide ratio of 1:7.5 rendered the material inactive.

Inasmuch as it had been postulated that the active component of poly A:U was poly A, with poly U acting merely as a carrier (Braun and Nakano, 1967), complexes composed of 2 proportions of poly A to 1 proportion of poly U were tested and found to be no more effective adjuvants than those having equimolar amounts of each. Similarly, uridine 5′ monophosphate could not substitute for poly U, nor could adenosine 5′ monophosphate, adenosine 5′ triphosphate, or oligoadenylic acid of chain lengths 3–6 substitute for poly A. Thus, it was concluded that helix formation between large molecular weight polynucleotides was a requirement for adjuvant action. Further evidence for the latter hypothesis was gained when the degree of hydrogen bonding was reduced by substituting poly A methylated at either the N-1 or N-6 position for poly A, and alkylated poly U for poly U. Following verification of a reduction in hydrogen bonding by showing a drop in Tm when complexing methylated poly A with poly U or alkylated poly U with poly A (Schmidtke, 1969), the biological activity was measured and the results are shown in Table 6. A diminution in activity was observed in the substituted preparations.

Table 6. Effect of Chemical Modification of Poly A and Poly U on the Adjuvant Action of the Resultant Complexes

Products injected BGG (0.05 mg) +	Antibody titer on day	
	6	9
...	40	80
Poly A:U, 150 μg each	1280	1280
Poly (N-1 Meth A[a] + ALK U), 150 μg each	320	160
Poly (N-6 Meth[b]A + ALK U[c]), 150 μg each	320	320
Poly (N-1 Meth A + U), 150 μg each	160	320
Poly (N-6 Meth A + U), 150 μg each	160	160
Poly (A + ALK U), 150 μg each	320	320
Poly (N-1 Meth A), 300 μg	80	80
Poly (N-6 Meth A), 300 μg	80	80
Poly (ALK U), 300 μg	80	160

[a] N-1 meth A = poly A methylated at the N-1 position.
[b] N-6 meth A + poly A methylated at the N-6 position.
[c] ALK U = Alkylated poly U.

EFFECT OF POLY A:U ON THE CELLS ENGAGED IN THE IMMUNE RESPONSE

Current knowledge permits the hypothesis that antibody synthesis to some antigens may result from (1) contact of antigen with macrophages, (2) transfer by macrophages to thymic lymphocytes of antigen attached to ribonucleic acid, (3) contact of these antigen reactive-thymic lymphocytes, after division, with lymphocytes originating in the bone marrow, and (4) production of antibody by the latter cell after expansion. Using this scheme as a working hypothesis, experiments were designed to test the cellular site(s) of action of poly A:U and poly I:C.

Table 7. Adjuvant Action of Macrophages and/or Poly A:U

Incubation mixture injected[a]	Reciprocal HA titer		
	6 days	8 days	10 days
Macrophages	0	0	0
BGG only	80	320	640
BGG + poly A:U	160	2560	10,240
Macrophages + BGG	320	5120	10,240
Macrophages + BGG + poly A:U	640	10,240	40,960
Macrophages + poly A:U	0	0	0

[a] Concentrations: Macrophages, 2.4×10^7/mouse; BGG, 0.5 mg; poly A:U, 150 μg each.

Table 8. Uptake of Aggregated I^{131} BGG by Macrophages Cultured as Monolayers

Time (hours)	Net uptake of BGG in counts/min		
	Antigen dose μg/ml	Addition of poly A:U	CPM/ 4.3×10^7 cells
1/2	500	—	7860
	500	+	5880
	50	—	970
	50	+	910
2.0	500	—	15,960
	500	+	17,810
	50	—	3180
	50	+	1550
4.0	500	—	25,110
	500	+	27,740
	50	—	4080
	50	+	1880
8.0	500	—	20,690
	500	+	15,840
	50	—	3590
	50	+	2100

Poly A:U added to incubation mixture at a concentration of 150 μg/ml of each polynucleotide.

Macrophage: In studying the effect of poly A:U on the macrophage, a system used by several others was employed (references in Johnson, 1969). Thus, macrophages were removed from mice stimulated with thioglycollate

4 days previously, incubated with BGG as antigen with and without the addition of poly A:U, washed 4 times, injected back into syngeneic mice, and antibody- or rosette-forming cells measured as a function of time. Representative results are shown in Table 7 and illustrate the increase in antibody synthesis when poly A:U was added to macrophages incubated with antigen. This enhancement was not associated with increased uptake of antigen, as is shown in Table 8. A lesser amount of antigen was retained in the presence of poly A:U than seen with antigen alone. When the ratio of retention of I^{131} labeled antigen to H^3 labeled poly A:U was measured, competition for receptor sites was revealed. These data are seen in Table 9, where it is evident that a 10-fold reduction in BGG resulted in a 4-fold increase in retention of poly A:U.

Addition of poly A:U to macrophages incubated with antigen was associated with an increase in uridine uptake and resulting RNA synthesis

Table 9. Retention by Macrophages of H^3-Poly A:U when Incubated in the Presence of Two Different Concentrations of Antigen

Materials incubated	Net uptake in CPM/5 × 10^7 Macrophages time (min)			
	10	20	30	60
Macrophages + poly A:U-H^3	1040	1380	1540	2410
Macrophages + BGG-I^{131}, 0.33 mg/ml	27,100	30,720	33,520	41,100
Macrophages + BGG-I^{131}, 0.033 mg/ml	5900	6290	7380	7560
Macrophages + poly A:U-H^3 + BGG-I^{131}, 0.33 mg/ml				
H^3	1090	1600	1450	2270
I^{131}	27,170	37,300	36,550	47,710
H^3/I^{131}	*0.04*	*0.04*	*0.04*	*0.05*
Macrophages + poly A:U-H^3 +BGG-I^{131}, 0.033 mg/ml				
H^3	980	1260	1480	2710
I^{131}	7590	9030	8610	12,610
H^3/I^{131}	*0.13*	*0.14*	*0.17*	*0.22*

Poly A:U-H^3 was used at a concentration of 150 μg each/ml (10.9 μg poly A/μC and 23.4 μg poly U/μC). The counts reflect absolute amounts of material taken up by macrophages; therefore, a change in ratio reflects a change in uptake of BGG in the presence of a constant amount of poly A:U.

Table 10. Reversal of Actinomycin D Inhibition of Macrophage Action

Incubation mixture injected[a]	Reciprocal HA titer	
	7 days	9 days
BGG only	320	1280
BGG + poly A:U	640	2560
Macrophages + BGG	1280	5120
Macrophages + BGG + poly A:U	1280	10,240
Macrophages + BGG + Act D	10	160
Macrophages + BGG + poly A:U + Act D	160	1280
Macrophages + ET[b] + BGG	2560	2560
Macrophages + ET + BGG + Act D	10	80

[a] Concentrations: Macrophages, 2×10^7/mouse; BGG, 0.5 mg; poly A:U, 150 µg each; actinomycin D, 15 µg, ET, 1 µg.
[b] ET = endotoxin.

Table 11. Induction of Antibody Synthesis by an RNA-Rich Fraction Isolated from Macrophages

Material injected	RFC/10^6 spleen cells	HA antibody titer on day	
		6	35
BGG	6400	40	20,480
RNA, 0.1 mg	2000	0	40
RNA, 0.1 mg + poly A:U	5250	2	40
RNA[a], 0.2 mg	5300	4	20
RNA[a], 0.2 mg + poly A:U	8700	16	320
...	495	0	ND

[a] RNA treated with Pronase.
Concentrations: BGG, 0.5 mg/mouse; poly A:U, 150 µg each/ml; RFC = rosette-forming cells measured on day 6 after injection. All mice injected with BGG 30 days after initial injection of BGG or RNA. ND = not done.

which could be correlated with increased antibody synthesis. Several different species of RNA were probably stimulated by poly A:U. One, messenger RNA, was made evident through use of actinomycin D which obliterated this enhancement when added with antigen (Table 10). Poly A:U was able to overcome this inhibition. However, when RNA was isolated from macrophages (Johnson, 1969) and demonstrated to have the capacity

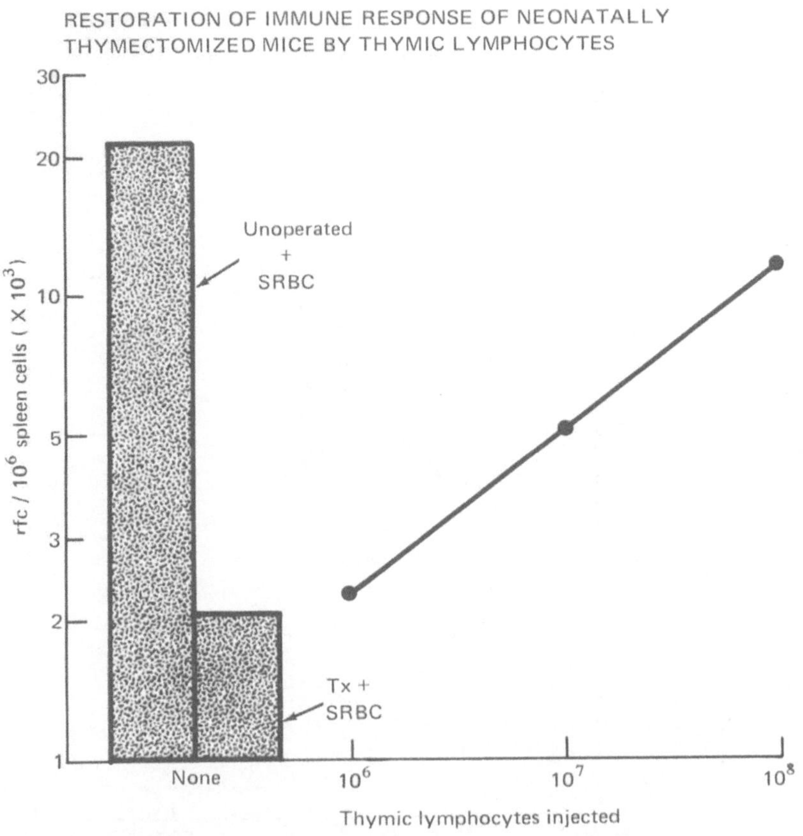

Mice received 4 X 10^8 SRBC ± thymic lymphocytes I. V. and spleens removed 6 days later and assayed for rfc. Each point represents the arithmetic mean of 3 separate experiments with 4 - 5 mice / group.

Fig. 1

to induce antibody synthesis, it proved to be too small in size (4S) to code for either a light or heavy chain of the immunoglobulin molecule. Thus, stimulation of a second type of RNA was indicated. The biological efficacy of this 4S RNA is documented in Table 11. It may be seen that injection of RNA did not give rise to significant levels of circulating antibody unless stimulated with poly A:U. Using a more sensitive test, splenic rosette-forming cells, RNA per se was immunogenic. Antigen "contamination" was undetectable, or below 10^{-10} g, as measured isotopically. However, a qualitative or quantitative difference from BGG existed when RNA was used as the immunizing preparation inasmuch as injection of BGG to these mice on day 30 did not give rise to an anamnestic response.

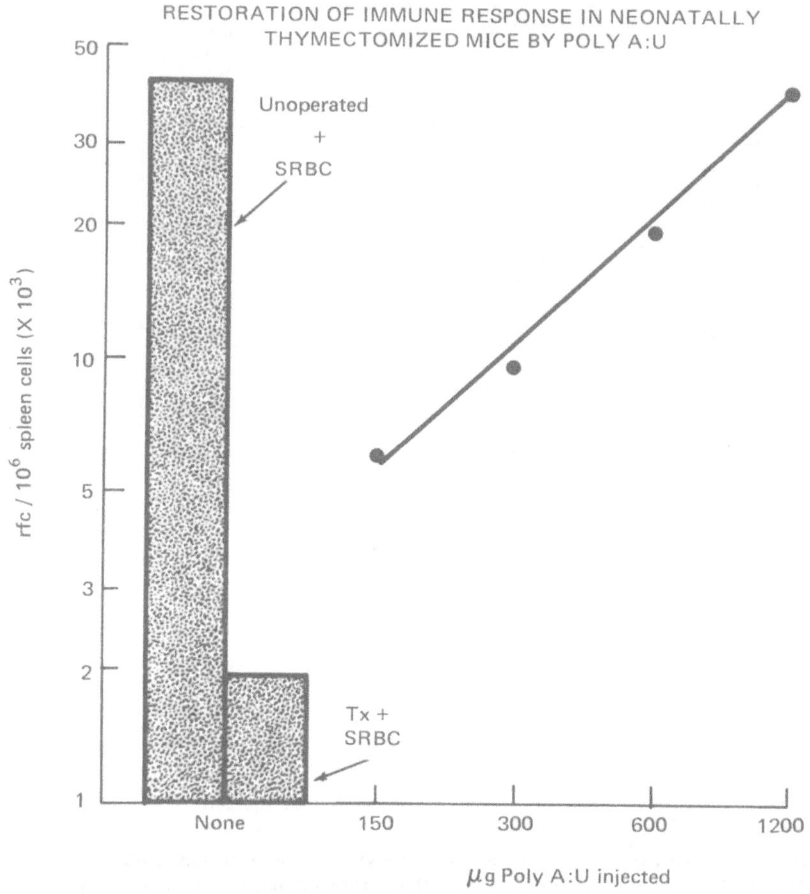

Mice received 4 X 10^8 SRBC ± Poly A:U I. V.,
Spleens removed 6 days later and pooled for rfc assay.
Each point represents the arithmetic mean of 3 separate experiments with 4–5 mice / group.

Fig. 2

Thymocyte: Thymectomy of the newborn mouse results in loss of immunologic responsiveness to sRBC at 4 weeks. This unresponsiveness can be overcome by injection of thymocytes with antigen (Miller and Mitchell, 1968) with the extent of the response being dependent on the number of lymphocytes injected. Confirmation of this is seen in Figure 1. However, in further experiments it was found that poly A:U per se, without thymocytes, was capable of restoring immunologic competence in a similar manner. The data are shown in Figure 2. Verification of the ability of poly A:U to restore

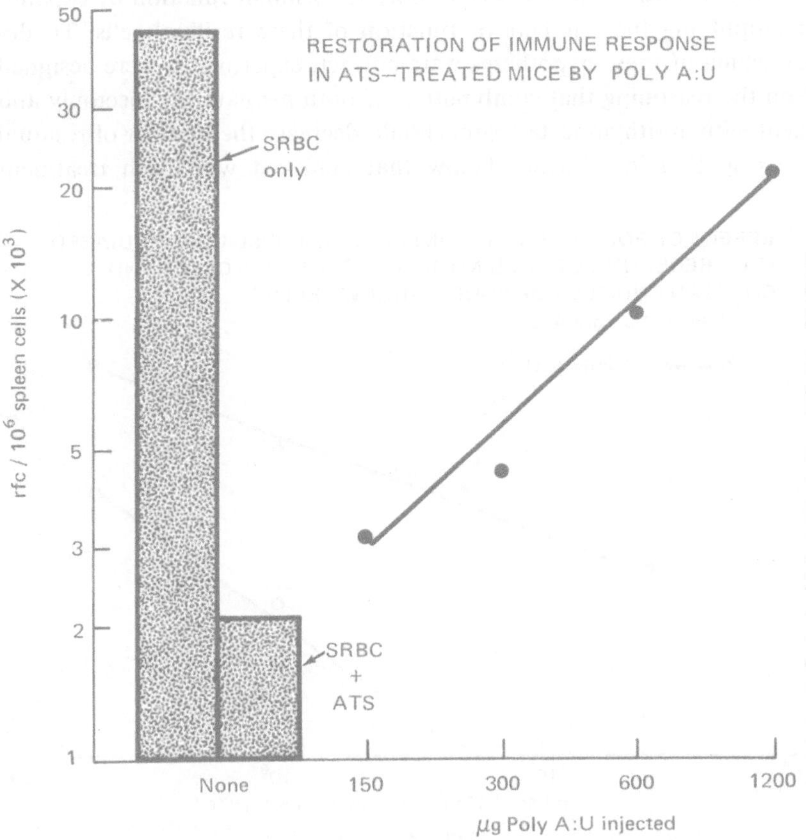

Mice received 4 X 10^8 SRBC ± Poly A:U 3 days after 0.4ml ATS I.P. Spleens were assayed for rfc 6 days after injection of antigen. Each point represents the arithmetic mean of 3 separate experiments with 3 - 4 mice / group.

Fig. 3

competency in mice depleted of lymphocytes was seen using a different approach: antithymocyte serum. Thus, injections of rabbit anti-mouse thymocyte serum 3 days prior to antigen reduced immunologic capacity to 4% that of normal mice; but once again, administration of poly A:U restored their competence (Figure 3).

These data might be explained by either of two hypotheses. First, the thymus could activate the peripheral immunologic system by secreting a hormone which is polynucleotide in nature and thus, poly A:U could be capable of substituting directly for thymic function. Second, since *complete* removal of all antigen-reactive cells is never fully achieved by either neonatal thymectomy or antithymocyte serum, poly A:U might function by expanding or amplifying the numbers or function of these residual cells. To determine which of these hypotheses were correct, experiments were designed based on the reasoning that combination of both neonatal thymectomy and treatment with antithymocyte serum should decrease the number of residual cells bearing thymic influence below that achieved with each treatment

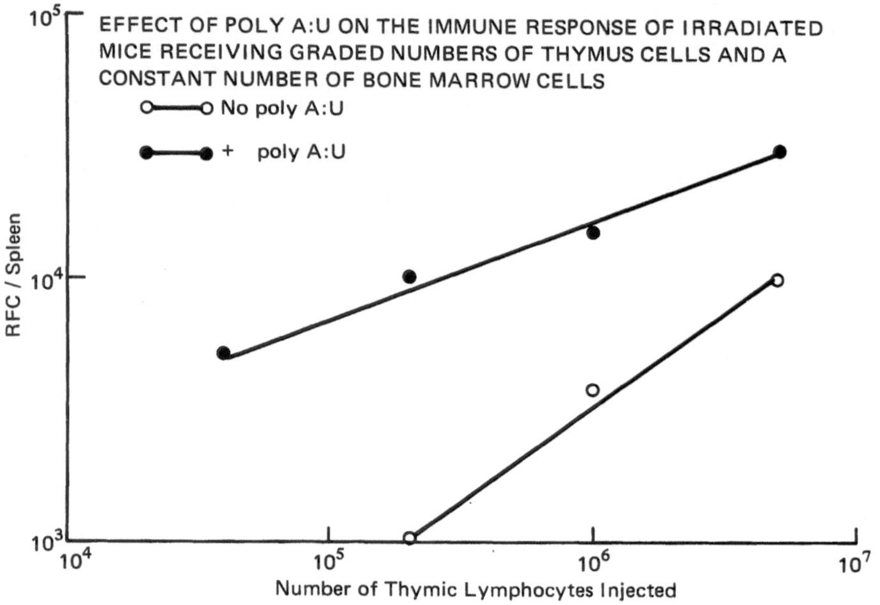

Fig. 4

Table 12. Restoration of Immune Response in Thymectomized and Thymectomized + ATS*-Treated Mice

Products injected	RFC[a]/10^6 spleen cells			
	Un-operated	Thym-ectomized only	ATS only	Thym-ectomized + ATS
sRBC	23,000	1600	2000	615
sRBC + poly A:U	ND	11,700	9700	2000

* ATS = antithymocyte serum.
[a] RFC = rosette-forming cells.
Mice received 4×10^8 sRBC ± 600 μg poly A:U 3 days after i.p. injection of 0.4 ml ATS. Values represent arithmetic mean of three separate experiments.

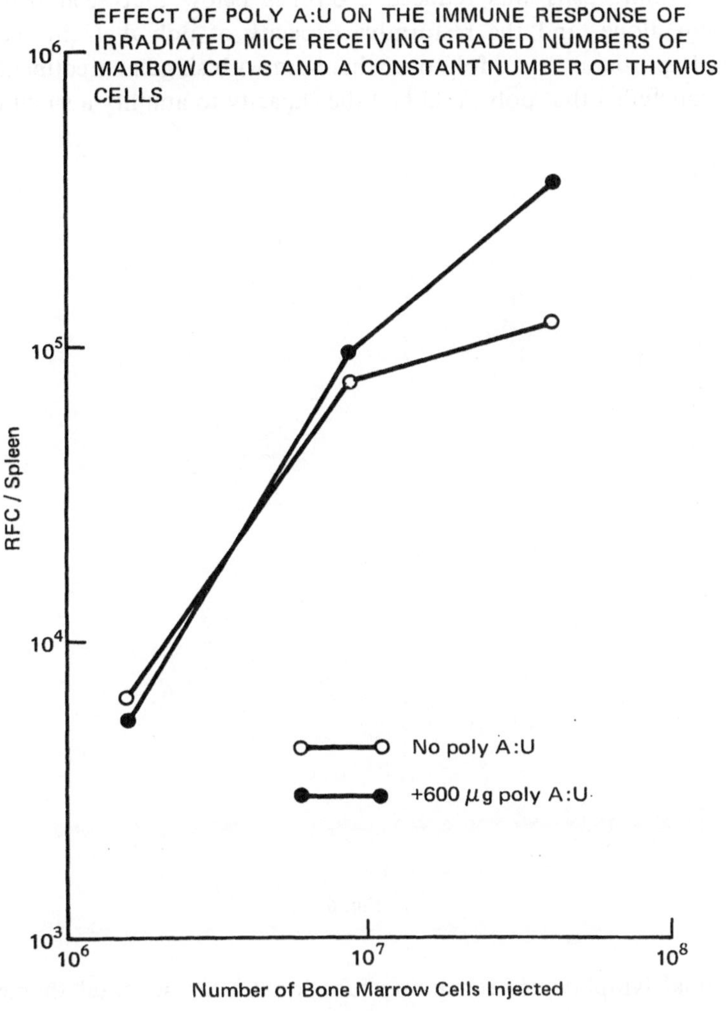

Fig. 5

alone. If poly A: U functioned by amplifying a small number of residual cells, its effect should be reduced if the number of cells to be amplified were reduced. On the other hand, if poly A: U were replacing thymic hormone action on a nonthymic cell, a decrease in the residual cell number should have no effect on the recovery induced by poly A: U. The results following injection of poly A: U into mice receiving both treatments is shown in Table 12. It may be seen that when both neonatal thymectomy and antithymocyte serum were administered to the same mice, the number of rosette-forming cells was reduced 2–3 times below that seen with either treatment alone, and the restorative capacity of poly A:U in such mice dropped approximately 5-fold over that seen with singular treatment. Thus, it was concluded that poly A:U had the capacity to amplify a small number

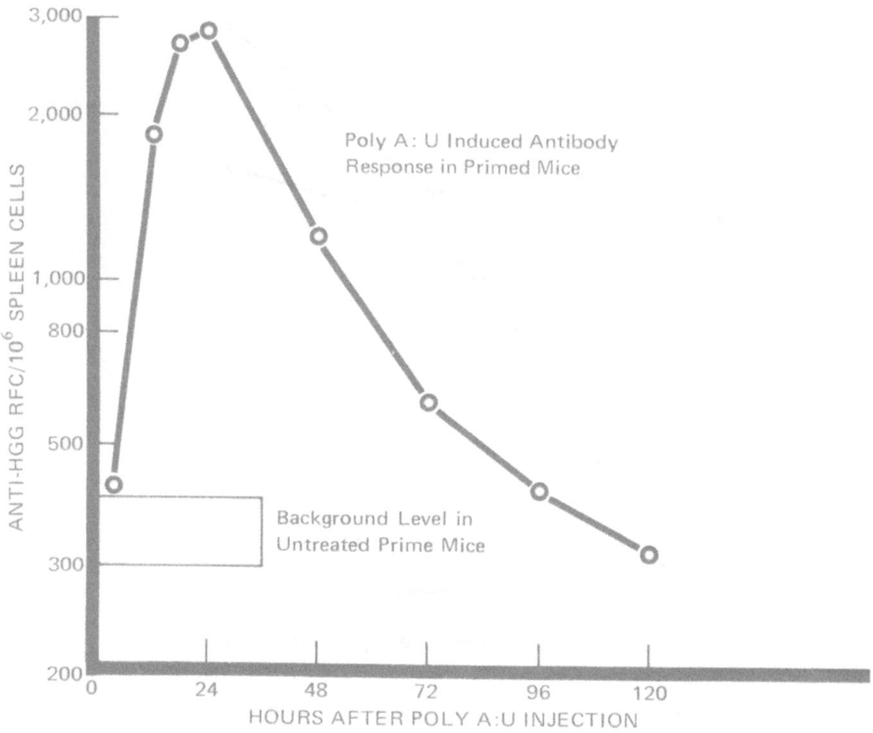

Fig. 6

of residual lymphocytes remaining functional after neonatal thymectomy. Since at least two different lymphocytic cells are involved in the production of antibody to sRBC in mice, for example, a thymic-influenced (antigen reactive cell) and a bone-marrow-derived cell (antibody producing cell), the model system of Miller and Mitchell (1968) was used to determine which lymphocytic cell was involved in poly A:U action. Further evidence documenting the conclusion that the thymic-influenced lymphocyte was a site of action of poly A:U was gained by injecting low-graded numbers of thymocytes with and without poly A:U into irradiated mice receiving bone marrow cells in excess. If poly A:U possessed the ability to amplify a small number of thymocytes, then antibody should be stimulated with doses of thymocytes too low to induce a response without poly A:U. This was shown to be the case, as is seen in Figure 4. In a reverse experiment, that is, where low-graded doses of bone marrow cells were injected into mice receiving an excess of thymocytes, no effect of poly A:U on bone marrow

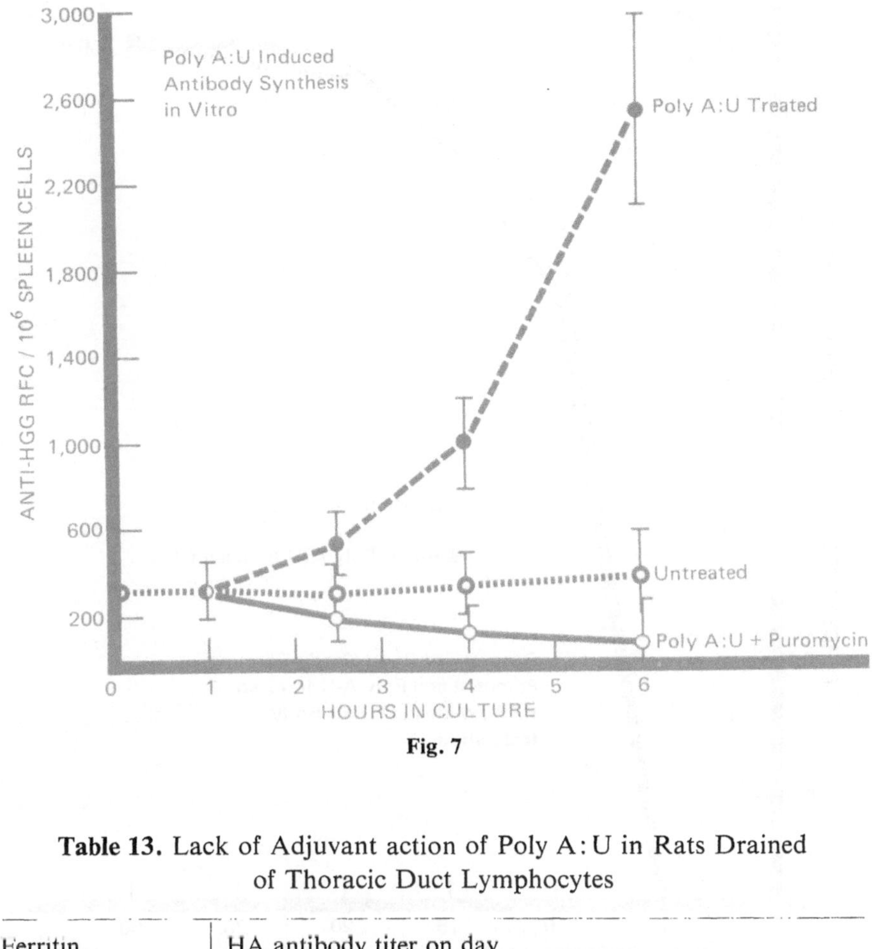

Fig. 7

Table 13. Lack of Adjuvant action of Poly A:U in Rats Drained of Thoracic Duct Lymphocytes

Ferritin, 0.1 mg +	HA antibody titer on day						
	Primary			Secondary			
	7	15	21	3	7	12	21
...	190	110	40	2150	4300	7240	8610
Poly A:U, 250 μg	170	110	20	260	2040	2030	4560
Endotoxin, 10 μg	510	2280	2280	10,200	84,000	186,000	92,000

Sprague-Dawley rats drained from the thoracic duct for 1 day.

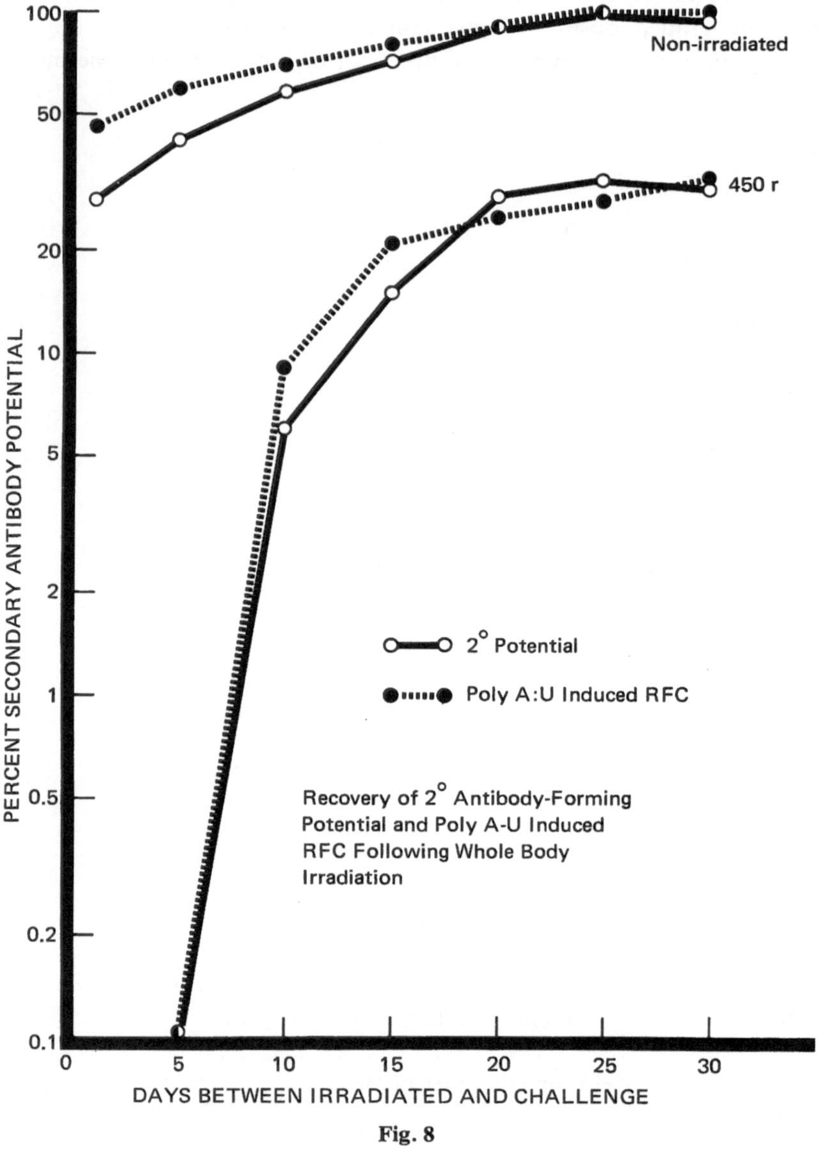

Fig. 8

cells was observed (Figure 5). The divergence of the lines when high numbers of bone marrow cells were injected was caused by the thymocytes now becoming limiting.

Evidence that poly A:U action was mediated through circulating lymphocytes was gained also by testing its effect in rats whose thoracic duct was

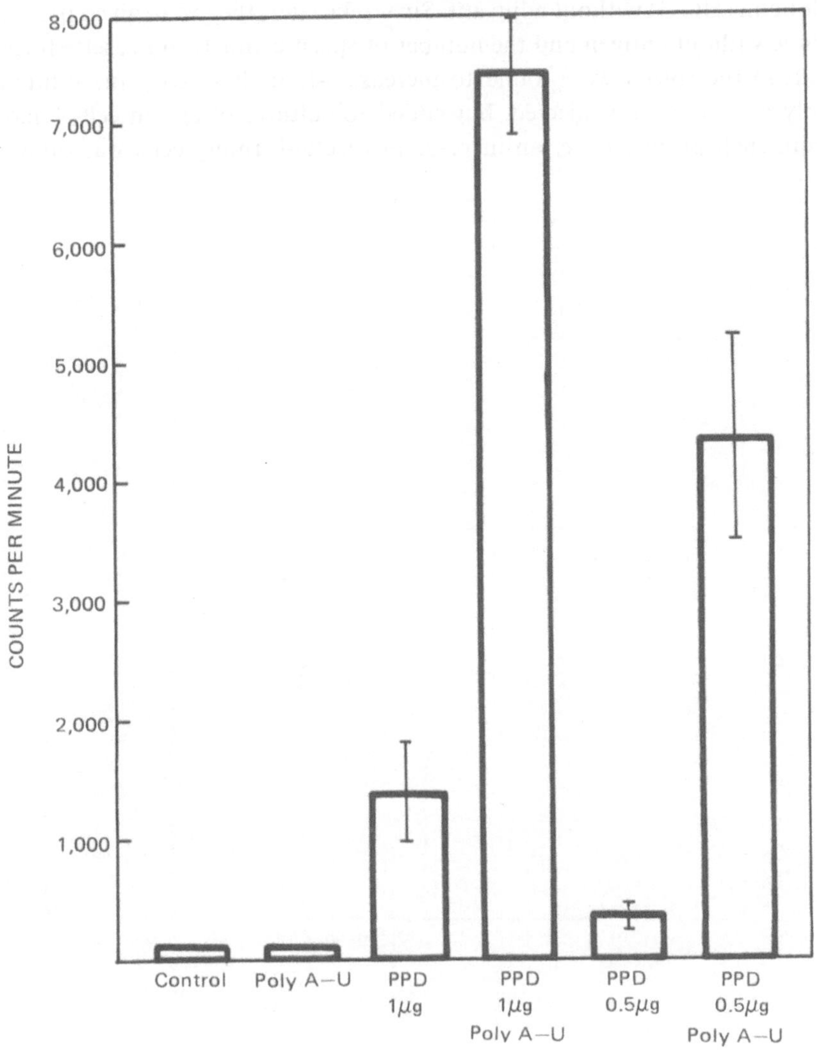

Fig. 9

drained of lymphocytes over a 24-hour period. It is seen in Table 13 that poly A:U was incapable of elevating either primary or secondary antibody levels to ferritin under these conditions. In contrast, endotoxin, which was used as a standard reference adjuvant throughout all these studies, was active in the drained rat, indicating a difference in capabilities of the two adjuvants.

Memory cells: The effect of poly A:U on immunologic memory was tested as follows: Mice were given a single injection of 100 μg human gamma globulin (HGG) without adjuvant. Six weeks later, they were given poly A:U alone without antigen and the number of specific anti-HGG rosette-forming cells in the spleen were found to increase 4-fold (Figure 6). In addition, if poly A:U were not injected, but added to cultures of spleen cells removed from such primed mice, an increase in rosette-forming cells was observed.

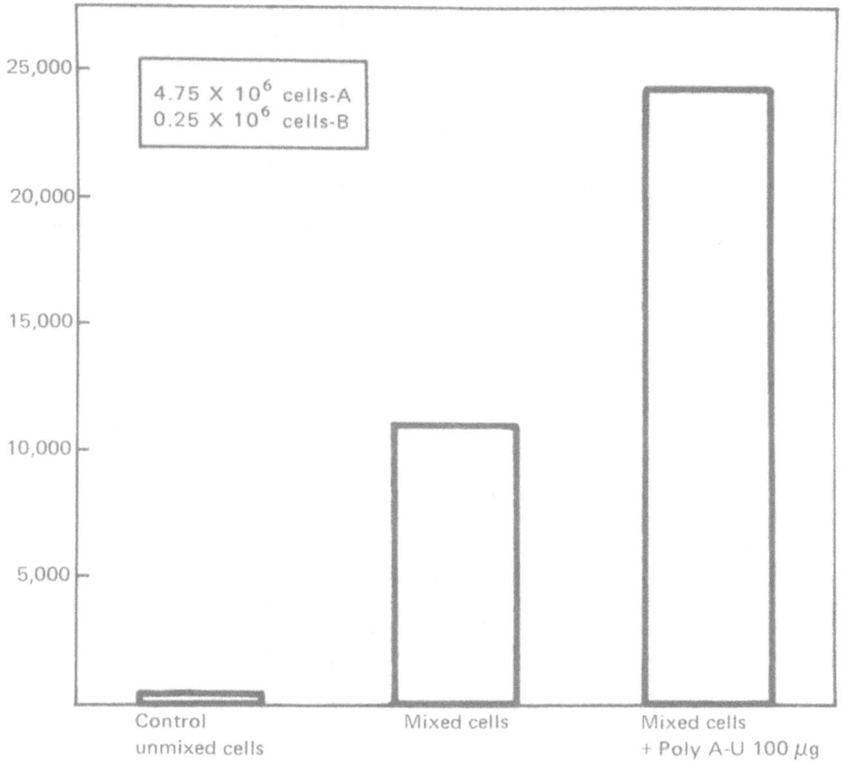

Fig. 10

This increase was inhibited by puromycin, indicating mediation by an active synthetic process (Figure 7). It was concluded that poly A:U had the capacity to cause the appearance or synthesis of antibody in spleen cells independent of antigenic stimulation. The functional nature of the cell involved was established as a memory cell by experiments employing the design of Nettesheim and Williams (1968). Here the recovery of secondary antibody potential after irradiation could be measured in a predictable fashion, and it was paralleled by the appearance of rosette-forming cells following poly A:U treatment (Figure 8). Thus, it was concluded that poly A:U can enhance greatly the number of memory cells capable of participating in an antigen-antibody reaction (rosette formation) independent of antigen. The practicality of this observation is now under investigation.

Delayed hypersensitivity: A study of the effect of poly A:U on cells expressing delayed hypersensitivity was conducted with several different systems (Friedman, Johnson, and Pan, 1969). First, a stimulation by poly A:U of antigen activation of leucocytes was noted in an enhancement of tritiated

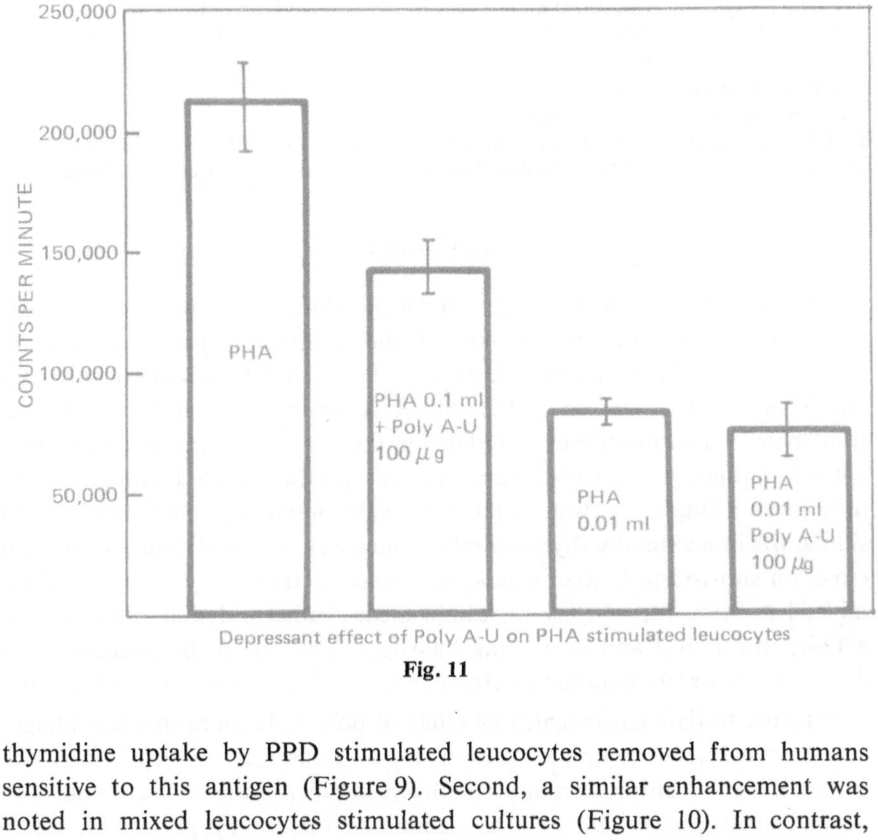

Depressant effect of Poly A-U on PHA stimulated leucocytes

Fig. 11

thymidine uptake by PPD stimulated leucocytes removed from humans sensitive to this antigen (Figure 9). Second, a similar enhancement was noted in mixed leucocytes stimulated cultures (Figure 10). In contrast, poly A:U depressed the increase in thymidine uptake induced by phytohemagglutinin when added to human leucocytes (Figure 11).

Lastly, the capability of poly A:U in hastening the immune processes involved in graft rejection is documented in Table 14. It may be seen that neonatal thymectomy prolonged to 24 days the retention of skin grafts transferred from C 57 B 1 donor mice to BALB recipients, which had a normal life of 13.5 days. Injection of poly A:U 1/2, 1, and 3 days after grafting, on the other hand, hastened the rejection time of the graft in thymectomized mice such that it equalled that seen in normal mice.

Table 14. Restoration of Homograft Rejection in NT_x* Mice by Poly A:U

Mice injected with:	Mean survival time (days)			
	Unoperated	No. of mice	T_x	No. of mice
...	13.5 (± 0.8)[a]	6	24.0 (± 3.0)	7
Poly A - U	12.0 (± 1.3)	6	13.0 (± 3.2)	8

* NT_x = neonatally thymectomized.
[a] ( ) indicates 95% confidence interval.
BALB/Aj recipients received whole thickness skin grafts from C57B1/6j donors. Mice receiving poly A:U were injected with 600 µg i.p. 12, 24, and 72 hr after grafting.

SUMMARY

An in-depth study was carried out on the stimulation by homoribopolynucleotides of various parameters of the immune response. Equimolar complexes of poly A and poly U or poly I and poly C increased both 19S and 7S antibody levels, as well as the secondary response in mice. In addition, increased antibody was stimulated in rabbits and guinea pigs. Poly A:U without antigen evoked the appearance of specific rosette-forming cells in mice given a single injection of HGG 6 weeks previously. Also, rejection of skin grafts in neonatally thymectomized mice was hastened, and an adjuvant effect on short-term human leucocyte cultures activated by antigen and on mixed leucocyte interactions was established. Neither polymer alone showed activity in any test and decreasing hydrogen bonding in the complex, heat denaturation or ribonuclease treatment resulted in a loss of adjuvant activity.

Evidence to date has revealed an effect of poly A:U on both macrophages and thymic-derived lymphocytes, but not on bone marrow cells. Thus, the enhancement of antibody synthesis observed when antigens were added to normal macrophages *in vitro* was stimulated further by poly A:U. RNA synthesis (4S) was increased and an RNA-rich fraction isolated from macrophages was rendered capable by this homoribopolymer complex of eliciting antibody formation when injected into syngeneic mice or added to spleen cells in culture. Antigenic uptake was not increased by poly A:U. In addition, poly A:U injected with sheep red blood cells at 5 weeks of age into neonatally thymectomized mice stimulated normal antibody synthesis. Using as a model system recovery of antibody response in irradiated mice by passive transfer of thymic- and bone-marrow-derived cells, poly A:U rendered 4×10^4 thymic-derived cells (a nonfunctional dose by itself) capable of initiating antibody synthesis. No effect on bone marrow cells was observed. An amplifying effect by homoribopolymers on antigen reactive cells was concluded.

References

Braun, W. and Nakano, M. (1967). Antibody formation: Stimulation by polyadenylic and polycytidylic acids. Science **157:** 819–821.

Friedman, H. M., Johnson, A. G., and Pan, P. (1969). Stimulatory effect of polynucleotides on short term leukocyte cultures. Proc. Soc. Exp. Biol. Med. **132:** 916–918.

Johnson, A. G., Schmidtke, J. R., Merritt, K., and Han, I. (1968). Enhancement of antibody formation by nucleic acids and their derivatives. *In* Nucleic Acids in Immunology. 379–385. Ed. by O. J. Plescia and W. Braun. Springer-Verlag, New York.

Johnson, A. G. and Hoekstra, G. (1967). Acceleration of the primary antibody response. *In* Ontogeny of Immunity. 187–190. Ed. by R. T. Smith, R. A. Good, P. A. Miescher. University of Florida Press, Gainesville.

Johnson, H. G. (1969). Factors affecting the enhancement of antibody synthesis by competent macrophages. Doctoral Dissertation, The University of Michigan, Ann Arbor.

Miller, J. F. A. P. and Mitchell, G. F. (1968). Cell to cell interaction in the immune response. I. Hemolysin forming cells in neonatally thymectomized mice reconstituted with thymus or thoracic duct lymphocytes. J. Exp. Med. **128:** 801–820.

Nettesheim, P. and Williams, M. L. (1968). Regenerative potential of immunocompetent cells. II. Factors influencing recovery of secondary antibody forming potential from X-irradiation. J. Immunol. **100:** 760–770.

Schmidtke, J. R. (1969). Enhancement of antibody formation by synthetic polynucleotides. Doctoral Dissertation, The University of Michigan, Ann Arbor.

EFFECT OF POLYNUCLEOTIDES (POLY I·C AND POLY A·U) ON PROTEIN, RNA, AND POLYSOMAL SYNTHESIS IN RABBIT AND HUMAN LUNG, AND RAT AND HUMAN PERITONEAL MACROPHAGES

ROGER M. MORRELL

Hypersensitivity Diseases Research
The Upjohn Company, Kalamazoo, Michigan

INTRODUCTION

The many functions of macrophages guarantee them a varied repertory of synthesized proteins. Examples are interferon (Finkelstein *et al.*, 1968; Smith and Wagner, 1967), endogenous pyrogens (Hahn *et al.*, 1967), proteins of the complement system (Stecher *et al.*, 1967), other serum proteins (Stecher and Thorbecke, 1967), and enzymes, of which Pearsall and Weiser (1970) have documented 13 classes for rabbit peritoneal macrophages and 16 classes for rabbit lung alveolar macrophages. Our interest in detailed mechanisms of protein synthesis in macrophages is based on their role in afferent immunity, which has been adequately reviewed (McMaster, 1953;

Abbreviations: A, absorbancy units; ATP (GTP, UTP, CTP), triphosphate of adenosine (guanosine, uridine, cytosine); BCG, bacillus Calmette-Guerin, tuberculovaccine; BSA (HSA), bovine (human) serum albumin; DNA, deoxyribonucleic acid; dpm, disintegrations per minute; Fc, FAb, complement-active and antigen-binding portions of antibody molecule; leu-, leucine; 2-ME, 2-mercaptoethanol; MgAc, magnesium acetate; P-105, 105,000 × g pellet; PEP, phosphoenolpyruvate; poly I · C (A · U), homoribopolynucleotide, double-stranded, of polyinosinic-polycytidylic (polyadenylic-polyuridylic) acids; polysome, polyribosome; PPD, purified tuberculoprotein derivative; RNA, ribonucleic acid; RNase, pancreatic ribonuclease; S-105 (S-10), 105,000 × g supernatant (10,000 × g supernatant); sp. act., specific activity; TCA, trichloroacetic acid; Tris, tris (hydroxymethyl) aminomethane.

Pearsall and Weiser, 1970). Other presentations at this Symposium will explore aspects of antigen-processing, the effects of polynucleotides on antibody-enhancing properties of macrophage-containing cell populations, and the role of the RNA-containing moieties from macrophages in antibody production. Our work on protein synthesis in rabbit and human lung macrophages has been published in abstract form (Morrell, 1969a; Morrell, 1970). The use of polynucleotides in our system was suggested by the observation (to be reviewed) that the immunizing antigen PPD stimulated RNA, polysomal, and protein synthesis *in vitro* in rabbit lung macrophages, and further, that labeled PPD bound to polysomes during their enhanced activity. Ample precedent exists for the evaluation of polynucleotides in macrophage-containing systems, either as antigen "substitutes" or as "adjuvants" (Braun and Nakano, 1967; Schmidtke and Johnson, 1968; Johnson and Johnson, 1968; Schmidtke *et al.*, 1969; Schmidtke and Johnson, 1969; Field *et al.*, 1968).

In this paper we summarize the effects of poly I·C and poly A·U on RNA, polysomal, and protein synthesis in four different macrophage populations.

MATERIAL AND METHODS

Procurement of macrophages

Rabbit lung and rat peritoneal macrophages were collected at room temperature by the methods described by Weir (1967), using thioglycollate stimulation of rat peritoneum 72 hours prior to collection, and using Hank's balanced salt solution (HBSS) in both cases for rinsing and suspension. Human lung macrophages were recovered from bronchoscopy washings or from biopsy specimens (kindness of Dr. John Neerken). Human peritoneal macrophages were obtained after centrifugation from peritoneal dialysates of uremic patients. Human cells were resuspended in HBSS for all operations after the initial cell buttons were obtained. The medical histories of patients were not taken into account in obtaining macrophages. "Sensitized" rabbit lung macrophages were obtained from animals immunized by the following schedule: purified protein derivative of tuberculoprotein (Parke-Davis) was suspended at 40 mg/ml in complete Freund's adjuvant and injected into adult New Zealand rabbits of both sexes at 0.25 ml subcutaneously (s.c.) in multiple sites every third day for 4 injections, and after the last s.c. injection, 0.25 ml of the suspension was injected intravenously. One week later, an intradermal test for delayed hypersensitivity was administered, and lungs were removed from positive animals after 72 hours. Macrophages from all sources were purified by dispersing them in glass dishes and allowing them to "climb" onto the glass surface for 2 hours, after which the glass was gently rinsed with HBSS and the cells removed by gentle scraping. Lung macrophages of 95–98% purity were obtained by this method, whereas peritoneal preparations were 92–95% pure. Macrophages are not the only glass-adherent cells, but no practical methods have been devised to attain purer preparations.

Assays for protein synthesis

Standard methods (Nirenberg, 1963; Morrell, 1969b) were used to determine incorporation of ^{14}C-leu (sp. act. 165 mc/mm) into TCA-precipitable protein in whole cells (10^6/0.5 ml) or in cell-free preparations. Incubations were 20 minutes at 30° for both preparations. Whole cells were in HBSS; cell-free preparations were in 1.0 ml containing 100 μ*M* Tris HCl, *p*H 7.8, 10 μ*M* MgAc, 60 μ*M* KCl, 3 μ*M* ATP, 0.1 μ*M* GTP, 5 μ*M* PEP, 20 μg pyruvate kinase, 15 μ*M* 2-ME, 0.15–2.0 mg washed ribosomes (RNA), 0.6 mg S-105 protein, 0.05 μ*M* unlabeled amino acids. Other additions for different experiments (see Results) were as follows: ^{14}C-leu 10 μC/ml (whole cells) or 0.5 μC per tube (cell-free), PPD 2 μg/ml, puromycin (PURO) 200 μg/ml, cycloheximide (CHX) 6 μg/ml, chloramphenicol (CAP) 2–4 mg/ml, actinomycin D (ACT.D) 10–40 μg/ml, RNase 10 μg/ml, horse antihuman thymocyte globulin (ATG) 5 μg/ml, homoribopolynucleotides (poly I·C, A·U) 100 μg/ml. Polyadenylic acid (K form) and polyuridylic acid (NH_4) were obtained from Miles Laboratories and stored in 0.15 *M* NaCl, —20° at 5 mg/ml. Prior to use they were complexed *in vitro* at 100 μg/0.2 ml by mixing. Poly I·C was prepared the same way, but was also obtained in double-stranded form (10% by weight, $S_{20,w} > 12$) from P. L. Laboratories. Labeled PPD (^{125}I) was prepared by the chloramine-T method (Hunter and Greenwood, 1962). At the conclusion of incubation, tube contents were precipitated by adding an equal volume of cold 10% TCA containing 1% serum albumin. The tube contents were washed onto filter circles (Whatman GF/C). The precipitate was washed 3 times with 5% TCA, twice with 95% 2-isopropanol, and, in some experiments, with 95% ethanol-ether (3:1). Pads were dried, placed in scintillation vials with 10 ml of scintillation fluid (5 g 2–5, diphenyloxazole plus 0.3 g 1,4-bis-2-(4-methyl-s-phenyloxazolyl)-benzene in 1 litre toluene), and after equilibration were counted by channels ratio with correction for quenching and background. Protein was determined by the method of Lowry (1951) but specific activities (dpm/mg protein) were also assayed by plating of radioactive TCA-insoluble material, dissolved in 88% formic acid, onto preweighed aluminum planchets and counted in a gas-flow counter. For protein determinations, such samples were treated first with hot 5% TCA (90° for 30 minutes) to remove nucleic acids, and then washed with isopropanol and ether.

Assay for DNA-dependent RNA polymerase

The method of Seifart and Sekeris (1969) was used. 0.3 μC ^{14}C-UTP (22 mC/mM) was incorporated over 20 minutes at 37° in a system containing 0.75 μ*M* each of ATP, GTP, and CTP, 4.5 μ*M* creatine phosphate, 15 μg creatine phosphokinase, 4.5 μ*M* 2-ME, 3.0 μ*M* $MnSO_4$, 30 μ*M* Tris HCl *p*H 7.8, 60 μ*M* $(NH_4)_2SO_4$, and 180 μg A grade calf thymus DNA in 0.45 ml. "Enzyme" to start the reaction was 0.2 ml of S-10 (containing 20–200 μg protein). Incorporation is reported (Tables 5 and 6) as *p*moles in RNA/20 min/mg protein.

Polysomal analysis (Wettstein *et al.*, 1963; Noll, 1969)

Polysomes from whole cells were obtained from the 0.5% DOC-treated S-10 after homogenization by 5–10 strokes of a glass pestle in a Potter-Elvejhem homogenizing tube. The cell-free incubation system was not rehomogenized prior to isolation of polysomes. The clarified P-105 was resuspended in buffer which had the same composition as the incubation medium. The RNase content of the samples to be layered on sucrose gradients was not measured, but satisfactory resolution was obtained using 10–34% gradients made in incubation buffer containing bentonite (W. H. Curtin No. 32854) 2 mg/ml. Polyvinylsulfonate (PVS, Eastman No. 8587) did not excel bentonite in protecting polysomes. The cold polysomal preparations were layered in volumes not exceeding 1.5 ml (about 20–30 A_{254} units) on 30 ml gradients and centrifuged at 4° (Beckman, L-2) in the SW 25.2 swinging bucket rotor for 2.5 hours at 24,500 rpm. Gradient tubes were placed in a specially designed tubeholder allowing airfree bottom puncture, and gradients were pushed out from below with 60% Na-K-tartrate solution, through a flowcell designed to minimize schlieren artefacts. The absorbancy at 254 nm was recorded by a Gilford recording spectrophotometer. One or 1.5 ml fractions were collected. Sedimentation coefficients ($S_{20,w}$) were calculated by the methods of Noll (1969), Martin and Ames (1961), and Staehlin and Meselson (1966), using marker 70S ribosomes.

Sources of reagents

Radiochemicals were obtained from the Radiochemical Center (Amersham) and New England Nuclear Corporation. Calf thymus DNA (highly polymerized, A grade) was obtained from Calbiochem, and RNase (pancreatic) from Worthington. Unlabeled amino acids were supplied by Sigma, PURO from Nutritional Biochemicals, CAP from Parke-Davis, CHX and ACT.D from Mann, phenol reagent from Scientific Products, Inc., and other analytical grade chemicals were purchased from Fisher. ATG (lot 16,138-2) was obtained from The Upjohn Company.

RESULTS

Rabbit alveolar macrophages (PPD-sensitized)

In whole cell incubations (Table 1), poly I·C and poly A·U produced about 5- and 2.5-fold increases in protein synthesis, respectively, while PPD was 33% more active in the cell-free than the whole-cell system, although it produced a 5–6-fold increase in whole cells. CHX and CAP inhibited protein synthesis in both systems to about the same extent (25 and 18%, respectively). This was significant, but much less than the great inhibition produced by PURO, ACT.D, and RNase. Poly A·U resulted in only a 3-fold increase in the cell-free system, whereas poly I·C stimulated as well in cell-free systems as in whole cells. The polynucleotides did not stimulate RNA synthesis in whole cells, nor the cell-free system, but PPD produced a 40–45-fold increase in whole cells and an 80-fold increase in the cell-free system. Negative controls were BCG, sheep red cells, BSA, and HSA. PURO, CHX, and CAP did not inhibit RNA synthesis, whereas ACT.D and RNase did. The gradient tracings (Figure 1) show that in whole cells poly I·C and poly A·U did not cause a change in the control profile, but PPD (10-minute aliquot) caused polysomal enhancement and differentiation,

Table 1. PPD-Sensitized Rabbit Alveolar Macrophages Vol.-Adjusted DPM at 10 Minutes ± S.D. (5–1)[a]

System	Whole cells		Cell-free (S-10)	
	Per mg prot.	RNA	Per mg prot.	RNA
CS	3817 ± 201	2112 ± 175	3500 ± 302	1850 ± 189
− COFACT.	...	...	28 ± 7	17 ± 3
− RIBO.	...	...	101 ± 9	1764 ± 163
− (S-105)	...	...	67 ± 8	42 ± 5
− MG^{++}	...	...	78 ± 7	54 ± 6
+ I·C	20,072 ± 1231	2298 ± 180	19,650 ± 1901	2029 ± 193
+ A·U	10,351 ± 631	1909 + 99	9872 + 801	1791 ± 165
+ PPD	18,330 ± 675	95,401 ± 3102	27,550 ± 1923	152,400 ± 11,750
+ PURO	96 ± 10	2021 ± 169	103 ± 9	1731 ± 161
+ CHX	2902 ± 172	2091 ± 143	2430 ± 224	1903 ± 187
+ CAP	3119 ± 212	2235 ± 217	3012 ± 351	1761 ± 190
+ ACT.D	31 ± 9	27 ± 11	17 ± 4	20 ± 5
+ RNase	9 ± 4	12 ± 6	11 ± 3	19 ± 4

[a] PPD-sensitized rabbit alveolar macrophages. DPM at 10 minutes ± S.D. 5 experiments, 1 animal per experiment. Counts incorporated into protein or RNA per mg protein, as described in the text according to methods for whole-cell or cell-free assays for protein and RNA synthesis.

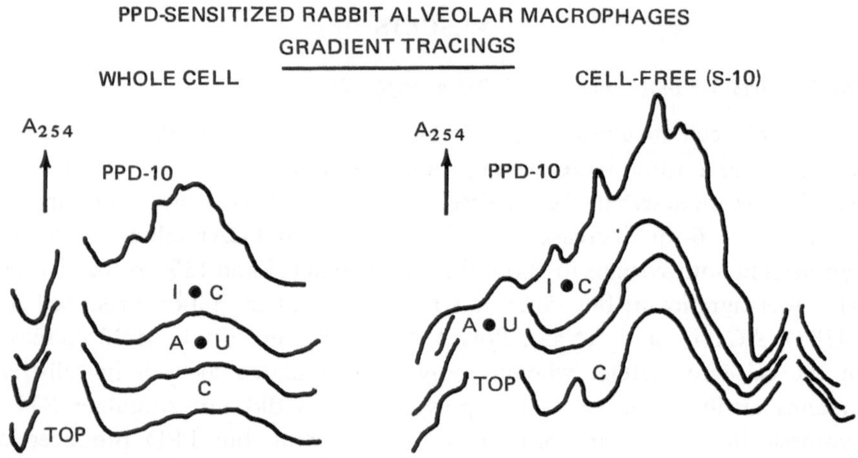

Fig. 1. Gradient tracings of PPD-sensitized rabbit alveolar macrophages. Polysomes were obtained from whole cells or from the cell-free system by homogenization (whole cells) and pelleting at 105,000 × g for 90 minutes as described in the text. Polysomes were then analyzed on sucrose density gradients as described in the text. There is no abbreviation of the 80S peak in the case of PPD-10 (10 minutes) for the cell-free system, because this peak was very small. Additions as described in text.

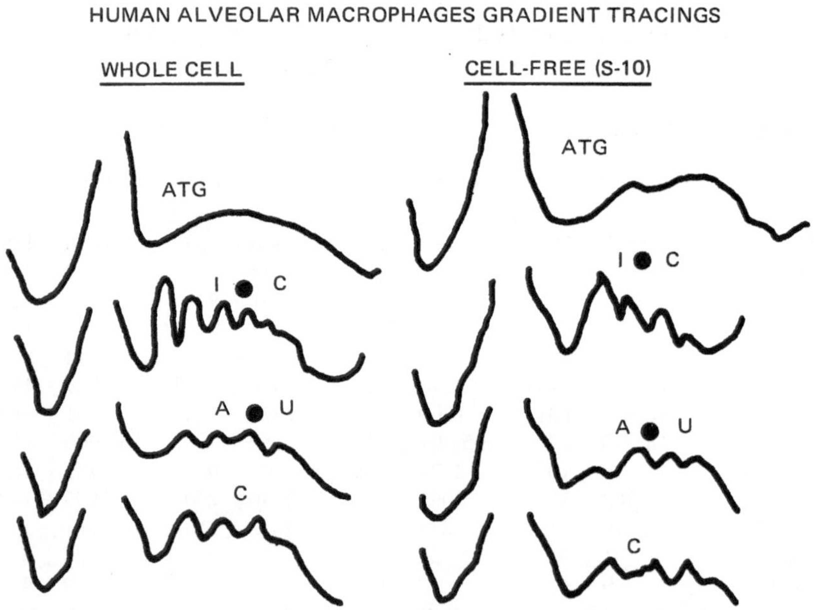

Fig. 2. Gradient tracings of human alveolar macrophages. Polysomes were obtained from whole cells or from the cell-free system by homogenization (whole cells) and pelleting at 105,000 × g for 90 minutes as described in text. Polysomes were then analyzed on sucrose gradients as described in text. Additions to systems were as described.

as seen also in the cell-free system. The control gradients in the cell-free system display a small peak between the 80S peak and the main polysomal peak. This was not present after poly I·C and poly A·U, but was present in the PPD-treated profiles.

Human alveolar macrophages

Whole cells incorporated more into protein (Table 2) than did rabbit alveolar macrophages, but the cell-free level was about the same as in the rabbit. Poly I·C caused a 25% increase in whole cell and almost 50% in cell-free protein synthesis, but poly A·U had no effect in either. ATG was markedly inhibitory to protein synthesis in the whole-cell system but did not affect the cell-free system. When poly I·C and ATG were both added to whole cells, there was partial protection from inhibition by ATG if poly I·C was added 5 minutes before; however, if ATG was added first it exerted virtually its full inhibitory effect. In the cell-free system, poly I·C followed by ATG resulted in greater than 50% increase in protein synthesis, whereas ATG followed by poly I·C did not inhibit. PURO inhibited both

Table 2. Human Alveolar Macrophages Vol.-Adjusted DPM at 5 Minutes ± S.D. (5)[a]

System	Whole cells		Cell-free (S-10)	
	Per mg prot.	RNA	Per mg prot.	RNA
CS	18,500 ± 1785	2550 ± 241	3450 ± 336	1812 + 170
– COFACT.	...	...	55 ± 6	75 ± 4
– RIBO.	...	...	17 ± 2	1921 ± 185
– (S-105)	...	...	92 ± 8	85 ± 7
– MG++	...	...	56 ± 4	27 ± 3
+ I·C	24,251 ± 2501	2431 ± 212	7502 ± 698	2923 ± 301
+ A·U	18,783 ± 1973	2319 ± 229	3391 ± 324	2207 ± 192
+ ATG	220 ± 43	3100 ± 298	3363 ± 329	1905 ± 205
I·C + ATG	6500 ± 701	39,000 ± 3706	8212 ± 841	10,110 ± 1120
ATG + I·C	500 ± 49	20,500 ± 1999	3318 ± 401	1756 ± 164
+ PURO	34 ± 4	2427 ± 225	11 ± 2	1503 ± 155
+ CHX	13,659 ± 1423	2630 ± 217	75 ± 8	1910 ± 186
+ CAP	16,101 ± 1509	2411 ± 252	3122 ± 302	1755 ± 166
+ ACT.D	1397 ± 140	382 ± 39	54 ± 7	24 ± 4
+ RNase	50 ± 6	31 ± 4	10 ± 3	13 ± 2

[a] DPM in TCA insoluble material (protein synthesis) or Schmidt-Thannhauser separated RNA (RNA synthesis) per mg protein, ± S.D. at 5 minutes of incubation. Five separate specimens. Counts obtained according to assays for protein or RNA synthesis as described in the text.

systems substantially, but CHX inhibited whole cells by only about 30%, whereas it inhibited the cell-free system almost completely. CAP inhibited both by only 10% or less. Although ACT.D inhibited both, it inhibited whole cells much less than the cell-free system as observed in the case of CHX. RNase inhibited both markedly. ATG caused a 19% increase in RNA synthesis in whole cells, but none in the cell-free system. When preceded by poly I·C, ATG caused a 19-fold increase in RNA synthesis in whole cells, and an 8-fold increase in the cell-free system. When ATG was added before poly I·C, there was a 10-fold increase in the whole cells, but none in the cell-free system. PURO, CHX, and CAP had little or no effect on RNA synthesis; ACT.D and RNase essentially abolished it. The gradient profiles (Figure 2) show relative differentiation of polysomes in controls compared with the rabbit, and this pattern (especially dimer and trimer) is further developed in the presence of poly I·C, but not poly A·U. ATG had no effect, and may actually have diminished the polysomal differentiation.

Rat peritoneal macrophages

Poly I·C caused about a 35% increase in whole-cell protein synthesis, poly A·U about 20%. In the cell-free system, the increases were 40% and none for the 2 polynucleotides. PURO, ACT.D, and RNase inhibited much

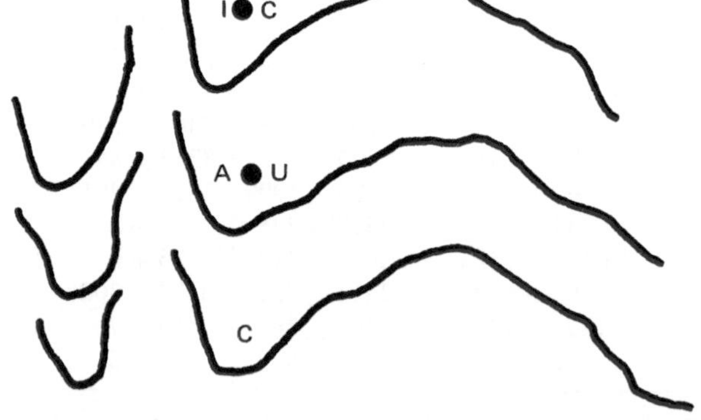

Fig. 3. Gradient tracings of rat peritoneal macrophages. Polysomes were obtained from cell-free system by pelleting at 105,000 × g for 90 minutes as described in the text. Polysomes were then analyzed on sucrose gradients with additions as described in text.

Table 3. Rat Peritoneal Macrophages Vol.-Adjusted DPM at 10 Minutes (3–8)[a]

System	Whole cells		Cell-free (S-10)	
	Per mg prot.	RNA	Per mg prot.	RNA
CS	5605 ± 549	3230 ± 311	3190 ± 308	2271 ± 219
− COFACT.	...	...	73 ± 7	22 ± 3
− RIBO.	...	...	40 ± 5	2130 ± 224
−(S-105)	...	...	23 ± 4	31 ± 3
− MG^{++}	...	...	27 ± 5	41 ± 7
+ I·C	8903 ± 797	3121 ± 390	5069 ± 519	2431 ± 247
+ A·U	7110 ± 603	3184 ± 321	3254 ± 333	2019 ± 199
+ PURO	25 ± 3	3056 ± 329	31 ± 4	2132 ± 217
+ CHX	2190 ± 221	3198 ± 320	1430 ± 141	2297 ± 246
+ CAP	4823 ± 479	3295 ± 335	2875 ± 250	2190 ± 270
+ ACT.D	76 ± 8	55 ± 6	49 ± 5	17 ± 2
+ RNase	12 ± 3	18 ± 3	11 ± 2	16 ± 3

[a] DPM in TCA insoluble material (protein synthesis) or Schmidt-Thannhauser-separated RNA (RNA synthesis) per mg protein, ± S.D. at 10 minutes of incubation. Three experiments, 8 rats per experiment. Counts obtained according to assays for protein or RNA synthesis as described in text.

(whole cells), but CHX and CAP inhibited less, although CHX was about twice as inhibitory as CAP. Essentially the same relationships occurred for the inhibitors in the cell-free system. As for RNA synthesis, it was not affected by polynucleotides in either system. PURO, CHX, and CAP did not inhibit, but ACT.D and RNase did. The polynucleotides did not alter the polysomal profile (Figure 3) in either system.

Human peritoneal macrophages

Poly I·C stimulated protein synthesis in whole cells by 27%, poly A·U by 13%. In the cell-free system, relative increases from these polynucleotides were 20 and 10%, respectively. ATG did not affect whole-cell protein synthesis, but had a 9% positive effect in the cell-free system. This is notably different from the human alveolar macrophage. PURO, ACT.D, and RNase inhibited both whole cell and cell-free systems markedly, CHX inhibited whole cells by 78% and the cell-free system almost totally. CAP inhibited whole cells by 15%, and the cell-free system by about 50%. RNA synthesis in whole cells was stimulated arithmetically by poly A·U and ATG and slightly but significantly by poly I·C. There was no stimulation in the cell-free system. PURO inhibited whole-cell RNA synthesis by 20% but CHX

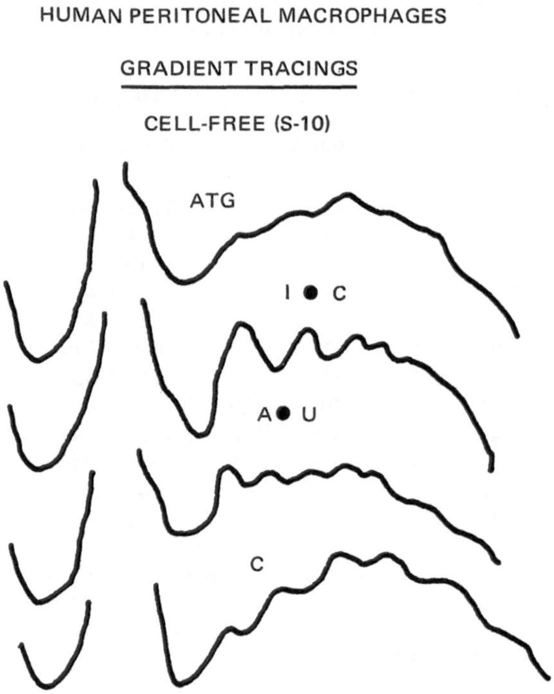

Fig. 4. Gradient tracings of human peritoneal macrophages. Polysomes were obtained from the cell-free system by pelleting at 105,000 × g for 90 minutes as described in the text. Polysomes were then analyzed on sucrose density gradients as described in the text. Additions as described in the text.

and CAP did not inhibit at all. In the cell-free system, PURO, CHX, and CAP were not inhibitory; ACT.D and RNase markedly inhibited RNA synthesis in both systems. The polysomal control profile (Figure 4) shows moderate differentiation into separate peaks, which appear to be enhanced (especially the dimer and trimer) by the polynucleotides, mostly by poly I·C, whereas ATG had no effect.

DNA-dependent RNA polymerase

In lung macrophages (Table 5) the resting values for the enzyme are increased 11-fold by addition of PPD to PPD-sensitized cells, and 2–3-fold by polynucleotides, poly I·C being slightly more effective. The polynucleotides cause from 2–3 times increase in human lung macrophages, whereas ATG causes a 3-fold increase. In peritoneal macrophages, poly A·U is 7% more stimulatory than poly I·C in rat cells, and 30% more stimulatory in human cells. ATG causes a 41% increase in human cells.

Table 4. Human Peritoneal Macrophages Vol.-Adjusted DPM at 10 Minutes (4–1)[a]

System	Whole cells		Cell-free (S-10)	
	Per mg prot.	RNA	Per mg prot.	RNA
CS	6902 ± 701	3538 ± 302	4391 ± 441	2972 ± 289
– COFACT.	···	···	33 ± 6	27 ± 5
– RIBO.	···	···	20 ± 4	2801 ± 292
– (S-105)	···	···	24 ± 5	19 ± 7
– MG	···	···	75 ± 12	35 ± 11
+ I·C	9413 ± 915	4382 ± 419	5311 ± 544	3033 ± 321
+ A·U	7914 ± 781	3780 ± 362	4703 ± 458	2831 ± 290
+ ATG	7279 ± 714	3618 ± 339	4943 ± 487	3101 ± 318
+ PURO	33 ± 4	2719 ± 268	42 ± 10	2899 ± 302
+ CHX	1210 ± 125	3490 ± 310	21 ± 7	2717 ± 283
+ CAP	5967 ± 601	3611 ± 363	2209 ± 214	3007 ± 322
+ ACT.D	22 ± 3	10 ± 4	23 ± 5	22 ± 9
+ RNase	14 ± 2	9 ± 5	16 ± 4	5 ± 8

[a] DPM in TCA-insoluble material (protein synthesis) or Schmidt-Thannhauser-separated RNA (RNA synthesis) per mg protein, ± S.D. Four experiments, 1 subject per experiment. Counts obtained according to assays for protein or RNA synthesis as described in the text.

Table 5. DNA-Dependent RNA Polymerase[a]

Preparation	Specific activity (mean) pmoles in RNA/20 min/mg prot.
Rabbit lung	
nonsensitized – PPD	(3.4)
nonsensitized + PPD	5.4
sensitized – PPD	2.8
sensitized + PPD	39.2*
sensitized + I·C	9.5*
sensitized + A·U	7.5*
Human lung	
control	(3.9)
+ I·C	8.3*
+ A·U	9.4*
+ ATG	10.3*

[a] Incorporation of ^{14}C-UTP into RNA in the presence of "enzyme" as described under Material and Methods. Specific activity given is the mean of the experiments for each cell type. Control levels are in parentheses. Numbers with asterisk indicate significant difference from control, $p = 0.05$.

Table 6. DNA-Dependent RNA Polymerase[a]

Preparation	Specific activity (mean) *p*moles in RNA/20 min/mg prot.
Rat peritoneum	
control	(3.5)
+I·C	7.6*
+A·U	8.1*
Human peritoneum	
control	(3.7)
+I·C	6.4*
+A·U	9.1*
+ATG	5.8*

[a] Incorporation of ^{14}C-UTP into RNA in the presence of "enzyme" as described under Material and Methods. Specific activity given is the mean of the experiments for each cell type. Control levels are in parentheses. Numbers with asterisk indicate significant difference from control, $p = 0.05$.

DISCUSSION

Of the 4 types of macrophages studied, the effects of polynucleotides are most evident in the lung cells, possibly because sensitized rabbit cells were studied, and because the human cells are metabolically very active on their own. Although one might be tempted to state that poly I·C substituted for antigen in the sensitized rabbit system, this is not the case since it did not stimulate RNA or polysome synthesis. Since its effect on protein synthesis is nonspecific, it is thereby also selective.

The differences between the effects of poly I·C and poly A·U are large enough to require explanation, but this will be impossible until more information is available on the intracellular reactions that these polynucleotides influence. Probably more than a relative messenger function is involved. Work is in progress to determine the effect, if any, of these polynucleotides on the synthesis of the acidic proteins which seems to result specifically from antigenic stimulation of PPD-sensitized rabbit lung macrophages. The role, if any, of these proteins in antibody formation is unknown. The human cell-free system appears to be more efficient than, and/or qualitatively different from that of the sensitized rabbit macrophages because of its markedly enhanced protein synthetic response to poly I·C. The inability of ATG to inhibit the cell-free system suggests that its inhibition of whole cells requires the plasma membrane. The partial protection from this inhibition afforded by prior treatment with poly I·C (but again, *not* poly A·U) is difficult to explain without postulating an interaction between poly I·C and ATG at the plasma membrane, or an intracellular interaction between events initiated by the early presence of poly I·C and the later influence of ATG. ATG is a 7S globulin, and would be expected to bind to the plasma membrane through its Fc (tail) piece, leaving its FAb fragments free in space like the pincers of a crab. It is clear that ATG still exerts part of its inhibitory effect when preceded by poly I·C; therefore, it is insufficient to say that prior poly I·C treatment prevents binding of ATG to the membrane. Further, since ATG *followed by* poly I·C is still inhibitory, one might say that ATG has occupied membrane sites available to poly I·C. If this were true, then how does one explain continued synergistic stimulation of RNA synthesis by poly I·C with ATG? Although RNA synthesis is only about half as much when ATG precedes poly I·C as in the other way around, it is still almost 10 times control with the combination, suggesting that both molecules still exert an effect, regardless of their order of appearance. The suggestion of a sole membrane action of ATG is contradicted by the ATG-stimulated RNA synthesis when ATG is preceded by poly I·C in the cell-free system. However, this response does not duplicate the situation in whole cells. In the latter, it will be recalled that ATG followed by poly I·C gave 50% of the incorporation into RNA attained by poly I·C followed by ATG. However, ATG followed by poly I·C in the cell-free system is *not* at all stimulatory. Therefore, a specific order of effect is required for stimulation of RNA synthesis in the cell-free system, as well as for inhibition of protein synthesis in whole cells. It is not clear how poly I·C might accentuate polysomal profiles in whole cells without actually providing a template, unless it simply makes the existing message more efficient in assembling ribosomes, possibly through its stimulatory effect on RNA polymerase (see below), or through a totally different mechanism such as complexing of inhibitory proteins. Again, poly A·U does not have this effect.

The enhancement of the dimer and trimer peaks of the polysomal profile by poly I·C in both human alveolar and peritoneal macrophages (Figures 2 and 4) might indicate selective stimulation of specific protein synthesis on polysomes having 2 or 3 ribosomes.

Both types of peritoneal macrophages were less active than lung cells in response to polynucleotides. Poly A·U stimulated protein synthesis in rat whole cells. In the human cell-free system (peritoneal macrophages) poly A·U stimulated protein synthesis slightly, whereas it did not stimulate protein synthesis in the rat cell-free system. In distinction to effects on whole human lung cells, ATG did not inhibit whole human peritoneal macrophage protein synthesis, and stimulated the cell-free system slightly. The effects of anti-lymphocyte serum on lymphocytes and macrophages are dose-related (Barth *et al.*, 1969); in these experiments the concentration was not varied. There may be specific ion requirements, not met in these experiments, for certain activities of ATG. Human peritoneal macrophages were more sensitive to CHX and CAP than the other cells tested, and displayed partial, definite inhibition of RNA synthesis by PURO. If the turnover of RNA in these cells were very rapid, then the dependence of RNA synthesis on new enzyme biosynthesis could be PURO-sensitive. On the same hypothesis, one might expect "endproduct inhibition" through the presence of the unrecognized competitor for RNA binding sites. If this were true, then why would not the resting level of RNA synthesis be higher? The human peritoneal macrophages were obtained from uremic patients. What effects this disorder would have on the metabolism of the cells cannot be stated.

The RNA polymerase levels show, in the rabbit lung alveolar cells, that the specific sensitizing antigen is much more efficient than poly I·C or poly A·U in inducing or stimulating biosynthesis of that enzyme. The polynucleotide effects are not insignificant, and ATG stimulates RNA polymerase synthesis in both human lung and peritoneal macrophages. Experiments with labeled polynucleotides and labeled ATG will be necessary to understand these findings. It would be of interest to assess the effect of exogenously added RNA polymerase in nonsensitized cells or sensitized without PPD. Genome transcription under such circumstances can be studied by the method of Kohl *et al.* (1968). If macrophage RNA polymerase is rifamycin-sensitive, the relationship between polymerase synthesis, specific RNA synthesis, and immunogenicity could be studied (Geiduschek and Sklar, 1969).

The increase of the polysome population suggests significant attachment of ribosomes to mRNA, but no evidence has been obtained concerning the type of RNA synthesized, except that in the rabbit macrophages most of the ^{3}H-Ur is attached to heavier polysomes. The short labeling period makes it unlikely that this radioactivity is in ribosomal RNA. A possibility not yet ruled out is that some free ribosomes may attach to inactive ("stored") mRNA (Spirin and Nemer, 1965). The use of labeled polynucleotides might allow one to determine whether free ribosomes attach to them as well. Polyribosomes in antibody-forming tissues have been described (Nakashima *et al.*, 1967; Vasil'chenko *et al.*, 1969).

Braun (1969) has suggested that poly I·C may be more injurious to macrophage membranes than poly A·U. Although we have not yet investigated the immunogenicity of polynucleotide-treated macrophages as obtained in our incubation systems, we would not have predicted the effects of poly A·U reported by Johnson (1969), particularly the reversal of ACT.D inhibition of RNA synthesis in mouse peritoneal macrophages following exposure to specific antigen. This may reflect a species or experimental difference. Chemically altered poly A·U was not as good an adjuvant as the "normal" complex (Schmidtke, 1969). In the experiments reported here, poly A·U stimulated whole-cell rat peritoneal macrophage RNA synthesis arithmetically, whereas poly I·C and poly A·U stimulated human peritoneal macrophage RNA synthesis significantly. This might suggest more than a releasing or permeability-enhancing role for the polynucleotides as adjuvants. Salas and Bollum (1969) have studied binding of some polynucleotides to ribosomes. Single-stranded poly U and poly A were unable to stimulate incorporation of amino acids in rat liver mitochondria (Hanninen and Alanen-Irjala, 1968). Margolis and Levy (1969) reported stimulation of rabbit kidney cell RNA and protein synthesis by poly I·C.

It is premature to speculate on the mechanism of sensitivity to the inhibitors of protein synthesis used, except that the responses to PURO, ACT.D, and RNase are as one would expect. The responses to CHX and CAP require more work to explain, and, in particular, we are investigating the possibility of a CHX-sensitive and CHX-resistant protein-synthesizing system (Lukins and Linnane, 1967) in the rabbit cells, and also the kinetics of PURO-induced release of peptides from tRNA (equivalent to the peptidyl transferase reaction) in the presence and absence of CAP (Weber and De Moss, 1969).

Phagocytosis has been shown to increase the synthesis of enzymes (Jansa and Kratky, 1968; Cohn, 1970) and incorporation of inorganic phosphate-^{32}P (Oren *et al.*, 1963) into macrophages. Although no studies have been reported on the functional significance of amino acid incorporation during exposure of macrophages to antigen, polynucleotides, and so on, biological studies will be required to distinguish between protein synthesis directed toward the cell's survival versus that which might alter the immune response.

References

Barth, R. F. *et al.* (1969). Effects of antilymphocyte sera on antibody forming cells and macrophages in the early immune response. Transplant. Proc. **i:** 1, 433–435.

Braun, W. (1969). Relationships between the effects of poly I·poly C and endotoxin. Nature **224:** 1024–1025.

——— and Nakano, M. (1967). Antibody formation: Stimulation by polyadenylic and polycytidylic acids. Science **137:** 819.

Cohn, Z. (1970). Physiology and biochemistry of macrophages. Proc. III Internat. Inflammation Club, Excerpta Med. (in press).

Field, A. K. *et al.* (1968). Induction of interferon and host resistance *in vitro* by complexed polynucleotides. Bacteriol. Proc. 145.

Finkelstein, M. S. *et al.* (1968). Interferon inducers *in vitro*: Difference in sensitivity to inhibitors of RNA and protein synthesis. Science **131**: 465–468.

Geiduschek, E. P. and Sklar, J. (1969). Role of host RNA polymerase in phage development. Nature **221**: 833–836.

Hahn, H. H. *et al.* (1967). Studies on the pathogenesis of fever. XV. The production of endogenous pyrogen by peritoneal macrophages. J. Exp. Med. **126**: 385–394.

Hanninen, O. and Alanen-Irjala, K. (1968). Effect of polyuridylic and polyadenylic acids on protein synthesis in isolated rat liver mitochondria. Acta Chem. Scand. **22**: 3072–3080.

Hunter, W. M. and Greenwood, F. C. (1962). Preparation of iodine-131-labeled human growth hormone of high specific activity. Nature **194**: 495.

Jansa, P. and Kratky, J. (1969). Enzyme pattern of macrophages in the course of inflammation. Blut **18**: 218–219.

Johnson, H. G. (1969). Factors affecting the enhancement of antibody synthesis by competent macrophages. Ph. D. Thesis, University of Michigan.

——— and Johnson, A. G. (1968). Enhancement of antibody synthesis in mice by macrophages stimulated *in vitro* with antigen and polyadenylic, polyuridylic acids. Bacteriol. Proc. 75.

Kohl, D. M. *et al.* (1969). The role of RNA polymerase in the control of RNA synthesis *in vitro* from *Rana pipiens* embryo chromatin. Biochim. Biophys. Acta **179**: 23–28.

Lowry, O. H. *et al.* (1951). Protein determination by color reaction with Folin's phenol reagent. J. Biol. Chem. **193**: 265.

Margolis, S. and Levy, H. B. (1970). The action of poly I·C on the RNA metabolism of primary rabbit kidney cells. Fed. Proc. (abstr. No. 2187).

Martin, R. G. and Ames, B. N. (1961). Analysis of sedimentation coefficients on sucrose density gradients. Application to protein mixtures. J. Biol. Chem. **236**: 1372.

McMaster, P. D. (1953). Sites of antibody formation. *In* The Nature and Significance of the Antibody Response. 13–45. Ed. by A. M. Pappenheimer, Jr. Columbia University Press, New York.

Morrell, R. M. (1970). Effect of horse anti-human thymocyte gamma globulin (ALG) on RNA-dependent protein synthesis in human lung macrophages. Clin. Res. **18**: 2, 429.

——— (1969a). RNA and protein synthesis and polysomal activation by specific antigen in PPD-sensitized rabbit alveolar macrophages. Abstrs. VI Ann. Mtg., Reticuloendothelial Soc.

——— (1969b). Biosynthesis of peptides in canine hypothalamus. Dissertation, University Microfilms, Inc., **29**: 9.

Nakashima, S. *et al.* (1967). Polyribosomes in antibody-forming tissues of hyperimmunized rabbits. Biochim. Biophys. Acta **145**: 671–678.

Nirenberg, M. W. (1963). *In* Methods in Enzymology **6**: 17. Ed. by S. P. Colowick and N. O. Caplan. Academic Press, New York.

Noll, H. (1969). *In* Techniques in Protein Synthesis (II). 101–178. Ed. by P. N. Campbell and J. R. Sargent. Academic Press, New York.

Oren, R. *et al.* (1963). Metabolic patterns in 3 types of phagocytic cells. J. Cell Biol. **17**: 487–501.

Pearsall, N. and Weiser, R. (1970). The Macrophage. Lea and Febiger, Philadelphia.

Salas, J. and Bollum, F. J. (1969). Binding of biosynthetic polyribo- and polydeoxynucleotides to ribosomes. J. Biol. Chem. **244**: 1152–1156.

Schmidtke, J. R. (1969). Enhancement of antibody synthesis by synthetic polynucleotides. Ph. D. Thesis, University of Michigan.

——— and Johnson, A. G. (1968). Stimulation of antibody formation by complexes of polyadenylic acid and polyuridylic acid. Bacteriol. Proc. 75.

——— ——— (1969). Factors affecting the stimulation of antibody formation by complexes of polyadenylic and polyuridylic acids. Bacteriol. Proc. 99.

Seifart, K. H. and Sekeris, C. E. (1969). Extraction and purification of DNA-dependent RNA polymerase from rat nuclei. European J. Biochem. **7**: 408–412.

Smith, T. J. and Wagner, R. R. (1967). Rabbit macrophage interferons. I. Conditions for biosynthesis by virus-infected and uninfected cells. J. Exp. Med. **125**: 559–577.

Spirin, A. S. and Nemer, M. (1965). Messenger RNA in early sea urchin embryos; cytoplasmic particles. Science **150**: 214–217.

Staehlin, T. and Meselson, M. (1966). *In vitro* recovery of ribosomes and of synthetic activity from synthetically inactive ribosomal subunits. J. Mol. Biol. **15**: 245.

Stecher, V. J. *et al.* (1967). Sites of production of primate serum proteins associated with complement system. Proc. Soc. Exp. Biol. Med. **124**: 433–438.

Stecher, V. J. and Thorbecke, G. J. (1967). Sites of synthesis of serum proteins. I. Serum proteins produced by macrophages *in vitro*. J. Immunol. **99**: 643–652.

Vasil'chenko, V. N. *et al.* (1969). Studies on spleen polysomes of immunized animals. Biokhim. **34**: 139–143.

Weber, M. J. and De Moss, J. A. (1969). Inhibition of the peptide bond synthesizing cycle by chloramphenicol. J. Bacteriol. **97**: 1099–1105.

CONVERSION WITH POLYNUCLEOTIDES OF A GENETICALLY CONTROLLED LOW IMMUNE RESPONSE TO A HIGH RESPONSE IN MICE IMMUNIZED WITH A SYNTHETIC POLYPEPTIDE ANTIGEN

EDNA MOZES, G. M. SHEARER,* MICHAEL SELA, and WERNER BRAUN

*Department of Chemical Immunology, The Weizmann Institute of Science, Rehovot, Israel***

and

The Institute of Microbiology, Rutgers University, New Brunswick, New Jersey

INTRODUCTION

During the past few years a number of investigations have demonstrated that genetics plays a significant role in the ability of several animal species to elicit immune responses to a variety of antigens (McDevitt and Benacerraf, 1969). Results of experiments using synthetic polypeptides of restricted heterogeneity indicated that antibody responses of mice to these types of immunogens were controlled by dominant, quantitative, autosomal genes (McDevitt and Sela, 1965, 1967; Pinchuck and Maurer, 1968; Mozes *et al.*, 1969a, 1969b). In view of the complexity of the cellular events involved in immune processes, it is possible that certain genetic controls are expressed at the cellular level. Hence, genetic controls could affect any of a number of cell-mediated steps.

* Postdoctoral Fellow of the American Cancer Society.

** Work performed in the Department of Chemical Immunology, The Weizmann Institute of Science, was supported in part by Agreement 06-035 with the National Institutes of Health, Public Health Service, U.S. The activities of one of us (W. B.) at the Weizmann Institute were supported by an NIH Special Fellowship.

Much of our knowledge concerning the cellular aspects of immunity has been obtained using the antigens of sheep erythrocytes as immunogens. The immune response of mice to sheep erythrocytes requires the cooperation of more than one functionally distinct cell type (Claman *et al.*, 1966; Mosier, 1967; Mosier and Coppleson, 1968; Raidt *et al.*, 1968; Mosier *et al.*, 1970; Shortman *et al.*, 1970). Certain of these cells, broadly referred to as immunocompetent precursors, are capable of recognizing antigen and/or of generating mature antibody-forming cells as a result of interaction, proliferation, and differentiation. Suspensions of a sufficient number of spleen cells contain all the cell types necessary for antibody production. These cells constitute a functional unit known as an antigen-sensitive unit (Shearer *et al.*, 1969a; Shearer and Cudkowicz, 1969a), which can be detected by transplanting graded and limiting cell numbers into heavily irradiated syngeneic recipient mice with antigen. The *donor-derived* immune responses are subsequently analyzed in the host animals several days later. Such limiting dilution experiments have been used to study the relative number and nature of the least frequent immunocompetent precursor cells which constitute antigen-sensitive units reactive with sheep erythrocyte antigens (Kennedy *et al.*, 1966; Shearer *et al.*, 1968, 1969a, 1969b; Cudkowicz *et al.*, 1969; Shearer and Cudkowicz, 1969b). In the present report the limiting dilution approach is utilized to study the cellular basis of genetic control of immunity to a synthetic polypeptide built on multichain polyproline in high and low responder inbred strains of mice.

Multiple-stranded homopolymers of ribonucleotides including poly A-poly U (poly A:U) have the capacity to enhance antibody formation in newborn (Winchurch and Braun, 1969), adult (Braun and Nakano, 1967), and aging (Braun, 1970) animals, as reported elsewhere in this Symposium. These phenomena appear to be a consequence of stimulatory effects on members of the macrophage cell population (Johnson and Johnson, 1968; Braun, 1969), as well as on lymphocytes (Braun *et al.*, 1970). It has been reported that polynucleotides increase the rate of appearance of antibody-forming cells (Braun and Nakano, 1967). This effect could be the result of more rapid proliferation of potentially immunocompetent cells, or it could be due to the stimulation of "silent" or "low affinity" precursors not ordinarily activated by the immunogen alone. One of the known effects of poly A:U and poly I:C is to evoke detectable immune responses to ordinarily poor immunogens, such as tumor cell antigens in syngeneic host animals (Braun, 1969; Levy *et al.*, 1969). It was of interest, therefore, to establish whether it is possible to enhance significantly the immune response of a genetic low responder mouse strain to a synthetic polypeptide by treatment with poly A:U.

RESULTS

A. *Genetic control of immune response to (T,G)-Pro—L.* Quantitative differences in the amount of antibodies produced to a synthetic polypeptide built on multichain polyproline, poly-L-(Tyr, Glu)-poly-L-Pro—poly-L-Lys, abbreviated (T, G)-Pro—L, have been demonstrated in several inbred strains of mice (Mozes *et al.*, 1969a). The SJL strain was the highest responder to this immunogen, whereas the DBA/1 strain was the lowest. Figure 1 illustrates the pattern of immune responses to (T, G)-Pro—L in parental, F_1, and backcross animals of these two strains (Mozes *et al.*, 1969b). The number of mice tested is plotted as a function of the percent of ^{125}I-labeled antigen bound to specific antibody. The response of (DBA/1 × SJL)F_1 mice was intermediate between those of the parental strains. (DBA/1 × SJL)F_1 × DBA/1 backcross offspring segregated as low and intermediate responders. The responses of F_1 × SJL progeny were intermediate and low, although they were expected to have been intermediate and high. There is no simple genetic explanation for these latter results, although genetic and possibly other factors of a complex nature may be involved. In general, the results presented in Figure 1 indicate that the ability of mice to respond to (T, G)-Pro—L is genetically controlled by dominant, quantitative, autosomal factor(s).

B. *Conversion of genetic low responder mice to high responders by treatment with poly A: U.* Since polynucleotides stimulate nonspecifically some facets of the immune system (Braun and Nakano, 1967; Braun, 1969), it was of interest to determine whether poly A: U affects the immune response of DBA/1 mice to (T, G)-Pro—L. Low responder DBA/1 and high responder SJL mice were immunized in the hind foot pads with 10 μg of (T, G)-Pro—L in complete Freund's adjuvant. Twenty-four hours later, 300 μg of poly A: U dissolved in phosphate-buffered saline (0.15 N NaCl, 0.01 M phosphate buffer, *p*H 7) was administered via the tail vein. Table 1 summarizes passive hemagglutination results obtained by titering low and high responder strains immunized and treated with poly A: U. DBA/1 mice injected with poly A: U after primary *and* secondary immunization responded as well as untreated high responder SJL mice (Table 1). About an 8-fold increase in hemagglutination titer was observed above the untreated DBA/1 mice. Treatment of the low responder animals only after primary immunization resulted in an enhancement almost as high as that obtained by giving poly A: U after primary and secondary antigenic stimulation. In contrast, injection of poly A: U only after secondary immunization resulted in a less dramatic increase in antibody titers. Administration of poly A: U to the high responder SJL strain after primary and secondary immunization resulted in a slight but insignificant increase in antibody titer. The results shown in Table 1 indicate

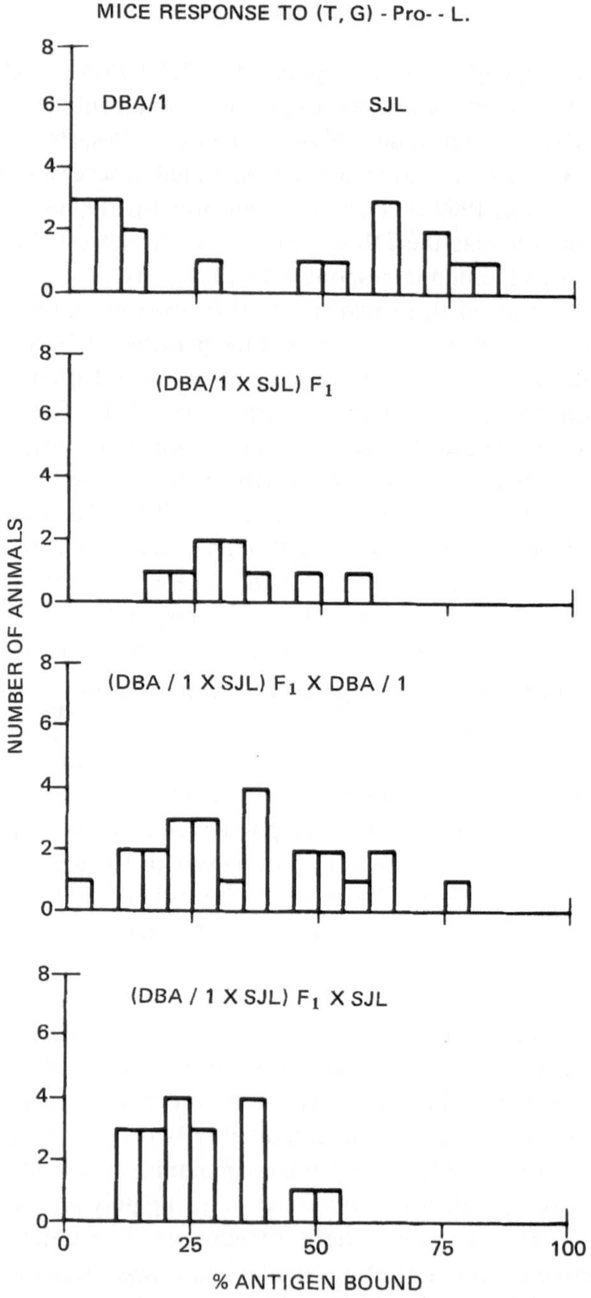

Fig. 1. Antibody responses of mice immunized with (T, G)-Pro—L assayed with ^{125}I labeled (T, G)-Pro—L. Horizontal axis plots percent antigen bound as a function of the number of mice exhibiting a given percent. All antisera were assayed at 1/50 dilution.

Table 1. Anti-(T, G)-Pro—L Titers in Intact High and Low Responder Mice 12–14 Days after Immunization[a]

Treatment[b] with poly A:U	Number of sera containing detectable antibody and range of hemagglutination titers	
	DBA/1	SJL
No treatment	10/10 1:8–1:16	5/5 1:64–1:128
Primary immunization	5/5 1:64	Not tested
Secondary immunization	10/10 1:16–1:32	Not tested
Primary and secondary immunization	10/10 1:64–1:128	5/5 1:128–1:256

[a] Mice were injected in hind foot pads with 10 µg (T, G)-Pro—L in complete Freund's adjuvant, followed 3 weeks later by 10 µg (T, G)-Pro—L in phosphate-buffered saline.
[b] Mice were injected intravenously with 300 µg poly A:U in phosphate-buffered saline 24 hours after (T, G)-Pro—L injection.

that the immune response of this genetic low responder strain can be "corrected" by poly A:U treatment. Although the experiment described here was not designed to establish at which level the polynucleotide was altering the immune response of DBA/1 low responders, the fact that poly A:U treatment had a more pronounced effect after primary than after secondary immunization suggested that a relatively early event in the immune response was affected.

C. *Estimation of immunocompetent precursor cell frequency by limiting dilution analysis.* A limiting dilution approach has been used to estimate relative numbers of the least frequent immunocompetent precursor cell type constituting splenic antigen-sensitive units reactive with sheep erythrocyte antigens (Kennedy *et al.*, 1966; Shearer *et al.*, 1968, 1969a). This analysis is based on the fact that a critically low number of cells transplanted will not generate a donor-derived immune response in recipients, since the least frequent precursor cell type relevant for the response is not contained in the inoculum. Conversely, a relatively high number of cells injected will contain *all* required precursor types, and a donor-derived response will result. Immune responses in the recipients are respectively classified as "negative" or "positive". By plotting the percentage of positive responses in recipients as a function of the number of cells transferred, a curve is obtained which conforms to the predictions of the Poisson model. Therefore, this statistical approach has been used to describe the theoretical probability that antigen-sensitive units contained in a given number of spleen cells generate an immune response. The procedure used for calculating this probability has been described elsewhere (Porter and Berry, 1963; Shearer *et al.*, 1968). Figure 2 gives an example of the limiting dilution approach for estimating frequency of antigen-sensitive units reactive with antigens of sheep erythrocytes in mouse spleens. Donor-derived responses were analyzed for direct (IgM) and indirect (IgG) hemolytic plaque-forming cells (Jerne *et al.*, 1963;

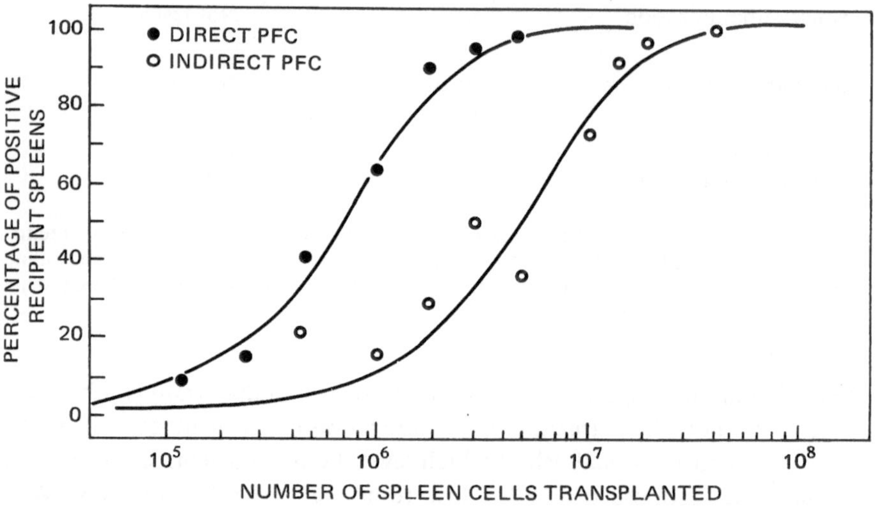

Fig. 2. Percentage of recipient spleens positive for direct (IgM) and indirect (IgG) plaque-forming cells after injection of graded numbers of spleen cells mixed with 5×10^8 sheep erythrocytes. Symbols indicate observed percentages and curves show expected percentages according to the Poisson model.

Wortis *et al.*, 1966). The inoculum size necessary to yield 63% positive recipient spleens, that is, the number of cells estimated to contain one relevant precursor on the average, was 1×10^6 and 7×10^6 spleen cells for direct and indirect plaque-forming cells, respectively.

D. *Frequency of splenic antigen-sensitive precursor cells reactive with (T,G)-Pro—L before and after poly A:U treatment.* In one series of experiments 51 SJL and 78 DBA/1 recipient mice were irradiated with 750–850 R and injected with graded numbers (5×10^5 to 6×10^7) of spleen cells from normal syngeneic donors. Twenty-four hours later, the recipients were immunized intraperitoneally with 10 μg (T,G)-Pro—L in complete Freund's adjuvant. In another group of experiments 87 SJL and 78 DBA/1 irradiated recipients were injected with similar numbers of spleen cells from syngeneic donors immunized 3 weeks earlier via the hind foot pads with 10 μg (T,G)-Pro—L in complete Freund's adjuvant. Recipients were injected intravenously with (T,G)-Pro—L in Eagle's medium at the time of spleen cell transfer. Recipients were bled from the retro-orbital plexus 12–16 days after immunization and the sera from individual mice were assayed by passive hemagglutination. The sera from 12 to 25 recipients were titered for each inoculum. The results are summarized in Table 2. As the number of injected spleen cells was

Table 2. Percentage of Positive Sera in Irradiated Syngeneic Recipient Mice 12–16 Days after Injection of (T,G)-Pro—L and Graded Numbers of DBA/1 or SJL Spleen Cells from Normal and Immunized Donors[a]

Number of cells injected ($\times 10^6$)	Percentage of positive sera in recipients			
	DBA/1 donors		SJL donors	
	Normal	Immunized	Normal	Immunized
0.5	...	...	0	26.7
1.0	...	0	22.1	61.1
2.0	...	...	...	77.0
4.0	25.0	...	50.0	89.0
10.0	33.3	46.2	63.6	93.4
20.0	50.0	53.7	100.0	...
40.0	60.0	72.0	...	100.0
60.0	100.0	100.0	...	...
	Limiting precursor cell frequency ($\times 10^{-6}$)			
	1/31 (1/22–1/46)	1/30 (1/22–1/46)	1/7.2 (1/4.8–1/12)	1/1.3 (1/1.1–1/1.8)

[a] Sera from 12–25 recipients were assayed for each inoculum. 95% confidence intervals shown in parentheses.

increased, the proportion of recipient sera containing a detectable amount of antibodies, that is, the fraction of positive recipient sera, also increased for both high and low responder strains. Approximately 63% of SJL sera were positive when 10^7 cells from nonimmunized syngeneic donors were transplanted, whereas 60% of DBA/1 sera were positive only after injection of 4×10^7 cells from nonimmunized donors. Statistical analysis of the data (Porter and Berry, 1963) indicated that one limiting precursor cell reactive with (T,G)-Pro—L was detected in 7×10^6 and one in 31×10^6 spleen cells for high responder SJL and low responder DBA/1 strains, respectively. The difference between precursor frequencies in the strains was more striking when spleen cells from preimmunized donors were compared. One relevant limiting precursor was detected per 1.3×10^6 and one per 30×10^6 for SJL and DBA/1 donors, respectively (Table 2). These strain differences were statistically significant at the 0.05% level. It is noteworthy that preimmunization of DBA/1 donors did not increase the detected precursor frequency, whereas similar injection of SJL donors resulted in a significant 6-fold increase in the relative number of precursors (Mozes *et al.*, 1970).

The results described above show that the low response of DBA/1 mice to (T,G)-Pro—L can be attributed to a reduced number of antigen-sensitive precursor cells as compared to SJL mice, and demonstrate (see Table 1) that treatment of intact DBA/1 mice with poly A:U enhanced the response to the level of SJL mice. Therefore, limiting dilution experiments were carried out using spleen cells from low responder donors that had been immunized and treated with poly A:U. In order to follow the experimental protocol that gave the maximum responses in intact DBA/1 mice, the polynucleotide was also injected into the recipients 24 hours after cell transfer and immunization. The results, presented in Table 3, indicate that poly A:U treatment increased the proportion of positive sera significantly. The inoculum that gave approximately 63% positive sera was found to range between 5×10^5 and 2×10^6 spleen cells. The data obtained using preimmunized DBA/1 do-

Table 3. Fraction and Percentage of Positive Sera in Irradiated Syngeneic Recipient Mice 12–16 Days after Injection of (T,G)-Pro—L, Poly A:U, and Graded Numbers of DBA/1 or SJL Spleen Cells from Immunized Donors[a]

Number of cells injected ($\times 10^6$)	DBA/1		SJL	
	Fraction of positive sera[a]	Percentage of positive sera	Fraction of positive sera[a]	Percentage of positive sera
0.12	6/12	50.0	...	...
0.5	7/11	63.6	5/11	45.4
1.0	9/15	60.0	4/7	57.1
2.0	11/17	64.4	4/6	66.7
4.0	...	...	7/9	77.7
5.0	6/8	75.0	...	...
10.0	8/10	80.0	...	...
20.0	6/7	85.0	...	...
60.0	5/5	100.0	...	...

[a] Donors were injected in hind footpads with 10 µg (T,G)-Pro—L in complete Freund's adjuvant 3 weeks before transfer, and 24 hours later with 300 µg poly A:U. Recipients were injected intravenously with 10 µg (T,G)-Pro—L at the time of spleen cell transfer, and 24 hours later with 300 µg poly A:U.

nors contained in Tables 2 and 3 are graphically illustrated in Figure 3. Graded numbers of cells from untreated donors follow the predictions of the Poisson model, whereas similar numbers of cells from poly A:U treated animals did not conform to these statistical predictions. In fact, the change in increments observed in percentage of positive sera with change in the number of injected spleen cells was less than that expected. Therefore, it was not possible by this approach to estimate the number of spleen cells

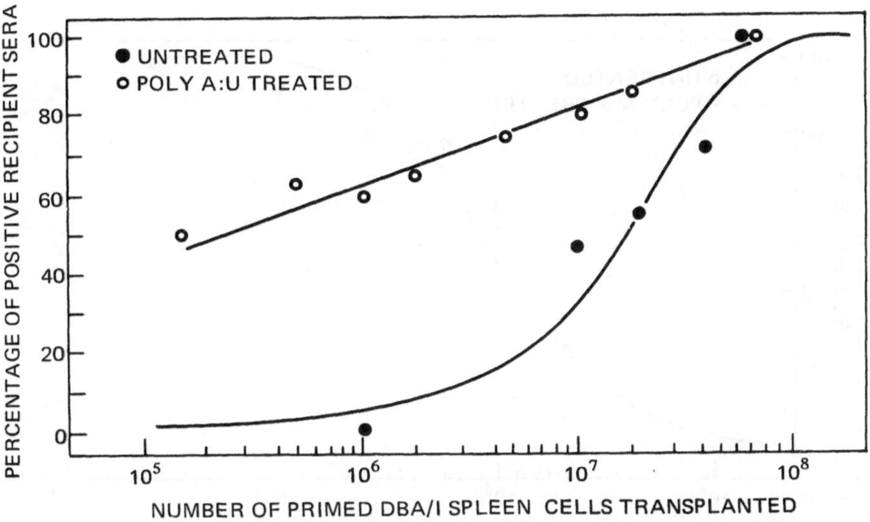

Fig. 3. Percentage of irradiated DBA/1 low responder mice with significant anti-(T,G)-Pro—L titers after injection of graded numbers of spleen cells from immunized donors and (T,G)-Pro—L. Symbols indicate observed percentages for untreated and poly A:U treated mice, curve gives expected percentages for untreated animals according to the Poisson model, and straight line shows eye-fitted data for poly A:U treated mice.

containing an average of one limiting immunocompetent precursor cell. Similar non-Poisson limiting dilution data have been reported for antisheep erythrocyte responses in mice using limiting numbers of bone marrow cells in the presence of excess thymocytes (Cudkowicz *et al.*, 1969, 1970).

In contrast to the significant effect of poly A:U on the DBA/1 response demonstrated by cell transfers, no detectable change was observed by similar treatment of SJL donors and recipients (Table 3). The number of grafted spleen cells in treated mice that yielded 63% positive sera was $1–2 \times 10^6$, indistinguishable from that obtained using SJL spleen cells (see Table 2). Transferred cells from untreated SJL mice followed the predictions of the Poisson model, as illustrated in Figure 4. The limited amount of available data obtained with SJL spleen cells suggested that there was no change in precursor frequency nor in the shape of the limiting dilution curve after injection of poly A:U.

E. *Limiting dilution experiments using mixtures of thymocytes and bone marrow cells.* In the mouse, for at least some immunogens, cooperating functional cell types are naturally separated between the organs of thymus and bone marrow (Claman *et al.*, 1966; Taylor, 1969; Chiller *et al.*, 1970). Under the proper experimental conditions, antibody formation results only

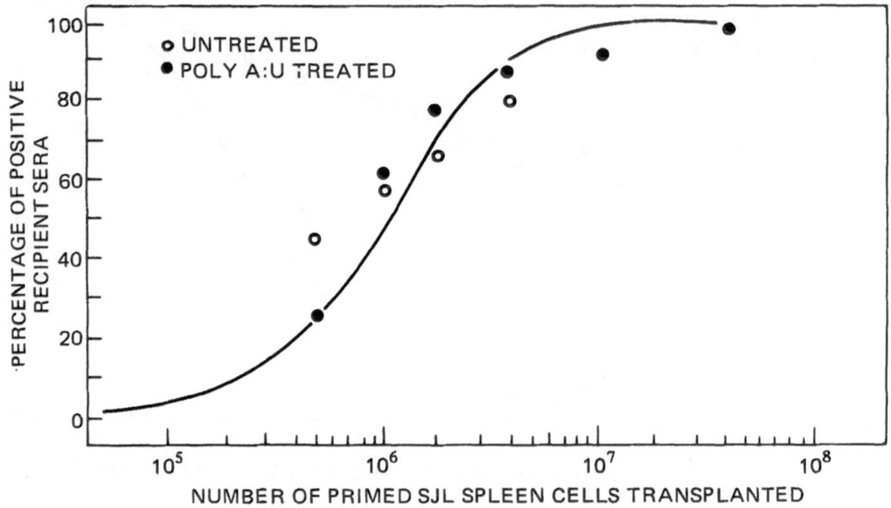

Fig. 4. Percentage of irradiated SJL high responder mice with significant anti-(T,G)-Pro—L titers after injection of graded numbers of spleen cells from immunized donors and (T,G)-Pro—L. Symbols indicate observed percentages for untreated and poly A:U treated mice, and curve shows expected percentages according to the Poisson model.

when cells from these two sources are mixed in the presence of antigen. The data shown in Table 4 demonstrate thymus-marrow cooperation for the (T,G)-Pro—L response in both DBA/1 and SJL mice. The transfer of 75×10^6 thymocytes gave no detectable response in 18 recipients, and 15×10^6 transferred marrow cells gave weak responses in only 4 of 35 mice tested. However, by transferring 75×10^6 thymocytes together with 15×10^6 marrow cells, all 16 SJL mice immunized yielded significant responses, whereas 8 of 11 DBA/1 mice gave positive hemagglutination titers. The fact that thymus-marrow cooperation is required for the (T,G)-Pro—L response and that not all DBA/1 recipients were positive suggests that one or more of the required cell types is limiting in the low responder strain and that the genetic defect can be demonstrated with cell mixtures from these two organs.

Table 4. Antibody Production in 750–850 R X-rayed Mice 14 Days after Injection of Thymus, Marrow, or Thymus and Marrow Cells and (T,G)-Pro—L[a]

Cells transplanted	Fraction of recipient sera containing antibodies to (T, G)-Pro—L	
	DBA/1	SJL
75×10^6 Thymus	0/9	0/9
15×10^6 Marrow	2/11	2/24
75×10^6 Thymus + 15×10^6 Marrow	8/11	16/16

[a] Recipients were immunized intraperitoneally with 10 µg (T,G)-Pro—L in complete Freund's adjuvant 24 hours after cell transfer.

Therefore, transfer experiments were carried out using both strains in which graded numbers of marrow cells were mixed with an excess of thymocytes, and in which graded numbers of thymocytes were mixed with a non-limiting number of marrow cells. An intraperitoneal injection of 10 µg of (T,G)-Pro—L in complete Freund's adjuvant was given to each recipient 24 hours later. The mice were bled and their sera were titered 14 days after transfer. The results are summarized in Table 5. As the number of marrow cells injected with 75×10^6 thymocytes was increased, there was a corresponding increase in the fraction of positive sera in recipients for both mouse

Table 5. Fraction and Percentage of Positive Sera in Irradiated Syngeneic Recipient Mice 14 Days after Injection of (T,G)-Pro—L and Varying Numbers of Thymocytes and Bone Marrow Cells from SJL and DBA/1 Donors

Number of thymocytes injected ($\times 10^6$)	Number of marrow cells injected ($\times 10^6$)	DBA/1		SJL	
		Fraction of positive sera	Percentage of positive sera	Fraction of positive sera	Percentage of positive sera
75.0	0.5	...	...	6/12	50.0
75.0	1.0	2/7	28.6	8/12	66.7
75.0	2.0	1/6	16.7	7/10	70.0
75.0	15.0	8/11	72.7	16/16	100.0
10.0	15.0	...	...	10/12	83.4
20.0	15.0	7/7	100.0	8/9	88.8
40.0	15.0	9/9	100.0	10/10	100.0
60.0	15.0	7/7	100.0	...	...

strains. The inoculum that generated 70% positive sera was 2×10^6 SJL marrow cells, whereas it was necessary to transfer 15×10^6 DBA/1 marrow cells in order to obtain comparable results. On the other hand, variable numbers of thymocytes injected with 15×10^6 marrow cells yielded 83–100% positive sera for the range of inocula tested in both strains. The results shown in Table 5 indicate that there are about 7-fold fewer cells in DBA/1 bone marrow, which are limiting and relevant for the (T,G)-Pro—L response, than in SJL marrow. The incomplete results for limiting numbers of thymocytes do not reflect any differences between the frequency of relevant thymocytes in high and low responder mice. However, this could be due to the fact that the critical limiting inocula are below 10^7 cells, the minimum used in this series of transfers. It is noteworthy that in high responder strains to sheep erythrocytes, one precursor cell was detected in 3–10 $\times 10^6$ thymocytes (Shearer *et al.*, 1969b; Shearer and Cudkowicz, 1969b). Although at the present time one cannot rule out the possibility that the thymocyte population carries a deficiency in cell frequency or perhaps in some other immunocompetent aspect, it is evident that the relative number of bone marrow cells is significantly lower in DBA/1 low responder than in SJL high responder mice.

If the deficient number of relevant precursors is found in the marrow population, then it might be possible to increase the percentage of positive sera for limiting numbers of low responder marrow cells by poly A:U treatment. Nonimmunized DBA/1 donors were injected with 300 μg poly A:U 4 days before cell transfer. Irradiated recipients were injected with 50×10^6

Table 6. Fraction and Percentage of Positive Sera in Irradiated Syngeneic Recipient Mice 14 Days after Injection of (T,G)-Pro—L, Poly A:U, Graded Numbers of Marrow Cells, and 50×10^6 Thymocytes from Nonimmunized, Poly A:U-treated DBA/1 Donors[a]

Number of marrow cells injected	Fraction of positive sera	Percentage of positive sera
1.0	8/12	66.7
2.0	7/12	58.3
4.0	8/11	72.7
6.0	7/8	87.5
8.0	8/9	88.8

[a] Donors were injected intravenously with 300 μg poly A:U 4 days before the cell transfer. Recipients were immunized intraperitoneally with 10 μg (T,G)-Pro—L in complete Freund's adjuvant 24 hours after transfer, followed by an intravenous injection of 300 μg poly A:U.

thymocytes and 1–8 × 10^6 marrow cells from the above donors. Twenty-four hours later, the recipients were immunized intraperitoneally with 10 μg (T,G)-Pro—L in complete Freund's adjuvant, followed by an intravenous injection of 300 μg poly A:U. The results, summarized in Table 6, suggest that a significant change occurred in the percentage of positive sera when compared with the critical bone marrow inocula from untreated DBA/1 donors (see Table 5). These data confirm the findings reported above that expression of the genetic defect is reflected in the population of relevant cells found in marrow, since a similar increase in percentage of positive sera was observed in spleens of low responders treated with poly A:U.

DISCUSSION

Synthetic polypeptides of limited heterogeneity have permitted meaningful studies of the genetic control of antibody production (McDevitt and Sela, 1965, 1967; Pinchuk and Maurer, 1968; Mozes *et al.*, 1969 a, 1969 b; McDevitt and Benacerraf, 1969). In order to understand the mechanisms responsible for control of immunity, it is necessary to establish whether the genetic factors involved have a cellular basis, and, if so, to determine their nature. In mice and guinea pigs immune response have been obtained by the injection of lymphoid cells from high responder donors into irradiated low responder recipients (McDevitt and Tyan, 1968; McDevitt and Benacerraf, 1969). Limiting dilution experiments using high responder SJL and low responder DBA/1 mice demonstrated that genetic control of immunity to (T, G)-Pro—L can be attributed to the relative number of limiting precursor cells reactive with the immunogen in spleens of the two mouse strains (see Table 2, Figures 3 and 4; Mozes *et al.*, 1970). To confirm the cellular basis of this genetic control, and to establish which cell population is deficient in precursor cell number, poly A:U was used to "correct" the response of DBA/1 mice to (T, G)-Pro—L. The results presented in this report demonstrate that it is possible to enhance significantly the response of the DBA/1 strain. The enhancement observed is correlated with changes in the limiting dilution analyses of spleen and bone marrow, since about 60% positive sera were obtained with transfers of 20–40 × 10^6 and 0.5–2 × 10^6 spleen cells and 73% with transfers of 15 × 10^6 and 4 × 10^6 marrow cells for untreated and poly A:U treated mice, respectively. Therefore, phenotypic expression of this genetic control in the low responder strain can be detected in the relative numbers of transplanted bone marrow cells. The graded numbers of thymocytes chosen were not optimal for this type of study, since the transfer of 2 × 10^7 DBA/1 thymocytes yielded 100% positive sera. Therefore, it is not possible at the present time to say with certainty that the thymus population does not also carry the defect. However, since an equal number of thymocytes from SJL donors resulted in 88.8% positive sera, it is unlikely that a marked reduction in the frequency of thymocytes will be found in the low responder strain. This would not rule out the possibility that antigen-dependent proliferation of individual relevant thymocytes could reflect the genetic defect.

The experiments reported here do not permit a detailed assessment of the cell-associated mechanism(s) by which polynucleotides enhance the immunocompetence of genetic low responder mice. Nevertheless, results of poly A:U treatment raise two points of particular interest. First, results of treatment of intact low responders was more striking after primary than after secondary immunization. This could indicate that poly A:U exerts its effects early in the immune process. Second, treatment of high responders resulted in only a marginal enhancement. One possible explanation for this could be that all available precursors in SJL mice were activated by the immunogen alone, whereas only a portion of potentially immunocompetent cells were stimulated in DBA/1 mice by (T, G)-Pro—L. If this genetic phenomenon can be attributed to differences in numbers of so-called "high affinity" and "low affinity" precursor cells, and if poly A:U "converts" low affinity into high affinity precursors, then one would have to conclude that SJL mice possess almost exclusively cells with a high affinity for the immunogen investigated.

The "non-Poisson" pattern of spleen cell limiting dilution experiments involving poly A:U treated animals is of interest, since this is the second case in which findings of this type have been reported (Cudkowicz *et al.*, 1969, 1970). In limiting dilution experiments the assumption is made that the least frequent cell type constituting an antigen-sensitive unit becomes immunologically activated in the recipient animal and that this activation is independent of the stimulation and frequency of other cooperating cell types. It is also assumed that only those cells that are immunologically functional *at the time* of transfer are activated by the immunogen administered to the recipient. In other words, if differentiation of immature donor-derived elements into antigen-sentitive cells does occur in the irradiated host during the 2-week period between injection and assay, it is assumed that any residual immunogen present during this period will not be capable of activating the newly formed antigen-sensitive units. The fact that poly A:U treatment resulted in shallow, non-Poisson, limiting dilution curves indicates that the predictions of the Poisson model are not being met under these conditions. Therefore, it is not possible by statistical approaches to interpret fully the biological significance of these results. Nevertheless, the fact that the shallow curves observed here in spleen and marrow was previously detected only in bone marrow (Cudkowicz *et al.*, 1969, 1970) supports the conclusion that the marrow population is altered by poly A:U, and that the limiting and genetically affected cell type in DBA/1 spleens is probably marrow-derived.

References

Braun, W. and Nakano, M. (1967). Antibody formation: Stimulation by polyadenylic and polycytidylic acids. Science **157:** 819–821.

Braun, W. (1969). Proc. VIth Inter. Congress of Chemotherapy, Tokyo.

——— (1970). *In* Aging, Autoimmunity and Tolerance. Charles C. Thomas, Springfield.

——— Yajima, Y., Jimenez, L., and Winchurch, R. (1970). *In* Developmental Aspects of Antibody Formation and Structure. Academic Press, New York.

Chiller, J. M., Habicht, G. S., and Weigle, W. O. (1970). Cellular sites of immunological unresponsiveness. Proc. Natl. Acad. Sci. U.S. **65:** 551–556.

Claman, H. N., Chaperon, E. A., and Triplett, R. F. (1966). Immunocompetence of transferred thymus-marrow cell combinations. J. Immunol. **97:** 828–832.

Cudkowicz, G., Shearer, G. M., and Priore, R. L. (1969). Cellular differentiation of the immune system of mice. V. Class differentiation in marrow precursors of plaque-forming cells. J. Exp. Med. **130:** 481–491.

Cudkowicz, G., Shearer, G. M., and Ito, T. (1970). Cellular differentiation of the immune system of mice. VI. Strain differences in class differentiation and other properties of marrow-cells. J. Exp. Med. **132:** 623–635,

Jerne, N. K., Nordin, A. A., and Henry, C. (1963). The agar plaque technique for recognizing antibody-producing cells. *In* Cell-bound Antibodies. 109–116. The Wistar Institute Press, Philadelphia.

Johnson, H. G. and Johnson, A. G. (1968). Bact. Proc. **75.**

Kennedy, J. C., Till, J. E., Simmovitch, L., and McCulloch, E. A. (1966). The proliferative capacity of antigen-sensitive precursors of hemolytic plaque-forming cells. J. Immunol. **96:** 973–980.

Levy, H. B., Law, L. W., and Rabson, A. S. (1969). Inhibition of tumor growth by polyinosinic-polycytidylic acid. Proc. Natl. Acad. Sci. U.S. **62:** 357–361.

McDevitt, H. O. and Sela, M. (1965). Genetic control of the antibody response. I. Demonstration of determinant-specific differences in response to synthetic polypeptide antigens in two strains of inbred mice. J. Exp. Med. **122:** 517–531.

——— (1967). Genetic control of the antibody response. II. Further analysis of the specificity of determinant-specific control and genetic analysis of the response to (H,G)-A—L in CBA mice. J. Exp. Med. **126:** 969–978.

——— and Tyan, M. L. (1968). Genetic control of the antibody response in inbred mice. Transfer of response by spleen cells and linkage to the major histocompatibility (H-2) locus. J. Exp. Med. **128:** 1–11.

——— and Benacerraf, B. (1969). Genetic control of specific immune responses. Advances in Immunol. **11:** 31–74.

Mitchell, G. F. and Miller, J. F. A. P. (1968). Cell to cell interaction in the immune response. II. The source of hemolysin-forming cells in irradiated mice given bone marrow and thymus or thoracic duct lymphocytes. J. Exp. Med. **128:** 821–838.

Mosier, D. E. (1967). A requirement for two cell types for antibody formation in vitro. Science **158:** 1573–1575.

——— and Coppleson, L. W. (1968). A three-cell interaction required for the induction of the primary immune response in vitro. Proc. Natl. Acad. Sci. U.S. **61:** 542–547.

——— Fitch, F. W., Rowley, D. A., and Davies, A. J. S. (1970). Cellular deficit in thymectomized mice. Nature **225:** 276–277.

Mozes, E., McDevitt, H. O., Jaton, J.-C., and Sela, M. (1969a). The nature of the antigenic determinant in a genetic control of the antibody response. J. Exp. Med. **130:** 493–503.

——— (1969b). The genetic control of antibody specificity. J. Exp. Med. **130:** 1263–1278.

Mozes, E., Shearer, G. M., and Sela, M. (1970). Cellular basis of the genetic control of immune responses to synthetic polypeptides. I. Differences in frequency of splenic precursor cells specific for a synthetic polypeptide derived from multichain polyproline, (T, G)-Pro—L, in high and low responder inbred mouse strains. J. Exp. Med. (In press).

Pinchuk, P. and Maurer, P. H. (1968). Genetic control of aspects of the immune response. *In* Regulation of the Antibody Response. 97–113. Charles C. Thomas, Springfield.

Porter, E. H. and Berry, R. J. (1963). The efficient design of transplantable tumor assays. British J. Cancer **17:** 583–595.

Raidt, D. J., Mishell, R. I., and Dutton, R. W. (1968). Cellular events in the immune response. Analysis and in vitro response of mouse spleen cell populations separated by differential flotation in albumin gradients. J. Exp. Med. **128:** 681–698.

Shearer, G. M., Cudkowicz, G., Connell, M. S. J., and Priore, R. L. (1968). Cellular differentiation of the immune system of mice. I. Separate splenic antigen-sensitive units of different types of anti-sheep antibody-forming cells. J. Exp. Med. **128:** 437–457.

Shearer, G. M., Cudkowicz, G., and Priore, R. L. (1969a). Cellular differentiation of the immune system of mice. II. Frequency of unipotent splenic antigen-sensitive units after immunization with sheep erythrocytes. J. Exp. Med. **129:** 185–199.

——— (1969b). Cellular differentiation of the immune system of mice. IV. Lack of class differentiation in thymic antigen-reactive cells. J. Exp. Med. **130:** 467–480.

Shearer, G. M. and Cudkowicz, G. (1969a). Cellular differentiation of the immune system of mice. III. Separate antigen-sensitive units for different types of anti-sheep immunocytes formed by marrow-thymus cell mixtures. J. Exp. Med. **129:** 935–951.

Shearer, G. M. (1969 b.) Distinct events in the immune response elicited by transferred marrow and thymus cells. I. Antigen requirements and proliferation of thymic antigen-reactive cells. J. Exp. Med. **130:** 1243–1261.

Shortman, K., Diener, E., Russell, P., and Armstrong, W. D. (1970). The role of nonlymphoid accessory cells in the immune response to different antigens. J. Exp. Med. **131:** 461–482.

Taylor, R. B. (1969). Immune paralysis of thymus cells by bovine serum albumin. Nature **220:** 611.

Winchurch, R. and Braun, W. (1969). Antibody formation: Premature initiation by endotoxin or synthetic polynucleotides in newborn mice. Nature **223:** 843–844.

Wortis, H. H., Taylor, R. B., and Dresser, D. W. (1966). Antibody production studied by means of the LHG assay. I. The splenic response of CBA mice to sheep erythrocytes. Immunology **11:** 603–616.

INFLUENCE OF SYNTHETIC POLYNUCLEOTIDES ON THE BACTERICIDAL POWER OF MOUSE MACROPHAGES AGAINST *LISTERIA MONOCYTOGENES*

ROBERT M. FAUVE

Service de Pathologie Experimentale
Institut Pasteur, Garches 92, France

In 1964, Braun and Kessel (1964) suggested that bacterial endotoxins may release stimulatory oligonucleotides from intracellular environments, the release mechanism being the capacity of endotoxins to alter or to damage mammalian cell membranes and cells. More recently, Braun and Nakano (1967) were able to support this assumption, and demonstrated that complexes of polyadenylic and polyuridylic acids or polycytidylic and polyguanylic acids enhanced the early rate of increase in numbers of antibody-forming spleen cells in mice immunized with sheep red blood cells or other particulate antigens. The same year, Lampson *et al.* (1967) reported that double-stranded ribonucleic acids derived from several sources were active inducers of interferon and of enhanced host resistance to viral infection. Since then, it has been shown that synthetic double-stranded polynucleotides, such as poly I:C and poly A:U, can increase the resistance of the host against several viral infections and experimental tumors as will be reported in this Symposium.

Among the mechanisms involved in this increased resistance against viral infections, it was shown that these synthetic polynucleotides are powerful interferon inducers and that they stimulate both humoral and cellular immune responses (Braun and Nakano, 1967; Turner *et al.*, 1970). The fact that poly I:C-treated mice exhibit an increased resistance to some viral infections, even when this compound is injected a relatively long time before a viral challenge, led us to consider that a stimulation of macrophages could also be responsible for this increased resistance.

In order to answer this question, we used, as an experimental model, *Listeria monocytogenes* infection in mice. It is indeed known that these bacteria do multiply in macrophages and that an increased resistance is mostly mediated, if not only mediated, by an increased bactericidal power of macrophages (Fauve *et al.*, 1966; Mackaness, 1964).

When N.C.S. (Fauve and DeLaunay, 1966) mice were injected subcutaneously with 100 γ of poly I:C in 0.2 ml of P.B.S., and 3 days later were challenged with 10 LD 50 of *Listeria monocytogenes*, all poly I:C-treated mice survived and all control mice died. When the challenge inoculum was 10-fold higher, 3 poly IC-treated mice out of 12 survived and all

Table 1. Effect of Poly I:C on Experimental Infection of Mice with *Listeria monocytogenes*

Inoculum	Control 0.2 ml saline	Poly I:C 100 γ s.c.
$2.8–3 \cdot 10^6$	12/12 D	12/12 S
$1.5–2 \cdot 10^7$	12/12 D	3/12 S

control mice died. Since poly IC has no effect on the growth of *Listeria monocytogenes*, as it is shown in Figure 1, it was evident that the host mechanisms of resistance to infection were stimulated. In order to know if poly I:C-treated mice have an increased blood clearance of *Listeria*, 8 control and 8 treated animals were injected intravenously, 3 days later, with $3–4 \times 10^7$ *Listeria*, It is shown in Figure 2 that only 2 minutes after infection, the number of *Listeria* found in the blood of treated mice is only one-half of the number of *Listeria* found in the blood of control mice. But, if one considers the slopes of the clearance curves in the two groups of mice, they are roughly similar.

In order to see if the liver and spleen macrophages from treated mice were more efficient than the macrophages from control mice in their ability to inhibit the intracellular multiplication of *Listeria*, the following experiment was done. Five groups of 4 mice were injected subcutaneously with 100 γ of poly I:C 6, 24, 48, or 72 hours before the infection with 3×10^6 *Listeria*. Two hours later the mice were killed by cervical exsanguination and their spleen and liver were ground. Spleen and liver homogenates after dilution were plated on tryptose broth agar and the number of colonies recorded. Figure 3 shows that as soon as 2 hours after infection the number

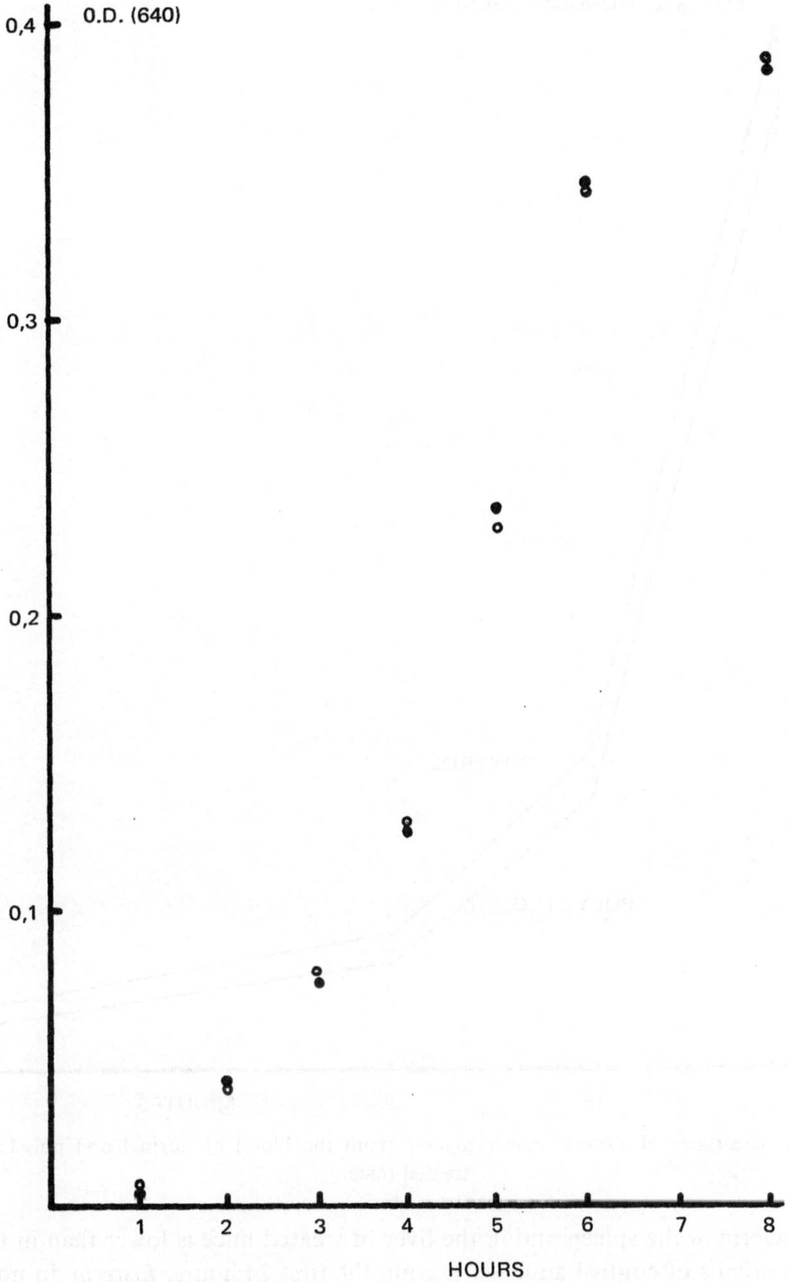

Fig. 1. Growth of *Listeria monocytogenes* in tryptose broth (●) and in tryptose broth containing 100 γ of poly I:C per ml (○).

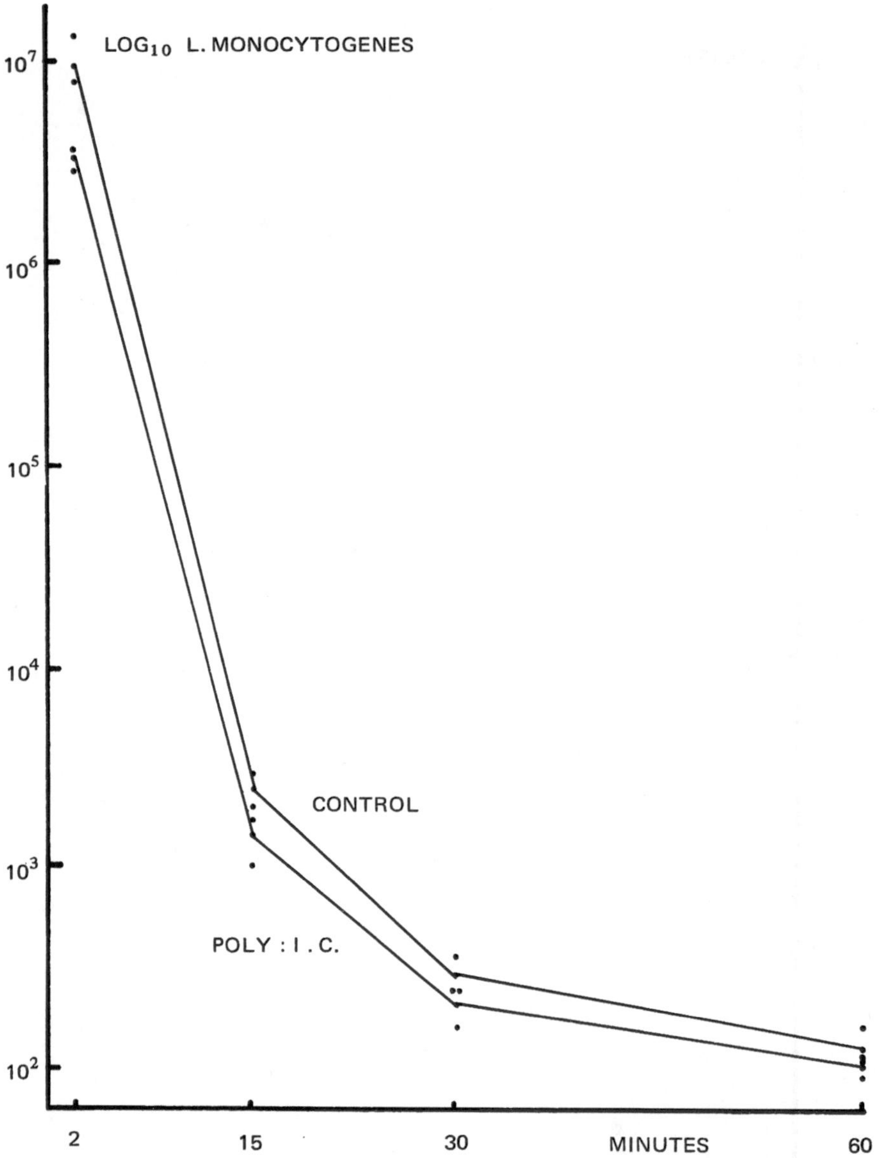

Fig. 2. Clearance of *Listeria monocytogenes* from the blood of normal and poly I:C-treated mice.

of bacteria in the spleen and in the liver of treated mice is lower than in the same organs of control animals. During the first 24 hours, *Listeria* do multiply in the liver and the spleen of all groups of mice and, during the last 24 hours *Listeria* multiplication is slower in mice that were injected 6, 24,

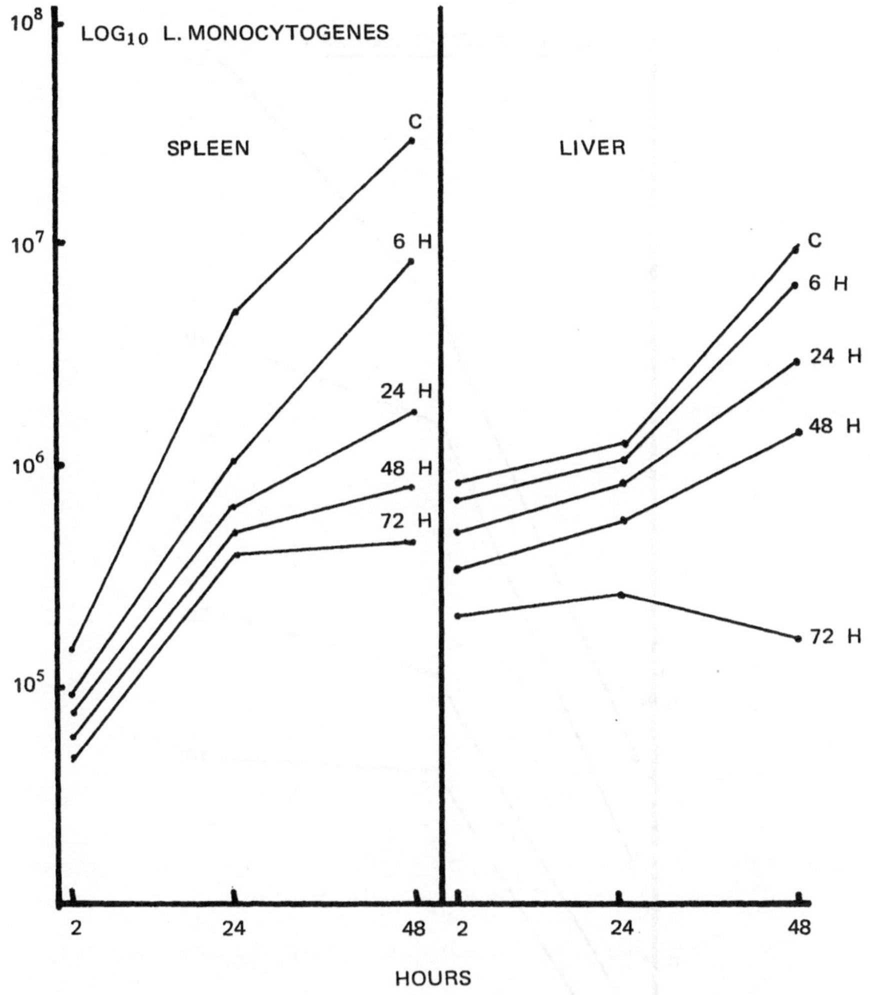

Fig. 3. Fate of *Listeria monocytogenes* in the liver and spleen of control mice (C) and of mice injected subcutaneously with 100 γ of poly I:C 6, 24, 48, or 72 hours before infection.

or 48 hours before challenge with poly I:C. Furthermore, when poly I:C was injected 72 hours before infection, the bacterial multiplication was inhibited both in the spleen and liver.

In order to know if this increased bactericidal power was dose-related, groups of 4 mice were injected subcutaneously with 10, 50, or 100 γ of poly I:C and 3 days later they were challenged intravenously with 2–3 $\times 10^6$ *Listeria*. It is again evident, as seen in Figure 4, that 2 hours after infection,

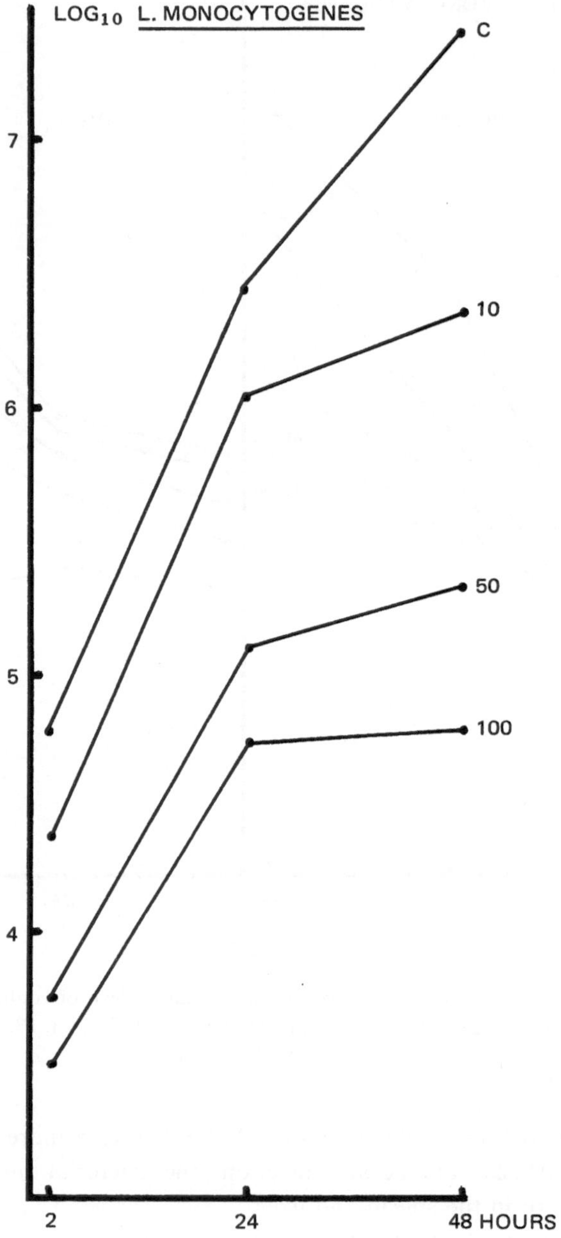

Fig. 4. Fate of *Listeria monocytogenes* in the spleen of control mice (C) and of mice injected subcutaneously with 10, 50, or 100 γ of poly I:C 72 hours before infection.

the number of bacteria in the spleen of all treated mice is lower than in the spleen of control animals, even in the group of mice injected with 10 γ of poly I:C, and that the bactericidal power of the macrophages from the spleen of mice injected with 100 γ of poly I:C is higher than in the spleen of mice injected with 10 or 50 γ of this compound. During the last 24 hours, the *Listeria* multiplication is slower in treated mice than in control animals and is even inhibited in animals that were treated with 100 γ of poly I:C.

Table 2

Time of infection (days after treatment)	Poly I:C (100 γ) Control (0.2 ml saline)	Increase of *L. monocytogenes* ($\log^{10}$) in spleen and liver 3 days after infection	
		Spleen	Liver
1	C	3	2.1
	P.I.C.	3	1
4	C	2.4	2.2
	P.I.C.	2.5	0.7
8	C	2.3	2.3
	P.I.C.	2.4	1.4
15	C	3	2.6
	P.I.C.	2.8	1.4
26	C	2.7	2.5
	P.I.C.	2.6	2.3

In order to know how long after poly I:C treatment this increased bactericidal power persists, 5 groups of 8 mice were injected subcutaneously, 4 with P.B.S. and 4 with 100 γ of poly I:C. One, 4, 8, 15, or 26 days later the animals were infected with 2–3 $\times 10^7$ *Listeria*. The results are reported in Table 2. It can be seen that even in the presence of this high inoculum, an increased bactericidal power of liver macrophages persisted until 15 days after poly I:C treatment.

Since poly A:U is also an inducer of interferon, but less so than poly I:C, and since it was suggested (Youngner and Hallum, 1968) that polynucleotides may be similar to endotoxin in their mode of action, we investigated the action of endotoxin and of poly A:U on the bactericidal activity of liver, spleen, and lung macrophages. For this purpose 3 groups of 8 mice were injected intravenously with 0.2 ml of saline, 0.2 ml of saline containing 100 γ endotoxin, or 0.2 ml of P.B.S. containing 100 γ of poly A:U. Forty-eight hours later all mice were injected with 5 $\times 10^7$ *Listeria*. Ninety minutes after

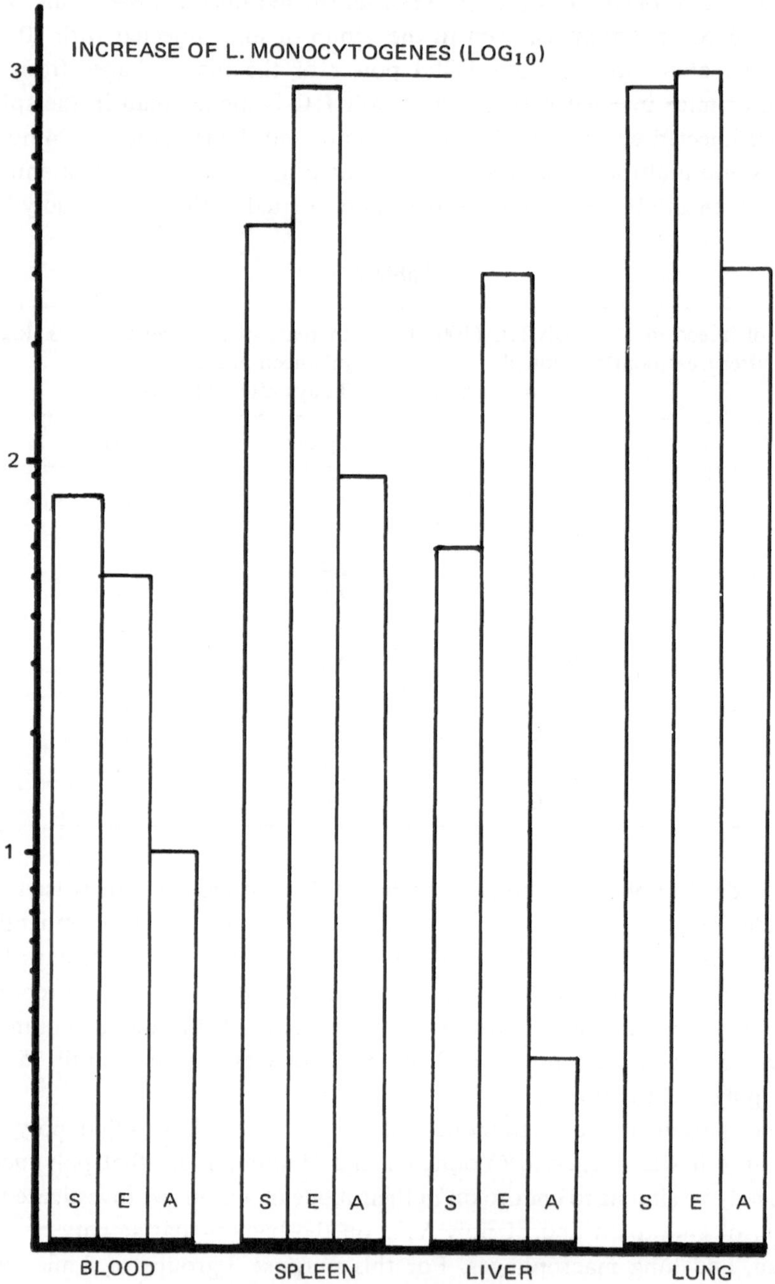

Fig. 5. Increase of *Listeria monocytogenes* ($\log_{10}$) in blood, spleen, liver, and lung of mice injected intravenously with saline (S), 100 γ of endotoxin (E) or 100 γ of poly A:U, 48 hours before infection.

infection, 4 mice in each group were anesthetized with ether, blood was withdrawn and the living *Listeria* were counted in the spleen, liver, and lung as previously described. Forty-eight hours after infection the same bacterial count was done with 4 other mice and the increase of bacteria is reported on Figure 5. In the blood of poly A:U-treated mice the number of bacteria is 8 or 5 times less than in saline- or endotoxin-treated mice. In the spleen of poly A:U-treated mice much less bacteria were found than in control and endotoxin-treated mice. More striking are the differences found in the liver of the 3 groups of mice. The liver of poly A:U-treated animals contained, respectively, 20 or 100 times less bacteria than the control and endotoxin-treated animals. Again, in the lung less *Listeria* were found among poly A:U-treated mice.

These results showed that both poly I:C and poly A:U given prior to a bacterial infection can protect mice against bacteria such as *Listeria monocytogenes*. These findings are in agreement with the work of Remington and Merigan (1970) who showed that prior treatment with poly I:C protected mice against *Listeria monocytogenes*. Our results showed that this increased resistance is reflected by an increased bactericidal power of spleen and liver macrophages and also, in the case of poly A:U, of lung macrophages. Since less bacteria were found in the blood, spleen, and liver of treated animals, as soon as 2 hours after injection, even when poly I:C was injected only 6 hours before injection, we can consider that the bactericidal activity of spleen and liver macrophages was increased following the injection of synthetic polynucleotides. It is not known yet if poly I:C increases the multiplication of liver or spleen macrophages and until now we were not able to demonstrate an increased bactericidal capacity of mouse peritoneal macrophages after these cells had been incubated *in vitro* with poly I:C.

The fact that poly I:C or poly A:U never decreased the resistance to infection is in contrast with the known fact that endotoxin decreases this resistance when injected a short time before the infectious challenge. This negative effect of endotoxin is well illustrated in our last experiment which shows that *Listeria* multiplication is less pronounced in poly A:U than in control animals and less extensive in control animals than in endotoxin-treated animals. It is also worth noting that we never observed splenomegaly and a decrease of macrophage spreading in mice treated with poly I:C or poly A:U, as it is the rule in endotoxin-treated mice.

It is still not possible to explain the mechanism of action of poly I:C or poly A:U on macrophages, but these results lead us to conclude that these synthetic polynucleotides are able to stimulate macrophages. Such stimulation can also explain in part the increased resistance of treated animals against bacterial, parasitic, and viral infections.

References

Braun, W. and Kessel, R. W. (1964). *In* Bacterial Endotoxins. Ed. by M. Landy and W. Braun. Rutgers University Press, New Brunswick, N. J.

——— and Nakano, M. (1967). Antibody formation: Stimulation by polyadenylic and polycytidylic acids. Science **157**: 819.

Fauve, R. M. and DeLaunay, A. (1966). Ann. Inst. Pasteur. **110**: 183.

——— Bouanchaud, D., and DeLaunay, A. (1966). Ann. Inst. Pasteur. **110**: 106.

Lampson, G. P., Tytell, A. A., Field, A. K., Nemes, M. M., and Hilleman, M. R. (1967). Inducers of interferon and host resistance. I. Double-stranded RNA from extracts of penicillium funiculosum. Proc. Natl. Acad. Sci. U.S. **58**: 782.

Mackaness, G. B. (1964). Symp. Soc. Gen. Microbiol. **14**: 213.

Turner, W., Chan, S. P., and Chiricos, M. S. (1970). Proc. Soc. Exptl. Biol. Med. **133**: 334.

Remington, J. S. and Merigan, T. C. (1970). Synthetic polyanions protect mice against intracellular bacterial infection. Nature **226**: 361.

Youngner, J. S. and Hallum, J. V. (1968). Interferon production in mice by double-stranded synthetic polynucleotides: Induction or release. Virology **34**: 177.

REMISSION INDUCTION WITH POLY IC IN PATIENTS WITH ACUTE LYMPHOBLASTIC LEUKEMIA (PRELIMINARY RESULTS)

G. MATHÉ, J. L. AMIEL, L. SCHWARZENBERG, M. SCHNEIDER, M. HAYAT, F. DE VASSAL, C. JASMIN, C. ROSENFELD, M. SAKOUHI, and J. CHOAY

Institut de Cancérologie et d'Immunogénétique, Hôpital Paul-Brousse; Service d'Hématologie de l'Institut Gustave-Roussy, 94-Villejuif; and Institut Choay, 75-Paris 16e

The oncostatic effect of polyinosinic-polycytidylic acid (poly IC) on several varieties of tumors has been reported previously (Levy *et al.*, 1969) and we have observed its effect on Walker's tumor. A number of clinical trials have been conducted (De Vita *et al.*, 1970; Krakoff *et al.*, 1970) but they merely confirmed tolerance to the product and did not yield therapeutical results. Although the mechanism of the oncostatic action exerted by polynucleotides on experimental tumors is not precisely defined, it is thought that it is probably multiple, for poly IC is cytotoxic (Braun, personal communication), is an inducer of interferon (Field *et al.*, 1967), is an adjuvant of immunity, as reported by Braun *et al.* (1968), and as observed by us by studying its effect on the multiplication of cells capable of forming antibodies in mice immunized with sheep red blood cells (Hayat and Mathe, in preparation). From this last-mentioned type of study, we have deduced the hypothesis according to which poly IC could have an action on acute leukemia only when the number of tumor cells is not very high. In fact, we have shown (Mathe, 1968; Mathe *et al.*, 1969) that the adjuvants of immunity in the animal exert a detectable oncostatic action only when the number of cancer cells is low.

Therefore, we decided to conduct a trial using poly IC in acute leukemia patients whose bone marrow contained less than 30% leukemia cells, and to administer it at weak doses, doses at which it is more an adjuvant of immunity, less cytotoxic, and less an inducer of interferon.

PATIENTS AND METHODS

Patients

Nineteen patients were included in this study, 15 patients with acute lymphoblastic leukemia and 4 with acute myeloblastic leukemia. There were 5 female and 14 male subjects. The patients' ages varied from 3 to 75 years. All patients were in the visible phase of the disease; 3 in phase I, the others in a first, second, third, or fourth relapse.

The preparation and characteristics of the poly IC employed

Poly IC is a complex formed by hydrogen bonds between poly I and poly C. The homopolymers are obtained enzymatically by polymerization of IDP 5′ phosphate on the one hand and CDP 5′ phosphate on the other hand, with the aid of polynucleotide phosphorylase of *A. vinelandii*. They were mixed in equimolar quantities in a medium of ionic concentration sufficient to form a bicatenary complex (0.15 *M* NaCl, 0.01 *M* phosphate buffer, *p*H 7.4) from each chain of poly I and poly C.

The bicatenary complex necessary for biological activity can be obtained in a reproducible manner, provided that the homopolymers are free from oligonucleotides, are deproteinized, and have high molecular weights. Table 1 shows that our preparations satisfy these conditions (ratio 286/260 weak; Ep at 260 nm high; M.W. calculated in the order of 10^6 and S of 8.9).

Table 1

P	280/260 in nm	Ep 260 nm (*p*H 7)
Poly I	0.790	6.2×10^3
Poly C	0.280	5.7×10^3
M.W. I: 1.4×10^6		
C: 1.3×10^6		
S: 8.9		

All preparations were carried out under sterile conditions. The homopolymers, preserved either in the lyophylized form or in solution at —20 °C, were put into solution and mixed just before injection. The formation of the double helix was controlled by spectrophotometric determinations at the temperature of the formation of the complex (25 °C) and at the temperature at which the product was used (37 °C).

Methods for the administration of poly IC and the surveillance of patients

Poly IC was administered daily at a dose of 1 mg/m^2 per day; the duration of treatment varied from 8 to 16 days.

The product was administered intravenously. The injection was made in saline using infusion tubing sets. The product was not diluted in the bottle.

All patients were hospitalized and submitted to twice daily clinical, and twice weekly haematological examinations. All patients were submitted to an immunological examination before, during, and after chemotherapy, comprising a series of tests exploring: delayed hypersensitivity to BCG, candidine, streptokinase, the level of small and large lymphocytes, monocytes and circulating "immunoblasts", the level of transformation of lymphocytes cultivated *in vitro* in the presence of phytohemagglutinin, the level of antisheep red cell, natural "rosettes", and the level of immunoglobulins (IgG, IgA, IgM).

RESULTS

Toxicity

The tolerance of the product at the doses used was excellent. No manifestations of toxicity were observed.

Table 2. Results in Acute Lymphoblastic Leukemia

Visible phase no.	Remissions		Failures	
	Apparently complete	Partial	Partial	Total
1				2
2	1			2
3	2		2	3
4	1			
5	1			1
Total	5		2	8

Remission induction

Out of 15 acute lymphoblastic leukemias treated (Table 2), 5 complete remissions (Figures 1 and 2), 2 nontotal failures, and 8 total failures were observed. The 4 cases of acute myeloblastic leukemia failed to respond.

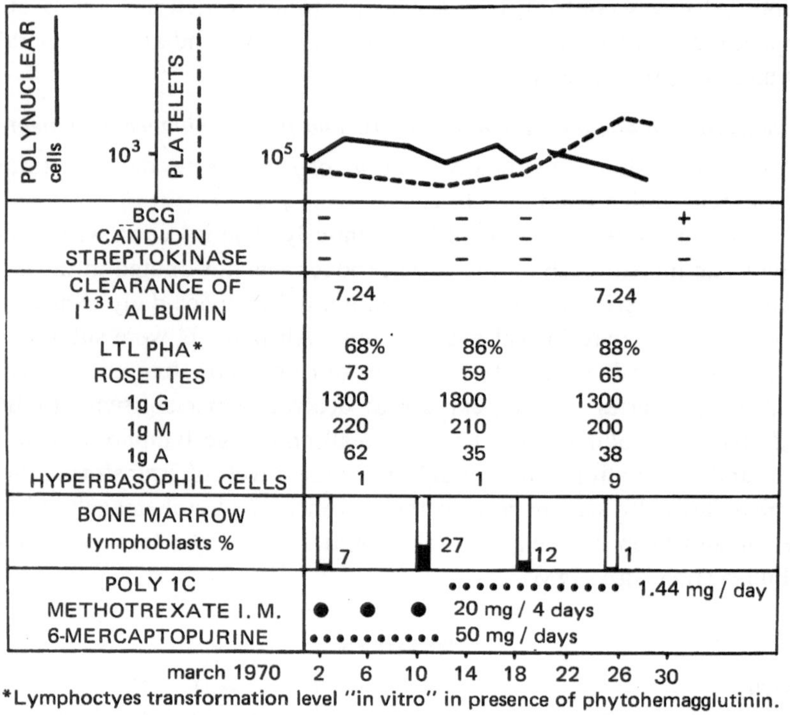

Fig. 1. Remission obtained with poly IC in a patient with acute lymphoblastic leukaemia.

Immunological effects

In 3 patients, delayed hypersensitivity reactions became positive; in 3 cases the level of transformation of lymphocytes *in vitro* in the presence of phytohemagglutinin increased significantly, from 68 to 88%, from 1 to 60% and from 22 to 77%, respectively. Finally, in 3 patients a significant increase in the circulating immunoblasts was observed (from 2 to 1760/mm^3, from 14 to 64, and from 0 to 22).

DISCUSSION

The results obtained in this trial, although preliminary, seem interesting to us because of the contrast with the negative results previously observed in patients with a great number of malignant cells (De Vita *et al.*, 1970; Krakoff *et al.*, 1970).

Our observation of remissions in patients with a moderate number of malignant cells is certainly not sufficient to assure that this is due to an immunological effect, and to eliminate the other actions of the product, such

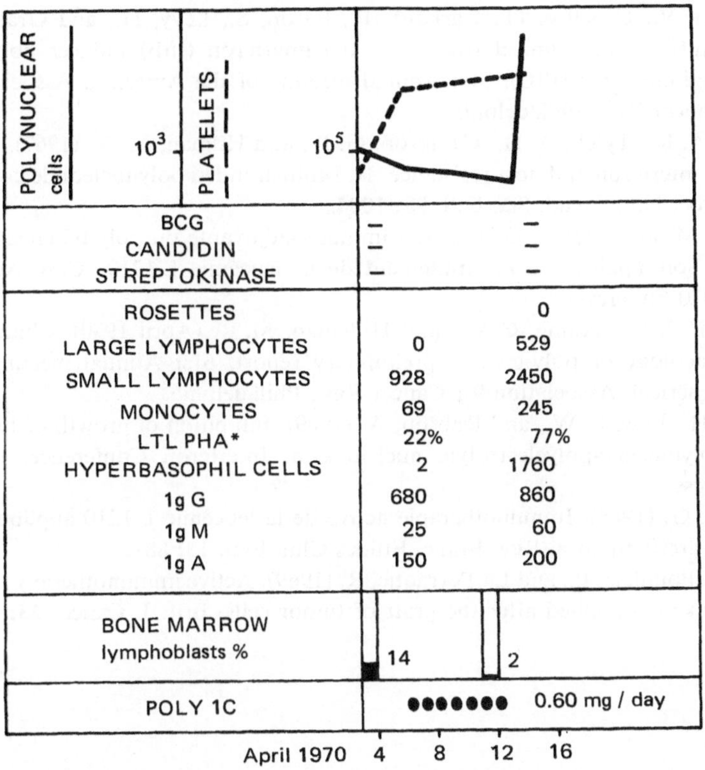
POLYNUCLEAR
cells
10^3
PLATELETS
10^5
BCG
CANDIDIN
STREPTOKINASE
ROSETTES
0
LARGE LYMPHOCYTES
0
529
SMALL LYMPHOCYTES
928
2450
MONOCYTES
69
245
LTL PHA*
22%
77%
HYPERBASOPHIL CELLS
2
1760
Ig G
680
860
Ig M
25
60
Ig A
150
200
BONE MARROW
lymphoblasts %
14
2
POLY 1C
0.60 mg / day
April 1970
4
8
12
16

*Lymphocytes transformation level "in vitro" in presence of phytohemagglutinin.

Fig. 2. Remission obtained with poly IC in a patient with acute lymphoblastic leukaemia.

as cytostatic effects and the induction of interferon. However, the modifications of immunological parameters that we observed support the hypothesis that immunological effects may have been responsible.

These trials will be followed not only by other trials with poly IC, but also by trials with poly AU, another polynucleotide with immunological stimulation, the action of which was observed in mice by us as well as by Braun *et al.* (1968).

References

Braun, W., Nakano, M., Jaraskova, L., Yajima, Y., and Jemenez, L. (1968). Stimulation of antibody-forming cells by oligonucleoproteides of known composition. *In* Nucleic Acids in Immunology. **1:** 347. Ed. by O. J. Plescia and W. Braun, Springer-Verlag, New York.

De Vita, V., Canellos, G., Carbone, P., Baron, S., Levy, H., and Gralnick, H. (April 1970). Clinical trials with the interferon (InF) inducer polyinosinic Cytidilic Acid (PIC). 61st Annual meeting of the American Association for Cancer Res., Philadelphia.

Field, A. K., Tytell, A. H., Campson, G. P., and Hilman, M. R. (1967). Inducers of interferon and host resistance. II. Multistranded polynucleotide complexes. Proc. Natl. Acad. Sci. U.S. **58:** 1004.

Hayat, M. and Mathe, G. L'action immuno-adjuvante du poly IC chez la souris et son application au traitement de la leucémie L 1210. C. R. Acad. Sci. 1970 (in press).

Krakoff, I. H., Young, C. W., and Hilleman, M. R. (April 1970). Clinical pharmacology of poly I:C. A preliminary report. 61st Annual meeting of the American Association for Cancer Res., Philadelphia.

Levy, H., Law, L. W., and Rabson, A. (1969). Inhibition of growth of tumors by polyinosinic-polyelectrolyte nucleic acid. Interferon Conference. Lyon, in press.

Mathe, G. (1968). Immunothérapie active de la leucémie L 1210 appliquée après la greffe tumora. Rev. Franc. Etudes Clin. Biol. **13:** 881.

——— Pouillart, P., and La Peyraque, R. (1969). Active immunotherapy of L 1210 leukemia applied after the graft of tumor cells. Brit. J. Cancer **23:** 814.

IMMUNITY AND TOLERANCE TO SYNTHETIC POLYNUCLEOTIDES IN NEW ZEALAND MICE

NORMAN TALAL and ALFRED D. STEINBERG

Arthritis and Rheumatism Branch,
National Institute of Arthritis and Metabolic Diseases,
National Institutes of Health, Bethesda, Maryland

Our laboratory has been studying the pathogenesis of the autoimmune disorder systemic lupus erythematosus as it occurs in humans and in the inbred New Zealand mice. The disease in both species is characterized by spontaneous formation of antibodies to double-stranded DNA and RNA which unite with antigen, fix complement, and deposit in the kidney where the immune complexes lead to a fatal glomerulonephritis. Three factors contribute to this disorder. These factors are genetic, immunologic, and viral.

Our concept of pathogenesis has been published elsewhere (Talal, 1970). Genetic factors, perhaps operating through a lysogenic or defective viral state, create a condition of immunologic imbalance in which cellular immunity is depressed and antibody responses are augmented. Depressed cellular immunity prevents normal control of virus infection, leading to the formation of double-stranded nucleic acids of viral or host origin. The hyperimmune antibody response results in the synthesis of antibody to nucleic acids, with consequent immune complex nephritis (Figure 1).

Genetic, Viral

↓

Immunologic Imbalance

↑Antibody response

↓Cellular response

↓

Anti-nucleic Acid Antibodies

Anti-viral

Anti-host

↓

Immune Complex Nephritis

Fig. 1. Pathogenesis of auto-immunity in NZB/NZW mice.

The excessive antibody response of New Zealand mice can be demonstrated with protein antigens, such as heterologous gamma globulin or bovine serum albumin, with particulate antigens such as sheep erythrocytes, and with the synthetic polynucleotides rI·rC or rA·rU. New Zealand mice produce more antibody to bovine gamma globulin than do three control strains of mice (Staples *et al.*, 1970). Continued intraperitoneal injection of poly I·poly C without carrier or adjuvant immunizes New Zealand mice but, once again, the three control strains fail to make antibody (Table 1). We might add that the ammonium sulfate binding method employed in these studies is the most sensitive of all assays for detecting antibodies to nucleic acids (Pincus *et al.*, 1969).

Table 1. Immunization with Polyinosinic·Polycytidylic Acid (No Freund's Adjuvant)

Strain	Number mice immunized/ Total number mice
NZB/NZW	18/22
C_3H/He	0/17
$C_{57}B$ 1/6	0/14
Balb/c	0/16

The immunogenicity of poly I·poly C explains the exacerbation of disease when New Zealand mice are treated with this potent interferon inducer. Figure 2 chronicles the effects of thrice weekly injections of 100–150 μg of poly I·poly C given in an effort to maintain a relatively steady concentration of interferon. Interferon is induced in concentration adequate to protect against murine leukemia virus which these mice harbor from birth. Despite interferon, antibodies to RNA and DNA are induced earlier, an RNA-containing immune complex nephritis develops, and the mice die 4–6 months before untreated controls (Steinberg *et al.*, 1969).

Induction of interferon.
Induction of antibodies to RNA and DNA.
Deposition of RNA-containing immune complexes in kidney.
Early death from nephritis.

Fig. 2. Effects of polyinosinic·polycytidylic acid in New Zealand mice.

Poly A·poly U, like poly I·poly C, is also immunogenic in New Zealand mice. The immunogenicity of both synthetic polynucleotides is greatly enhanced when they are emulsified in complete Freund's adjuvant. Poly A·poly U is the better immunogen of the two (Steinberg, Pincus, and Talal, 1971, in press).

Table 2. Immunization with Polyadenylic·Polyuridylic Acid in Complete Freund's Adjuvant

Strain	Antigen binding capacity
NZB/NZW	38.5
ALN	78.5
DBA/$_2$	18.7
C_3H/He	2.2
C_{57}B 1/6	1.7
Balb/c	1.4
Unimmunized controls	0.3

Poly A·poly U and poly I·poly C are also immunogenic in normal mice when given with adjuvant (Table 2). This result should caution those investigators contemplating the systemic administration of poly I·poly C to humans.

These antibodies to double-stranded RNA occur during the course of lupus in patients and in mice. Table 3 illustrates the incidence in several

Table 3. Antibodies to Polyinosinic·Polycytidylic Acid in Human Diseases

Disease	Number of patients	Percent positive
Systemic lupus erythematosus	78	45
Rheumatoid arthritis	27	15
Scleroderma	21	10
Sjogren's syndrome	28	7
Drug-induced lupus	28	7

human autoimmune disorders. Antibodies to RNA and DNA appear spontaneously in New Zealand hybrid mice (NZB/NZW F_1). Infection of these mice with known viruses such as lymphocytic choriomeningitis, polyoma, or Moloney leukemia will, like poly I·poly C, accelerate the onset of these presumed "autoantibodies" which may be antiviral nucleic acid antibodies.

Let us now turn to the question of tolerance. The administration of an antigen followed by the immunosuppressive drug cyclophosphamide 1 day later will induce a specific state of immunologic tolerance. Presumably, the antigen causes clones of responding cells to proliferate, which in turn, renders them more susceptible to the killing effect of cyclophosphamide. These cells are thereby eliminated and no longer present to respond to a subsequent challenge of antigen in Freund's adjuvant. Poly I·poly C and cyclophosphamide will induce tolerance in this way, as shown in Table 4. We have used this technique therapeutically to reduce anti-RNA antibody concentrations in older sick mice (Steinberg and Talal, 1970). We feel that it deserves continued study and possible therapeutic trial in human lupus, for it should avoid the often fatal infectious complications of long-term nonspecific immunosuppression, a current standard form of therapy.

We have also performed a cross-tolerance experiment, in which we could employ either poly I·poly C or poly A·poly U in each of three steps: (1) the initial tolerogen; (2) the challenge in adjuvant; and (3) the final radioactive

Table 4. Cyclophosphamide-Induced Tolerance in New Zealand Mice

Treatment	Antigen binding capacity
None	11.5
Poly I·Poly C	12.4
Cyclophosphamide	11.0
Poly I·poly C and cyclophosphamide	<0.1

antigen in the assay system. The results indicate that poly I·poly C is the better tolerogen and poly A·poly U is the more potent immunogen. The most complete tolerance occurred when poly I·poly C both preceded the cyclophosphamide and was used as the challenge. Poly A·poly U could partially overcome tolerance induced by poly I·poly C, and even more when poly A·poly U was also used as the tolerogen. Incidentally, these tolerance experiments add further definitive evidence that the synthetic polynucleotides are acting as true antigens and not simply as adjuvants.

In summary, we have briefly presented a concept of pathogenesis for systemic lupus erythematosus that accounts for the genetic, immunologic, and virologic contributions to this disease. We have shown that rI·rC and rA·rU are immunogenic in New Zealand mice, and also in normal mice if emulsified in adjuvant. We have induced specific immunologic tolerance to these synthetic polyribonucleotides, and are exploring the use of such immunologic tolerance as a therapeutic procedure in lupus. This approach may also have application in the area of transplantation immunology.

References

Pincus, T., Schur, P. H., Rose, J. A., Decker, J. L., and Talal, N. (1969). Measurement of serum DNA-binding activity in systemic lupus erythematosus. New England J. Med. **281**: 701–705.

Staples, P. J., Steinberg, A. D., and Talal, N. (1970). Induction of immunologic tolerance in older New Zealand mice repopulated with young spleen, bone marrow, or thymus. J. Exp. Med. **131**: 1223–1238.

Steinberg, A. D., Baron, S. H., and Talal, N. (1969). The pathogenesis of autoimmunity in New Zealand mice. I. Induction of anti-nucleic acid antibodies by polyinosinic·polycytidylic acid. Proc. Natl. Acad. Sci. U.S. **63**: 1102–1107.

——— Pincus, T., and Talal, N. (1971). The pathogenesis of autoimmunity in New Zealand mice. III. Factors influencing the formation of anti-nucleic acid antibodies. Immunology (in press).

——— and Talal, N. (1970). Suppression of antibodies to nucleic acids with polyinosinic·polycytidylic acid and cyclophosphamide in murine lupus. Clin. and Exp. Immunol. **7**: 687–691.

Talal, N. (1970). Immunologic and viral factors in the pathogenesis of systemic lupus erythematosus. Arth. and Rheum. **13**: 887–894.

DISCUSSION

DR. R. F. BEERS: In the foregoing one of the obvious problems facing the field today has been emphasized. When one attempts to understand the biological effects of polynucleotides on a molecular basis, many of them are interrelated. We heard discussions of the effects of polynucleotides on the production of interferon, of cytotoxic and adjuvant effects, and effects on cellular proliferation as well as migration. Later we will hear about toxological and pharmacological effects. Therefore, some of our discussion should be directed to the interrelationship of these various phenomena and I will ask Dr. Braun to make a few comments about that.

DR. W. BRAUN: I have tried to point out in my more formal remarks that the variety of phenomena that we are witnessing is due to the fact that we are dealing first of all with highly polymerized polyanions, which have membrane effects and probably interact with specific sites on the membrane, as has been brought out in many of the talks here. In addition, we have changes that occur in the molecule in an *in vivo* environment and give rise to derivatives that may have intracellular effects. I think the complexity is due to the fact that we are not dealing with one particular factor but with a variety of factors that may act both on the surface of the cell as well as in the interior of the cell.

DR. N. TALAL: Along the same lines, I would like to ask Dr. Braun for his thoughts about the following comment. We think about immunization as involving an interaction of antigens with specific receptors on the surface of sensitized lymphocytes. These receptors may be akin to antibodies. For example, in the situation I described, where we studied cells making antibody to poly IC, one could imagine that antibody being a receptor on the surface of the cells with a specific capability to interact with poly IC. We also think of poly IC inducing interferon by interacting with receptors on many cells. Some of these are lymphocytes and others are not. Should we think about the latter receptors, even when they are on cells other than lymphocytes, as also resembling antibodies in some way?

DR. W. BRAUN: I think we have to keep in mind the possibility that we are dealing with specific receptors, but also with reactive sites on the membrane that may interact with these materials in a far less specific way, perhaps in a manner reminiscent of the phytohemagglutin effects on cells, where we can mimic, in a nonspecific way, the consequences of a specific interaction between an antibody-like recognition site and an antigen.

I am wondering whether polynucleotides may not have two functions on the surface of the cell, namely (1) that of a specific interaction, which causes the phenomena that you have studied, that is, the immunogenicity of polynucleotides, and in addition (2) that of magnifying signals for effects elicited by signals on other sites, for example, by those produced as a result of specific interactions between an antigen and the antibody-like recognition site. I think the data that have now become available regarding the involvement of cyclic AMP on the magnification of the response to antigen, and the enhancement of polynucleotide effects by stabilizers of cyclic AMP suggest very strongly that we are dealing with a system that has the capacity to magnify signals. Let me add one thing, which I think is important, namely that these effects are not merely indirect effects due to the complexity of the *in vivo* environment. The same effects can be demonstrated in an *in vitro* system. Thus, the effects are direct effects on the cells involved in the responses.

DR. R. M. MORRELL: I would like to make a comment relating to the interaction between the polynucleotides and the antilymphocyte preparation or antithymus preparation. I think Dr. Braun's observations on the enhancement of antibody formation by ALS in combination with poly AU must in some way be correlated with our observations on the inhibition of globulin protein synthesis by antithymocyte sera. This was offset partially by poly IC. I think, at the present time, we need more information on how to integrate such data, but I think they move us in the same direction. 7S globulin, the antithymocyte globulin, probably binds to the membrane through its Fc portion, or tail piece, leaving the fragment hanging loose in the environment. How can this relatively small molecule interact with the huge poly IC molecules? I think this is something we have to solve.

DR. W. BRAUN: I am not sure it interacts with the polynucleotide molecules at all. I think what it may do is to get more than one membrane site involved. I don't see any reason why one has to assume an interaction between the polynucleotide and the antithymocyte globulin. I think they may produce combined effects.

Another aspect that may be important in any attempt to integrate our synergistic effects of antilymphocyte sera with your observations is the doses involved. I think we must be very careful in distinguishing between very strong effects, which may be inhibitory, and the effects of low concentration which serve to enhance.

DR. R. M. MORRELL: Yes, this is a very important point, and I was going to ask what you feel the relationship is of the variations in amounts of polynucleotides which have been utilized in different systems. For example, we have seen here variations in parenteral administration of polynucleotides to animals ranging from 100 μg i.v. to 1 mg i.v. Do you have any thoughts how this might relate?

DR. W. BRAUN: I think it may relate to the number of cells involved as well as to the cell types; it may also relate to cell membrane difference in different species, and it may relate to the point that we discussed earlier, namely that one is dealing with a multiplicity of effects, and that perhaps a large amount may superimpose a damaging effect on a stimulatory effect that is due to some derivative. In other words, I think that the complexity may be due to the involvement of a multicomponent system.

May I ask two questions. One is to Dr. Talal.

Do you know anything about the capacity of the polynucleotides to be immunogenic in incomplete Freund's adjuvant compared to complete Freund's adjuvant? It has been established for a long time now that polynucleotides may be highly immunogenic if they are complexed with an appropriate stabilizer or a carrier. Therefore, I am wondering if perhaps an interaction with the components of complete Freund's adjuvant rather than the uncomplexed antigen itself may be responsible for the immunogenic effects that you observed. Do you have any information on this?

DR. N. TALAL: Not yet.

DR. W. BRAUN: I am raising this question because we must keep in mind that if polynucleotides are going to be used for generalized stimulation, one does not have to fear immunogenic effects unless one gives them in complete Freund's adjuvant, and I don't know anybody who would want to use complete Freund's adjuvant together with polynucleotides.

DR. A. D. STEINBERG: Dr. Talal might point out that we have been able to induce immunological tolerance in normal strains of mice in addition to New Zealand mice. This indicates that normal strains of mice are capable of immunologically recognizing polynucleotides, specifically poly IC, without either protein carrier or adjuvants. This immunological recognition is specific.

DR. W. BRAUN: Yes, but I think that this is a different problem. It involves other steps than an activation, even though some of the same cells may be involved. In fact, I would anticipate that you would get stronger tolerance effects without the carrier molecule than you would with a carrier-containing molecule.

DR. A. D. STEINBERG: That is true of most forms of tolerance, but this is cyclophosphamide-induced tolerance which depends upon stimulation of cells that respond to the antigen and then are killed subsequently by the cyclophosphamide. We believe that this is evidence that the naked nucleic acids are being recognized immunologically.

DR. W. BRAUN: Why, then, do they not produce a positive response in terms of a measurable antibody response by the cells?

DR. A. D. STEINBERG: I think our assay is just not sufficiently sensitive to pick up the very low level of antibody that might be produced by this very weak antigenic stimulation.

DR. W. BRAUN: But if it is a very low level in the absence of complete Freund's adjuvant, we don't have to worry about complications either.

DR. A. D. STEINBERG: There are individual differences in these responses. The weak response without adjuvant is also found with some protein antigens which may be very weak and require Freund's adjuvant to produce detectable antibody. For example, Dr. Mozes can point out that her antigen is only an immunogen in complete Freund's adjuvant.

DR. W. BRAUN: Yes, and I am trying to stress that to my knowledge nobody who wants to use polynucleotides as stimulators of immune responses and antiviral effects is thinking of providing them in a form in which they would be highly immunogenic.

DR. N. TALAL: Could I make a comment about what we should worry about and what we need not worry about. Lupus seems to be a genetic disorder. Family studies and twin studies would indicate that one is genetically disposed to develop some or all of the manifestations of lupus. We have no idea of how many among us in the human population might be susceptible to the development of lupus, if given a proper stimulus. During my discourse I brought up the example of procaine amide-induced lupus. The incidence of antinuclear factor formation in procaine amide-induced lypus is 50–70%. The antinuclear factor goes away when medication is stopped. The people showing such responses are generally in advanced years, and there is evidence that aging itself has a profound effect on the immune system and may predispose toward autoimmunity. I think drugs such as poly IC deserve careful clinical trial. The point I want to make is that the high incidence of antinuclear factor in procaine amide-induced lupus suggests that we may be exposing some patients who are receiving poly IC to a similar risk. At the very least, we should do careful serological studies, employing the best assays possible, to gain as much information as we can of the possible immunologic consequences. Those of us who are interested in poly IC don't want to wake up 10 years from now and find ourselves observing problems such as other people have observed too late with other types of medication.

DR. W. BRAUN: I have one other question. It seems to me that among all the data that were discussed this morning, there is only one set of observations on which there is not complete agreement, and I wonder if we might not be able to achieve some agreement. It has been demonstrated that polynucleotides can stimulate the function of so-called macrophage cells. Dr. Johnson has pointed out that this stimulation is not associated with any increased uptake of antigen. In contrast, the work by Mr. Winchurch, in our laboratory, has quite clearly shown that if you use the proper interval to measure the effect between stimulation and uptake of foreign material you can show an enhancement. I therefore wonder whether the prior observations by Dr. Johnson of an inhibition might not involve a difference in timing, rather than a difference in effect.

DR. A. G. JOHNSON: First, you are involving an awful lot of cells. We measured the specific uptake of the antigen by macrophages in culture. I think there is a vast difference in what is measured *in vivo* and *in vitro*. We have done it as a function of timing and number of macrophages and we don't see an altered uptake. We have also done it in respect to differences in dosages of antigen added and at higher dosages we do not see any inhibition by poly AU. But you don't see any increase either. We have tested both macrophages and thymocytes with no increased uptake of antigen by the cells. We thought we might be concentrating it on the surface but this, too, does not seem to be the case.

DR. W. BRAUN: Are you sure that you have used appropriate amounts? If you remember, we found that you had to use truly minute amounts in order to demonstrate stimulatory effects *in vitro* and it may well be that the less drastic alteration of the polynucleotides in the *in vitro* system may produce effects *in vitro* that are quite different from what is encountered *in vivo*.

DR. A. G. JOHNSON: We used mainly 2 doses, 50 μg and 500 μg.

DR. W. BRAUN: That is too much.

DR. R. M. MORRELL: I would like to ask Dr. Braun or Dr. Johnson if there is any information on the effect of polynucleotides on lymphocytes as opposed to macrophages.

DR. W. BRAUN: I cited a considerable number of such effects. There are effects on lymphocyte transformation, effects on the rate of multiplication of activated lymphocytes, and there are effects on the lymphocytes in the interaction of thymus-dependent and bone-marrow-derived cells, and I think Dr. Johnson's work showed the same thing. Quite clearly, there are lymphocyte and macrophage effects.

DR. R. M. MORRELL: Let me restate my question more specifically. Have experiments been done with the objective of separating lymphocytes from mixed cell populations?

DR. W. BRAUN: Yes, such experiments have been done. We can separate spleen cell populations into macrophage and lymphocyte populations and these can then be treated separately with polynucleotides. Neither one of these subpopulations will yield antibody formation, but if you bring them together again, you get antibody formation *in vitro*. In such systems it has been demonstrated that there is a major effect of the polynucleotides on the lymphocytes.

DR. A. G. JOHNSON: I am struck with the similarity of the effects of poly AU and poly IC. I do not look at them from the complex angle of a variety of diverse effects in different cells by homopolyribonucleotides. I have been looking at it from the angle that only one major effect is exerted which results from increased protein synthesis. I am wondering if you have the same sort of effect on membranes of many cells so that the protein which that cell is going to make is going to be made in larger amounts in the presence of poly AU.

DR. W. BRAUN: There are some unpublished data supporting your view that the polynucleotides may enhance protein synthesis in all sorts of cells. I know some investigators have found an increased synthesis of enzymes involved in steroid transformation after treatment of an animal with polynucleotides.

DR. W. TALAL: I would predict that under the right circumstances, poly IC or poly AU would interfere with the induction of immunologic tolerance to antigens, such as bovine gamma globulin. I wonder if this experiment has been done.

DR. A. G. JOHNSON: Yes. Supernatant from centrifuged gamma globulin will produce tolerance. Poly AU will interfere with this.

DR. H. LEVY: I would like to make two comments relative to the present discussion. We have looked at the effects of varying doses of poly I and poly C on RNA and protein synthesis in a wide variety of tissue culture cells. You can get anything ranging from a strong stimulation to a strong inhibition of protein synthesis as measured by incorporation of amino acids depending on the concentration of drugs that you use with a given cell line. The level that will give you inhibition or stimulation varies quite a bit, depending on the concentration of drugs that you use with a given cell line. The level that will give you inhibition or stimulation varies quite a bit, depending on which cell you use, so that one has to look at a wide variety of concentrations to know whether one is going to get inhibition or stimulation.

The other comment deals with the nature of whether or not we are dealing with a specific messenger-like function. We did some experiments with L cells and with other cells, in which we tested to see whether poly IC has some messenger-like activity by seeing whether polyproline synthesis is specifically stimulated as measured by the incorporation of radioactive proline into the cell. Incorporation of radioactive proline was, indeed, stimulated by poly IC, but so was that of histamine and phenylalanine. Poly AU, which presumably would code for polyphenylalanine, showed about the same pattern. Polyproline incorporation was the most strongly stimulated, and the others were stimulated to a lesser extent. So I think the kind of stimulation we see when we talk about amino acid incorporation may not really be related to any coding effect on specific protein formation.

DR. J. W. GOODMAN: Dr. Johnson mentioned that the polynucleotides help to overcome tolerance to protein antigens, but what about the converse?

Is anything known about the effect of tolerance to the polynucleotides on the general adjuvant effect on other antigens?

DR. N. TALAL: When you induce immunologic tolerance to poly IC, you in no way interfere with the ability to have poly IC induce interferon.

DR. A. G. JOHNSON: If poly AU is injected prior to antigens, you can inhibit antibody responses, and if you give two injections of poly AU, one the day before and one with the antigen, then you will overcome the inhibition that is set up by early poly AU administration. I don't know anything about the nature of the tolerance to poly AU that can be produced, but it is an interesting question. Incidently, in such experiments, one has to be very careful that the preparations are not contaminated with endotoxin. Poly A and poly U are not pyrogenic nor is the poly A-poly U complex pyrogenic.

DR. R. C. SEEGAR: My question is directed to Dr. Mozes. You made the interesting suggestion that the defect in the low responder mice may be due to a low affinity receptor on the lymphocyte. Drs. Benacerraf, Siskind, and co-workers have suggested that serum antibodies reflect the affinity of the cellular receptor. Have you been able to compare the affinity of the serum antibody of the low responders and the high responders, and also have you been able to examine the affinity in the low responders following treatment with poly AU?

DR. E. MOZES: We have not done any studies on the affinity of the antibodies in low responders and high responders. I have been doing something similar to this in studies with PGAL, with the low responders and high responders to PGAL. According to preliminary results there was no difference in the antibodies in the low and high responders. However, I don't know if we can compare the results of PGAL with other antigens.

DR. O. J. PLESCIA: I should like to comment on the observations by Dr. Talal, especially regarding the antigenic specificity of poly A, poly U, and poly I-poly C, and the poly A-poly U complex. The latter can be quite different in terms of the number of strands which these helical polystrands contain. We have just completed an analysis of the specificity of antibodies to poly A + 2U, and multistranded poly CG. The antibodies made in response to poly AU react neither with poly A nor poly U, but react with multistranded poly A-poly U. However, these antibodies do not react at all with multistranded poly CG, and conversely, the antibodies that react with poly CG do not react with poly A-poly U. Therefore, I am somewhat surprised that there is an apparent antigenic cross-reactivity with poly AU and poly IC. Perhaps the answer may be that the poly AU used was predominantly double-stranded, although I am not so sure that this is the dominant configuration in the poly AU that was used.

Dr. Miller has made available to our laboratory a number of sera from NZB mice that have been injected with poly IC, beginning at different times from 2 weeks to 3 months. Poly IC was injected with or without DEAE-dextran. We have examined the sera of these mice for globulin that reacts specifically with native and denatured calf thymus DNA, soluble RNA, synthetic poly I, synthetic poly C, and double-stranded poly IC. These sera uniformly show a high level of gamma globulin reacting with DNA, somewhat less with denatured DNA and soluble RNA, and practically not at all with single-stranded poly I, poly C, and double-stranded poly IC. I also wonder if the early appearance of antinuclear factors may simply be an early reflection of a process that occurs with aging.

DR. N. TALAL: With respect to your first point, whether IC and AU can be distinguished immunologically, I can mention a series of experiments that we have conducted. We are looking at the ability of IC or AU to inhibit specifically the reaction with antibody that has been labeled with C-14 poly IC. When we examine sera from mice immunized either with IC or AU in Freund's adjuvant, both IC and AU can inhibit binding. Thus, under these circumstances, employing a very sensitive inhibition assay, we have not been able to distinguish IC from AU. These antisera seem to cross-react broadly.

I would like to know what type of assay you employ, particularly in the experiments with the New Zealand mice. You may not be finding antibodies to poly IC because your assay system is not sufficiently sensitive. With respect to your point about aging, I think this is quite true. One can look upon the problem of New Zealand mice as the problem of premature senility. Nevertheless, we have to know more about aging and about autoimmunity.

Has anyone looked to see whether antibodies to polynucleotides cross-react with teichoic acid or with pneumococcal polysaccharides? These are very ubiquitous, and perhaps study in germ-free animals might be of interest here.

DR. W. BRAUN: I think it may be important to determine whether the immunogenicity of the polynucleotides in strains of mice like the NZB strain is due to a uniqueness of the target of stimulation, or is due to certain materials that may be available in the circulation. I have long wondered why people involved in these studies have not done a very simple experiment based on the well-established fact that basic proteins when complexed to polynucleotides produce high immunogenicity of nucleic acids. I would like to know whether anybody has taken the serum from an NZB mouse, treated a polynucleotide preparation with it, and then injected the treated nucleic acid into another strain of mice that ordinarily will respond very poorly to polynucleotides to see whether you convert polynucleotides into a highly immunogenic complex by a basic protein that may be prevalent in strains that show a high response to polynucleotides.

DR. N. TALAL: We will do the experiment you suggest. I think it is a very worthwhile experiment.

DR. W. REGELSON: I would like to address a brief question to Dr. Talal in regard to the concept that he talked about in relation to aging. The whole story reminds me of the slow virus phenomenon that Dr. Hotchin has talked about in viral meningitis. With respect to the aging of mice involving an immune disease, one would wonder whether or not chronic administration of poly IC would accelerate the aging process.

DR. N. TALAL: I think the analogy with slow virus infection is a very good one. I talked about the depression of cellular immunity as being a very important factor in the pathogenesis of the New Zealand mouse disease. One way to produce a slow virus infection is to take normal animals and make them deficient in cellular immunity by exposure to antilymphocyte serum. I think that the New Zealand mice may already have a deficiency of cellular immunity because of genetic factors.

Part IV

Pharmacological Aspects of Polynucleotide Effects

THE DOUBLE-BITTED AXE: A STUDY OF TOXICITY OF INTERFERON RELEASERS

WARREN R. STINEBRING and MARLENE ABSHER
The University of Vermont College of Medicine
Burlington, Vermont

INTRODUCTION

The simplistic belief concerning the interferon system may be stated as: (1) Interferon, an inducible protein, specifically acts on cells to render them resistant to virus; (2) interferon itself is not antiviral but acts as a trigger for the synthesis of a protein by the affected cell which inhibits replication of virus; (3) viruses induce interferon production because the double-stranded replicative form of RNA, being foreign, "alerts" the cell to produce interferon; (4) nonviral inducers of interferon stimulate cells to produce interferon through some particular chemical grouping or property; (5) interferon inducers or interferon itself offer hope as prophylaxis or therapy against diseases caused by viruses, oncogenic or otherwise, tumors, virus-induced or otherwise, protozoa, fungi, and so on, and indeed any change in cellular homeostasis mediated by any foreign nucleic acid; and (6) there is enough knowledge extant concerning interferon induction so that these inducers may be used in human clinical trials. There are, of course, many sets of data that partially substantiate many of these beliefs but there are also results of studies that do not fit into this simplistic scheme. We believe, as do many workers, that much more basic work on all aspects of the interferon problem is needed. This paper is concerned with one of the most pressing problems in clinical application of the interferon system, that is, mode of action of interferon inducers in the intact animal.

The very nature of *in vivo* interferon inducers, ranging as they do from viruses through intact bacterial organisms, pathogenic or nonpathogenic (*Brucella abortus*, *Salmonella typhimurium*, *Serratia marcescens*, *Francisella tularensis, Listeria monocytogenes*), rickettsiae, chlamydiae, endotoxins, synthetic or natural RNA's, including poly I:C, to simple substances such as cycloheximide (see Finter, 1966; Rita, 1968; Vilcek, 1969) and bis-DEAE-fluorenone (Mayer and Fink, 1970; Krueger and Yoshimura, 1970), makes a single, specific chemical grouping or configuration as inducer seem unlikely. The induction of interferon by purified protein derivative (PPD) of *Mycobacterium tuberculosis* in animals rendered sensitive to those proteins by infection with Bacille-Calmette-Guerin (BCG) points to possibilities of relating specific immune toxic reactions to interferon production (Green *et al.*, 1969; Stinebring and Absher, 1969). In this regard, one is struck by the fact that substances that are interferon inducers seem to be cell toxins or stimulators.

EVIDENCE OF TOXICITY OF INTERFERON INDUCERS

There are a number of substances that have been seriously considered for clinical trial in humans because of interferon inducing properties, and indeed, some human trials have already started. These substances include the synthetic polymer, poly I:C (polyinosinic:polycytidylic acids) (Hilleman, 1970), pyran copolymer (divinyl ether maleic anhydride) (Merigan and Regelson, 1967; Regelson and Munson, 1969), polyacrylic acids (Niblack, 1969), and other polyanionic materials. Recently an inducer of interferon that can be administered orally has been reported and must be considered—bis-DEAE-fluorenone (Mayer and Fink, 1970; Krueger and Yoshimura, 1970).

A. Poly I:C

Hilleman (1970), in a masterly review of the work of his group and others on poly I:C, the most promising candidate, has discussed the toxicity of this material. A summary of the Merck group's findings is presented in Table 1. Absher and Stinebring (1969) have recorded lethal effects of poly I:C in mice at similar dose levels. Lindsay *et al.* (1969) have found poly I:C to be pyrogenic for rabbits even by the intranasal route. Poly I:C and other double-stranded RNA's cause abnormalities of the lymphoid system in adult mice and cause thymic atrophy, splenic hypoplasia, runting, and lymphopenia in neonatal mice (Leonard *et al.*, 1969). Poly I:C induces antibodies to DNA and RNA in young NZB/NZW, F, (B/W) females which suggests that this double-stranded RNA can act as an antigen, particularly in mice with a genetically hyperreactive immune response (Steinberg *et al.*, 1969). Intravenous administration of poly I:C to rabbits results in ocular

Table 1. Toxicology of Poly I:C (from Hilleman, 1970)

Mouse:	LD_{50} (i.v.) 16.5 mg/kg; (s.c.) > 50 mg/kg
Rabbit:	LD_{50} (i.v.) 0.58 mg/kg
Dog:	Retching, emesis, diarrhea with or without detectable blood; tremors, convulsions; fever; necrotic changes in liver, bone marrow; leucopenia; toxicity noted at 1 mg/kg (i.v.) as single dose, 0.125 mg/kg at weekly intervals for 3 weeks, 0.01 mg/kg daily for 8 or 28 days
Monkey:	Only effect noted at 1.0 mg/kg (i.v.) was elevation of alkaline phosphatase

toxicity characterized by hyperemia of the iris, flare in the anterior chamber, cataract, and so on (Burt *et al.*, 1969). Poly I:C is embryotoxic in the rabbit at a dose of 2 mg/kg for 2 days which has led Adamson and Fabro (1969) to comment, "Until the nature and species specificity of the embryotoxic effect of poly I:poly C are clarified, this compound should not therefore be used in women of childbearing age." In a trial made by Krakoff, Young, and Hilleman (1970) on 5 humans (cancer patients), poly I:C produced fever in one and no other drug-related effects were observed at various dosages administered intravenously. It should be noted that 3 of 5 patients showed no demonstrable rise in interferon titers after treatment. De Vita *et al.* (1970) had a somewhat different experience in 8 patients with widespread cancer. At an intravenous dosage of 0.3 mg/kg on day 1, and 8 through 35, no side effects were noted and no circulating interferon was detected. At higher doses, induction of interferon activity was noted in all patients but the appearance of fibrin monomers and/or prolonged clotting time was also noted, several patients had fever, nausea, and vomiting, one had anemia, and one had a severe bronchospastic reaction. Hill *et al.* (1969) have not noted toxic effects of intranasally instilled poly I:C at doses of 0.01 and 0.1 mg/kg/day for 7 days in human volunteers. Niblack *et al.* (1970) reported no toxicity following deep inhalation of 7 or 14 mg of poly I:C by normal humans although low levels of circulating antiviral substance were induced.

B. Pyran Copolymer

Pyran copolymer has been administered to over 60 patients by W. Regelson (see Regelson and Munson, 1969). His statements at the Second Conference on Antiviral Substances of the New York Academy of Sciences are excerpted as follows: "... we see sarcomas in hamsters after a prolonged treatment and a long latent period. ... Clearly pyran is carcinogenic. Pyran is also a very potent inhibitor of the polymerization of fibrinogen to fibrin.

... We get a few other 'little' side effects including the precipitation of coronary insufficiency in patients who have such a history. One runs the risk, for reasons we do not understand, of myocardial infarction in the patient population. Thrombocytopenia is also a problem. ... Of course as we reported ..., fever still results at higher dose levels and we do not quite understand why."

C. Polyacrylic Acids

Niblack (1969) has explored some of the properties of polyacrylic acids (PAA). He was able to correlate molecular weight, toxicity, and interferon-inducing capacity. Data were quite clear that PAA of between 5000 and 20,000 molecular weight was a good interferon inducer but the LD_{50} for mice was about 25 and 20 mg/kg, respectively. Polymers below 5000 in molecular weight lost interferon-inducing capacity and toxicity in parallel. In any case, this group has abandoned efforts to dissociate toxicity and interferon inducing capacity of PAA.

SIMILARITIES OF EFFECTS OF THESE MATERIALS

All in all, perusal of the data leads one to the conclusion that the interferon inducers mentioned above are not innocuous materials. Effects are quite broad in regard to toxic response and in some cases remind one of bacterial endotoxins. Indeed, Absher and Stinebring (1969), Braun (1969), and Regelson and Munson (1969) have called attention to these similarities. Effects of poly I:C or other polyanions on resistance to bacterial or protozoan infections are similar to effects of endotoxin on such experimental infections as has been pointed out by Herman and Baron (1970) and Weinstein *et al* (1970). Thus, although there is little doubt that interferon is induced by these substances, many of the effects may be attributable to reticuloendothelial system stimulation with concommitant release of active materials such as endogenous pyrogen, enzymes, oligonucleotides, and so on, from cells, and which can in turn modify the host in many ways. Table 2 compares similarities of effects among these materials.

NONTOXIC MATERIALS MADE TOXIC WITH ACQUIRED ABILITY TO RELEASE INTERFERON

Another line of evidence for the relationship between induction of interferon and toxicity or cell-stimulating effects of the inducer is immunologic. Glasgow (1966) has noted leukocytes from mice immunized with chikungunya virus (CV) released 2–10-fold more interferon than controls when

Table 2. Comparison of Activities of Endotoxin, Poly I:C, Pyran Copolymer

	Endotoxin	Poly I:C	Pyran
Fever induction	Yes	Yes (1)	Yes (2)
Induction of thymic atrophy in neonates	Yes (3)	Yes (4)	
Damage resulting in death of fetus	Yes (5)	Yes (6)	
Carbon clearance—depression followed by stimulation	Yes	Yes (2)	Yes (2)
Antitumor effect (animals or man)	Yes (7)	Yes (8)	Yes (2)
Nonspecific resistance to:			
Cryptococcus neoformans		Yes (2)	Yes (2)
Escherichia coli	Yes (9)	Yes (9)	
Production of tolerance to itself and to other stimulators	Yes (10)	Yes (10)	

(1) Lindsay *et al.* (1969).
(2) Regelson and Munson (1969).
(3) Rowlands *et al.* (1965).
(4) Leonard *et al.* (1969).
(5) Rieder and Thomas (1960).
(6) Adamson and Fabro (1969).
(7) Nauts *et al.* (1946).
(8) Levy *et al.* (1968).
(9) Weinstein *et al.* (1969).
(10) Absher and Stinebring (1969).

stimulated with homologous antigen. Green, Cooperband, Kleineman, and Kibrick (1969) have demonstrated, *in vitro*, that lymphocytes from tuberculin sensitive humans produced interferon upon challenge with PPD and other antigens if the lymphocytes came from a specifically sensitized person. Studies of the effect of an existing mycobacterial infection on the ability of the intact mouse to respond to nonspecific endotoxin stimulus have demonstrated an enhanced production of interferon. This enhancement of interferon production parallels the enhanced lethality of endotoxin in BCG infected mice studied by Suter. We have recently reported on the release of cycloheximide-resistant[1] interferon in BCG infected mice following intravenous administration of PPD (Stinebring and Absher, 1969; Absher and Stinebring, 1970). Of course, PPD is not a specific, purified protein but a

[1] Cycloheximide, in doses capable of suppressing protein synthesis, does not inhibit appearance of interferon in mice stimulated with endotoxin (Youngner *et al.*, 1965), statolon (Youngner and Stinebring, 1966), or poly I:C (Youngner and Hallum, 1968). This is considered evidence for the preformed state of interferon *releasable* by these substances.

mixture. The lack of ability of this specific batch to sensitize or provoke the Schwartzman phenomenon in rabbits or to induce interferon in normal mice militates against contamination with endotoxin being the cause of reactivity. The time pattern of appearance of the peak circulating levels of interferon in mice also was different from that seen with endotoxin induction. The specific immunologic reactivity to the PPD itself is the most logical explanation for the results. Thus, essentially, a nontoxic substance that is not an interferon inducer has become an interferon releaser when toxicity has become manifest through modification of host reactivity.

CONCLUSION

It comes as no surprise that materials such as those we have discussed can, through their cytotoxic effects, affect cells of the reticuloendothelial system or the lymphoid system and, through signal substances, other cells throughout the animal. That the cell-parasite-host relationships in protozoan, bacterial, or neoplastic diseases can be modified by such substances which can, in addition to other effects, produce interferon, seems reasonable. In many cases, data do not indicate interferon *per se* is involved in these effects. Indeed, when one deals with such substances one must be prepared to extend the studies of mechanisms of action to consider cell regulator systems mediated through products of affected cells, as in delayed hypersensitivity or in the case of homeostatic mechanisms controlled by cell products. [The reader is referred to papers by Dumonde *et al.* (1969), Maini *et al.* (1969), Granger and colleagues (1968) and others which discuss substances variously termed lymphokines, lymphocyte toxic factor, lymphocyte mitogenic factor, and so on, which are released from cells during delayed hypersensitivity reactions, and to papers by Rytömaa (1969) and Bullough (1965) for references to cell regulation apropos of the above discussion.]

We believe, and we stress *believe*, that a number of candidate materials for induction of the interferon system probably have similar modes of action. We believe that associated with interferon induction is a toxic action and that probably, toxicity and interferon induction are not separable. We feel that Gledhill was quite prophetic and correct in this statement made in 1964 (Gledhill, 1964): "The sparing substances in mouse sera ... are also produced in conditions of stress; like interferons they do not directly inactivate virus particles but their site of action is unknown. Possibly they serve to lessen the dissemination of infectious virus within the host, by stimulation of phagocytosis, by alterations of the vascular system, or in some other way. Equally, they may reduce the amount of virus produced by cells, either by increase of cellular resistance to infection, or by reduction of virus synthesis within infected cells. In the latter case, they would define biologically a class of sparing substances produced by stressed cells that render other cells less able to synthesize viruses. *The interferons wouldbelong to that class*" (italics ours). The reader can judge for himself why the title of this presentation is "The Double-Bitted Axe."

References

Absher, M. and Stinebring, W. R. (1969). Endotoxin-like properties of poly I · poly C, an interferon stimulator. Nature **223:** 715–717.

——— (1970). Delayed hypersensitivity and interferon release. Bact. Proc. p. 165 (abst.).

Adamson, R. H. and Fabro, S. (1969). Embryotoxic effect of poly I · poly C. Nature **223:** 718.

Braun, W. (1969). Relationships between the effects of poly I · poly C and endotoxin. Nature **224:** 1024–1025.

Bullough, W. S. (1965). Mitotic and functional homeostasis: A speculative review. Cancer Res. **25:** 1683–1727.

Burt, W. L., Dawson, C. R., Ostler, H. B., and Oh, J. O. (1969). Ocular toxicity of poly I · C. Personal communication.

Devita, V., Canellos, G., Carbone, P., Baron, S., Levy, H., and Gralnick, H. (1970). Clinical trials with the interferon (InF) inducer polyinosinic-cytidylic acid (PIC). Proc. Am. Ass. Cancer Res. **11:** 21 (abst.).

Dumonde, D. C., Wolstencroft, R. A., Panayi, G. S., Matthew, M., Morley, J., and Howson, W. T. (1969). "Lymphokines": Non-antibody mediators of cellular immunity generated by lymphocyte activation. Nature **224:** 38–42.

Finter, N. B. Ed. (1966). Interferons. Frontiers of Biology, **2**. North-Holland Publishing Company, Amsterdam.

Glasgow, L. A. (1966). Leukocytes and interferon in host response to viral infection. II. Enhanced interferon response of leukocytes from immune animals. J. Bact. **91:** 1285–1291.

Gledhill, A. W. (1964). Influence of endotoxin upon the pathogenesis of viral infections. *In* Bacterial Endotoxin, 418. Ed. by M. Landy and W. Braun. Institute of Microbiology, Rutgers, The State University, New Brunswick, New Jersey.

Granger, C. A. and Kolb, W. (1968). Ylmphocyte (sic) in vitro cytotoxicity: Mechanisms of immune and nonimmune small lymphocytes mediated target L cell destruction. J. Immunol. **101:** 111–120.

Green, J. A., Cooperband, S. R., and Kibrick, S. (1969). Immune specific induction of interferon production in cultures of human blood lymphocytes. Science **164:** 1415–1417.

Green, J. A., Cooperband, S. R., Kleinman, L. F., and Kibrick, S. (1969). Immune stimulation of interferon by non-viral antigens. Ann. N.Y. Acad. Sci. **173** (in press).

Herman, R. and Baron, S. (1970). Effects of interferon inducers on the intracellular growth of the protozoan parasite, *Leishmania donovani*. Nature **226:** 168–170.

Hill, D. A., Perkins, J. C., Worthington, M., Kapikian, A. Z., Chanock, R. M., and Baron, S. (1969). Preliminary study of the effect of poly I·poly C on experimental adenovirus infections in volunteers. Proc. Third Intl. Symp. on Medical and Applied Virol. (In press).

Hilleman, M. (1970). Prospects for the use of double-stranded ribonucleic acid (poly I:C) inducers in man. J. Inf. Dis. **121:** 196–211.

Krakoff, I. H., Young, C. W., and Hilleman, M. R. (1970). Clinical pharmacology of poly I:C—A preliminary report. Proc. Am. Ass. Cancer Res. **11:** 45 (abst.).

Krueger, R. F. and Yoshimura, S. (1970). Antiviral activity of bis-DEAE-fluorenone, an oral interferon inducer. Fed. Proc. **29:** 635.

Leonard, B. J., Eccleston, E., and Jones, D. (1969). Toxicity of interferon inducers of the double-stranded RNA type. Nature **224:** 1023–1024.

Levy, H. B., Law, L. W., and Rabson, A. S. (1969). Inhibition of tumor growth by polyinosinic·polycytidylic acid. Proc. Natl. Acad. Sci. U.S. **62:** 357–361.

Lindsay, H. L., Trown, P. W., Brandt, J., and Forbes, M. (1969). Pyrogenicity of poly I·poly C in rabbits. Nature **223:** 717–718.

Maini, R. N., Bryceson, A. D. M., Wolstencroft, R. A., and Dumonde, D. C. (1969). Lymphocyte mitogenic factor in man. Nature **224:** 43–44.

Mayer, G. D. and Fink, B. A. (1970). Bis-DEAE-fluorenone, an oral inducer of interferon. Fed. Proc. **29:** 635.

Merigan, T. C. and Regelson, W. (1967). Interferon induction in man by a synthetic polyanion of defined composition. New Engl. J. Med. **277:** 1283–1287.

Nauts, H. C., Swift, W. E., and Coley, B. L. (1946). The treatment of malignant tumors by bacterial toxins as developed by the late William B. Coley, M.D., reviewed in the light of modern research. Cancer Res. **6:** 205–216.

Niblack, J. F. (1969). Interferon stimulation by low molecular weight polyacrylic acids. Ann. N. Y. Acad. Sci. **173** (In press).

——— Knersch, A. K., and Vora, K. R. M. (1970). Personal communication.

Regelson, W. and Munson, A. E. (1969). The reticuloendothelial effects of interferon inducers: Polyanionic and nonpolyanionic phylaxis against microorganisms. Ann. N. Y. Acad. Sci. **173** (in press). See also discussion session, "Will interferon be clinically useful?" in same annal.

Rieder, R. F. and Thomas, L. (1960). Studies on the mechanisms involved in the production of abortion by endotoxin. J. Immunol. **84:** 189–193.

Rita, G. Ed. (1968). The Interferons. Academic Press, New York.

Rowlands, D. T., Claman, H. N., and Kind, P. D. (1965). The effect of endotoxin on the thymus of young mice. Am. J. Path. **46:** 165–176.

Rytömaa, T. (1969). Granulocytic chalone and antichalone. In Vitro. **4:** 47–58.

Steinberg, A. D., Baron, S., and Talal, N. (1969). The pathogenesis of autoimmunity in New Zealand mice. I. Induction of antinucleic acid antibodies by polyinosinic·polycytidylic acid. Proc. Natl. Acad. Sci. U.S. **63:** 1102–1107.

Stinebring, W. R. and Absher, P. M. (1969). Production of interferon following an immune response. Ann. N. Y. Acad. Sci. **173** (In press).

Vilcěk, J. (1969). Interferon. Virology Monographs, **6**. Springer-Verlag, New York.

Weinstein, M. J., Waitz, J. A., and Came, P. E. (1970). Induction of resistance to bacterial infections of mice with poly I·poly C. Nature **226:** 170.

Youngner, J. S. and Hallum, J. V. (1968). Interferon production in mice by double-stranded synthetic polynucleotides: Induction or release? Virology **35:** 177–179.

Youngner, J. S. and Stinebring, W. R. (1966). Comparison of interferon production in mice by bacterial endotoxin and statolon. Virology **29:** 310–316.

——— and Taube, S. E. (1965). Influence of inhibitors of protein synthesis on interferon formation in mice. Virology **27:** 541–550.

POLYINOSINIC-POLYCYTIDYLIC ACID TOXICITY*

FREDERICK S. PHILIPS, MARTIN FLEISHER, L. D. HAMILTON, MORTON K. SCHWARTZ, and STEPHEN S. STERNBERG

Sloan-Kettering Institute for Cancer Research, Memorial Hospital for Cancer and Allied Diseases, New York, New York
and
Brookhaven National Laboratory, Upton, New York

This report deals mainly with the pathological effects of polyinosinic-polycytidylic acid, poly (rI·rC), given in lethal or near lethal doses. The work was done to obtain information about the potential of the complexes for inducing dangerous toxicity in patients who might receive the agent in clinical trials against cancer. The experimental species studied included mice, rats, rabbits, and dogs.

Unless otherwise stated all tests were done with a single, uniform batch of dry, alcohol-precipitated poly (rI·rC), batch H 1, which had been prepared by Hamilton and described elsewhere in this Symposium. For injection samples were dissolved in sterile, pyrogen-free 5% dextrose in water. The animals employed were 5-week-old male mice (CD®-1 line derived from HA/ICR Swiss albinos), 4-week-old male rats (CD®line derived from Sprague-Dawley rats), male New Zealand albino rabbits, 3–4 kg, and adult mongrel dogs of both sexes. The mice and rats were obtained from the Charles River Breeding Laboratories, Wilmington, Mass.

Mice. Table 1 shows that the median lethal dose of poly (rI·rC), given in single i.v. injections, was about 20 mg/kg; single injections given i.p. were about $^1/_4$ as potent. When i.p. injections were repeated for a total of 5 successive daily doses, the lethality increased by only a factor of 2 over that of single i.p. doses.

* Supported in part by Grant CA 08748 from the National Cancer Institute, USPHS, and by the U.S. Atomic Energy Commission.

Table 1. Toxicity in Mice, Rats, and Rabbits

Species	Route of injection	Dose (mg/kg/day)	Number of successive daily doses	Mortality[a]
Mice	i.v.	50	1	21/21
		25	1	17/22
		12.5	1	4/19
		6.3	1	0/10
	i.p.	100	1	8/10
		50	1	2/16
		25	1	0/10
		50	5	9/10
		25	5	0/10
		12.5	5	1/10
Rats	i.v.	100	1	1/9
		50	1	0/9
		25	1	1/6
	i.p.	100	1	1/6
		50	1	0/6
		100	5	2/6
		50	5	0/6
Rabbits	i.v.	1.0	10	3/3
		0.3	10	3/4
		0.1	10	1/4
		0.03	10	0/4

[a] Surviving animals were observed for 4 weeks after injection.

Single i.v. doses of 50 mg/kg killed most mice between 3 and 9 hours after injection. A few animals receiving this dose had immediate, brief convulsions; these mice recovered promptly and were normal within 5 minutes. By 1 hour all were dishevelled and had diarrhea. The diarrhea increased and the animals became steadily more hypoactive until they died with a brief terminal convulsion. About half of the fatalities after 25 mg/kg i.v. occurred in less than 9 hours; the remainder and all fatalities in mice given 12.5 mg/kg took place between 9 and 20 hours. All mice surviving single i.v. doses of 6.3, 12.5, and 25 mg/kg lost 10–20% of initial body weight during the first day after injection. Recovery of weight began during the second or third day and was complete by 7 days when the animals were growing again like controls. The lethal effect of poly (rI·rC) was more delayed in the mice receiving the single intraperitoneal doses listed in Table 1. None died earlier than 9 hours after injection and a few at times later than 1 day. In those receiving repeated daily i.p. doses, deaths occurred between 1 and 7 days after the first injection.

Groups of 3 mice were killed for histopathological study at 6 hours and at 1 day after single i.v. doses of 25 and 12.5 mg/kg, respectively, and at 1 day after 2 successive daily i.p. doses of 50 mg/kg/day. At autopsy the 6-hour animals were found to have fluid distending the stomach and small intestine and microscopic study revealed a severe necrotizing lesion in the epithelium of the duodenum sufficiently advanced to have resulted in marked blunting of villi (Plate 1—Figures 1 and 2). Flattening and loss of epithelium of duodenal villi were also seen in the two other groups of mice; although necrosis was not present at the later times, the epithelial nuclei were enlarged and irregularly shaped. Changes were also present in the mucosa of stomach and colon; however, these were less marked than in duodenum and consisted primarily of scattered necrotic cells in the surface epithelium in a few of the mice.

In addition to the intestinal lesion, all the mice showed pathologic changes in lymphoid tissues. Karyorrhexis was present in follicles of nodes and spleen and in thymic cortex as early as 6 hours after the single i.v. dose of 25 mg/kg. Atrophy of the hematopoietic elements in the red pulp of the spleen was detected at 1 day after 12.5 mg/kg i.v. and was severe in the mice killed after the 2 i.p. injections. In the latter animals the thymic cortex was atrophic and replaced by large, eosinophilic histiocytes. These results confirm earlier descriptions of the lymphocytotoxic effects of poly (rI·rC) by Leonard *et al.* (1969).

Most of the above mice had abnormally enlarged cells in the germinal epithelium of testicular tubules; a few had scattered karyorrhectic cells in adrenal cortex, in ventral prostatic epithelium, and in parotid and submaxillary salivary glands. Finally, the lungs of the mice killed at 6 hours after 25 mg/kg i.v. had large numbers of leucocytes, mainly mononuclear cells, in alveoli and attached to the intima of small blood vessels.

Rats. This species was less susceptible to poly (rI·rC); only a minority died after receiving the doses listed in Table 1. Nevertheless, during the first 24 hours after injection, all rats, which survived the single i.v. doses, lost 6–15% of initial body weight. Moreover, rats killed at either 6 or 24 hours after 100 mg/kg i.v. had pathologic changes in duodenal villar epithelium similar to those described above in mice. Small numbers of karyorrhectic cells and of atypical nuclei were present in the surface epithelium of stomach and colon. Hemorrhages and karyorrhexis were prominent in lymphoid follicles of nodes and spleen and in the thymic cortex.

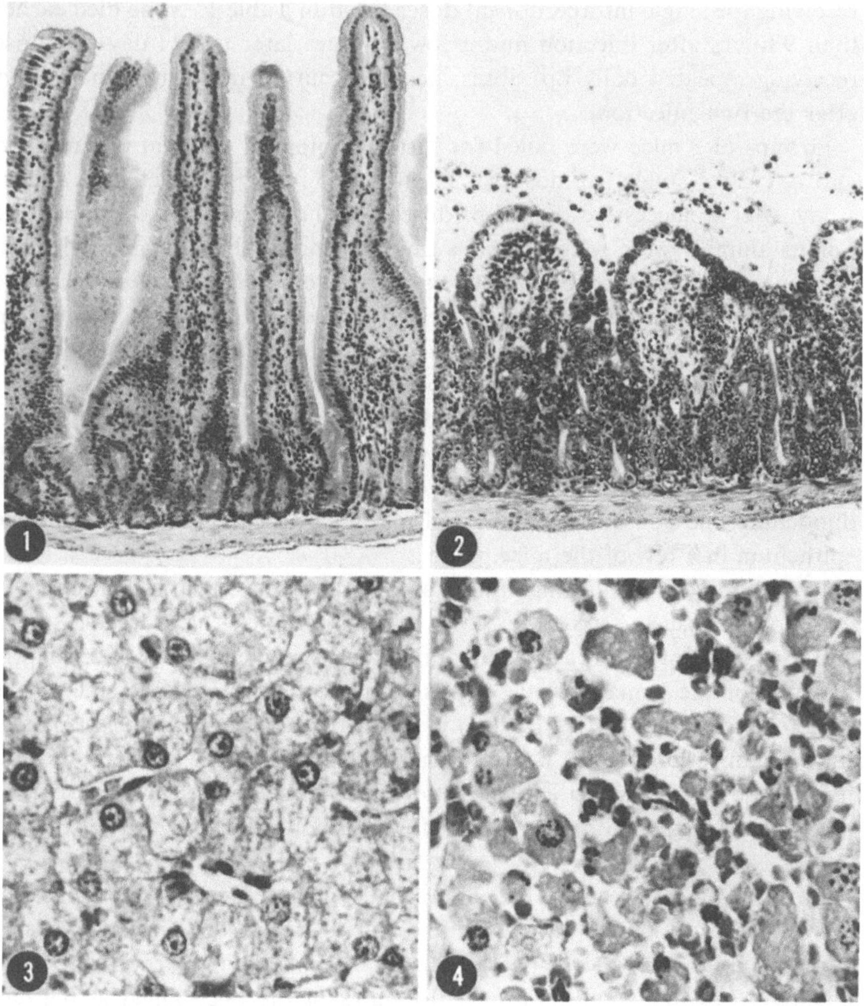

Plate 1—Fig. 1. Duodenum of a control mouse (× 180).

Plate 1—Fig. 2. Duodenum of a mouse killed at 6 hours after a single intravenous injection of poly (rI · rC), 25 mg/kg (× 180). The villi are shortened and blunted. Karyorrhectic debris is present in the lamina propria and epithelial cells of the villi. Necrotic cells are present in the surface epithelium and in the intestinal lumen. The crypts are less affected but have swollen and irregular nuclei in epithelial cells.

Plate 1—Fig. 3. Liver from a control dog (× 780).

Plate 1—Fig. 4. Liver from a dog killed at 2 hours after a single intravenous injection of poly (rI · rC), 1 mg/kg (× 780). This field, representative of the major portion of the organ, shows extensive disorganization of sinusoidal structure, hemorrhage, separation of hepatocytes, and punctate aggregation of chromatin in hepatocyte nuclei.

At 6 hours basophilic (nuclear?) debris was found in liver sinusoids and in interstitial spaces in hepatic portal areas. By 24 hours hepatic Kupffer cells were laden with basophilic debris. As in mice at 6 hours after i.v. injections, leucocytes were attached in large numbers to the intima of the small pulmonary blood vessels.

Rats were also killed for histopathological study at 24 hours after the last of 4 successive daily i.p. injections of 100 mg/kg/day and in these lymphoid tissue damage was severe. Thymi were involuted, lymph nodes and lymphoid follicles in spleen were atrophic, and the splenic red pulp was contracted. Trabecular vessels in spleen had inflammatory changes. The same animals had a few foci of karyorrhexis in epithelia of bladder, renal pelvis, and ventral prostate. Peritoneal inflammation was noted in the serosa of stomach and colon and there was edema and interstitial inflammation in pancreas and testis.

Rabbits. Of the four species tested herein rabbits proved to be the most susceptible to the lethal actions of poly (rI·rC). Table 1 shows that the median lethal dose i.v. was 0.1–0.3 mg/kg/day. One of the animals died within 8 hours after the first dose of 1 mg/kg; another died in less than 1 day after a single dose of 0.3 mg/kg. The remaining fatalities occurred by the end of the fourth day of daily treatment. Beginning at various times between 2 and 5 hours after injection, rabbits receiving lethal doses became increasingly dyspenic and hypoactive until they were moribund. The survivors in Table 1 were not significantly affected by treatment with 10 successive doses except for transient losses in weight of less than 10% of initial values.

Two rabbits, receiving doses of 1 mg/kg i.v., were killed for pathological study at 6 hours after injection: one, after a single dose and the other, after a second daily dose. Both animals had red-purple discoloration of the trachea and bronchi; microscopically the capillaries of the submucosa of the trachea were dilated and congested. The liver of the first animal had scattered foci of necrosis in hepatocytes; in the second rabbit 10–20% of the parenchyma was necrotic. In the latter animal karyorrhectic cells were present in the red pulp of the spleen, the thymic cortex, and in the surface epithelium of small and large intestine.

In confirmation of the study reported by Lindsay *et al.* the present sample of poly (rI·rC) proved to be highly pyrogenic in rabbits with significant fever induced by doses as low as 1 μg/kg (Figure 1). It was at least a thousand-fold more pyrogenic than either of the constituent homopolymers (Figure 2). Since it has been suggested that there are similarities in the actions of poly (rI·rC) and endotoxins, it is noteworthy that the course of the temperature increase after the polynucleotide complex was not identical with that caused by endotoxin. Minimally effective doses of the latter induce in rabbits maximal responses within less than 1 hour; larger doses cause an early peak followed by a second rise of greater magnitude some hours later (Bennett and Cluff, 1957). With poly (rI·rC), as seen in Figure 1, the lowest dose tested had no effect at 1 hour and maximal responses were delayed until 4 hours after injection.

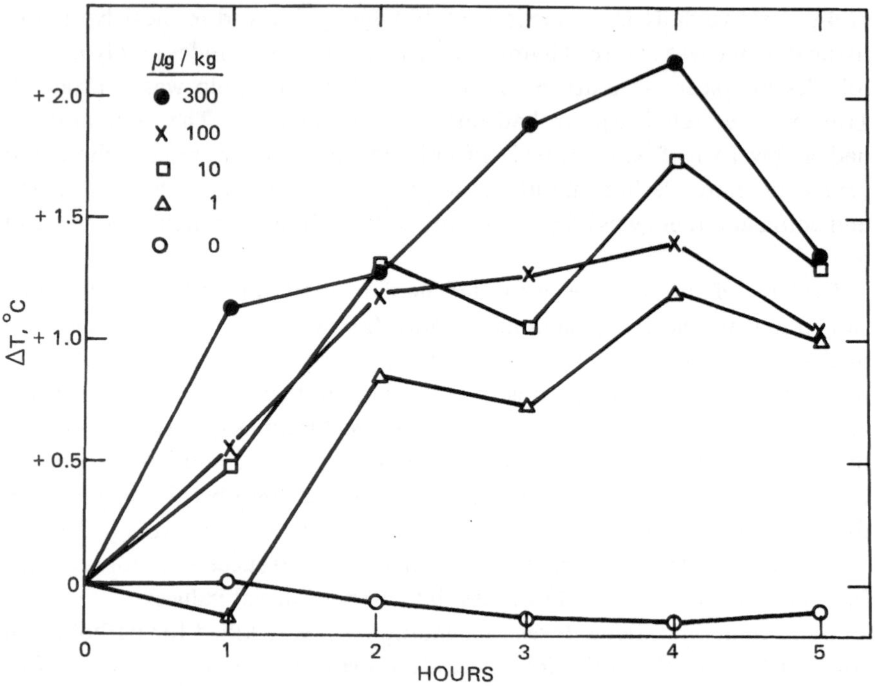

Fig. 1. Temperature changes in rabbits after different intravenous doses of poly (rI·rC). Assays were done according to standard U.S.P. procedures for pyrogen tests. Each symbol is the average of 4 animals receiving 1, 10, or 300 µg/kg, of 8 receiving 100 µg/kg, and of 19 controls given 5% dextrose.

Figure 3 illustrates another endotoxin-like effect of poly (rI·rC). All rabbits tested had a significant leucopenia due initially to decreasing numbers of circulating granulocytes. Heterophil counts remained low for 1 to 3 hours and then increased toward control levels. Lymphocyte counts, however, fell more slowly and were still depressed at 5 hours after injection. With endotoxin leucopenia develops more rapidly, being evident within a few minutes after injection and followed within 2 hours by leucocytosis, and it is mainly due to granulocytopenia (Bennett and Cluff, 1957). Interestingly, the decrease in circulating leucocytes after endotoxin has been attributed to their sequestration within pulmonary blood vessels and spleen and, as mentioned above, large number of leucocytes can be detected in pulmonary vessels and alveoli in rats and mice killed at 6 hours after i.v. injection of poly (rI·rC). The fact that leucopenic and pyrogenic activities are associated properties of both poly (rI·rC) and endotoxins suggests that the fevers due to the polynucleotide complex are mediated mainly by release of endogenous

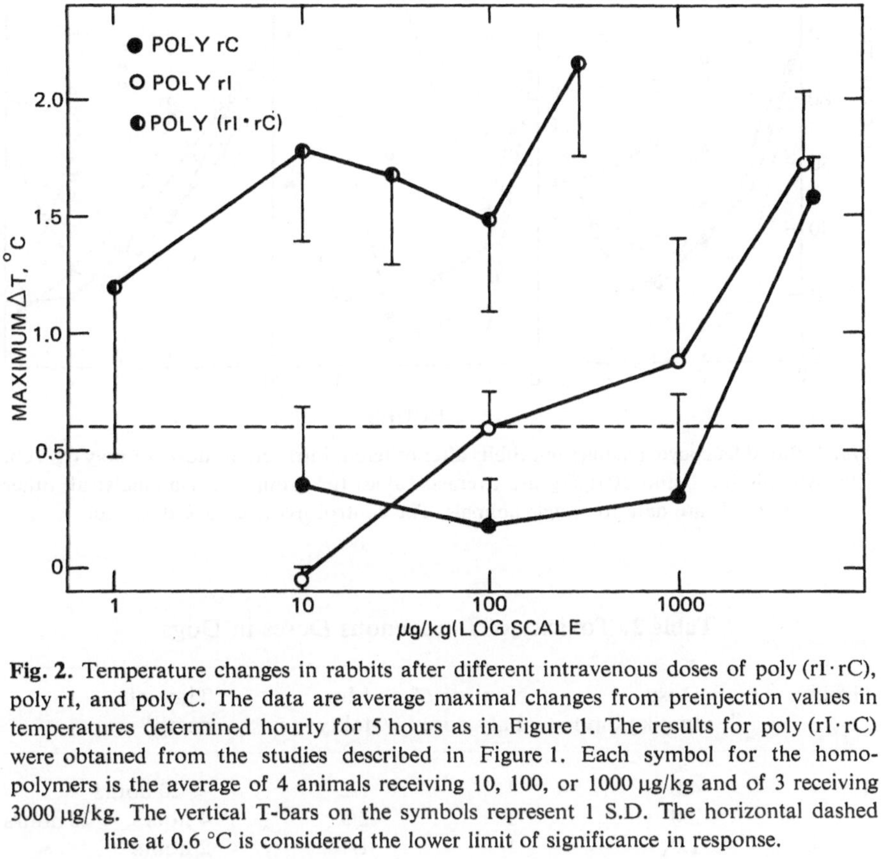

Fig. 2. Temperature changes in rabbits after different intravenous doses of poly (rI·rC), poly rI, and poly C. The data are average maximal changes from preinjection values in temperatures determined hourly for 5 hours as in Figure 1. The results for poly (rI·rC) were obtained from the studies described in Figure 1. Each symbol for the homopolymers is the average of 4 animals receiving 10, 100, or 1000 μg/kg and of 3 receiving 3000 μg/kg. The vertical T-bars on the symbols represent 1 S.D. The horizontal dashed line at 0.6 °C is considered the lower limit of significance in response.

pyrogen from granulocytes—a mechanism amply substantiated for endotoxins (Bennett and Cluff, 1957; Snell and Atkins, 1969).

Dogs. Preliminary studies in dogs were done with the animals of Group 1, in Table 2, which were tested with different, early batches of poly (rI·rC). The first 4 dogs, receiving single doses of batch CW-7, died within less than 20 hours after injection. Two survived repeated daily injections of the same preparation in doses of 0.5 mg/kg/day. The remaining 10 dogs of Group 1 were treated with 4 other small samples and all but one survived. (Elsewhere in

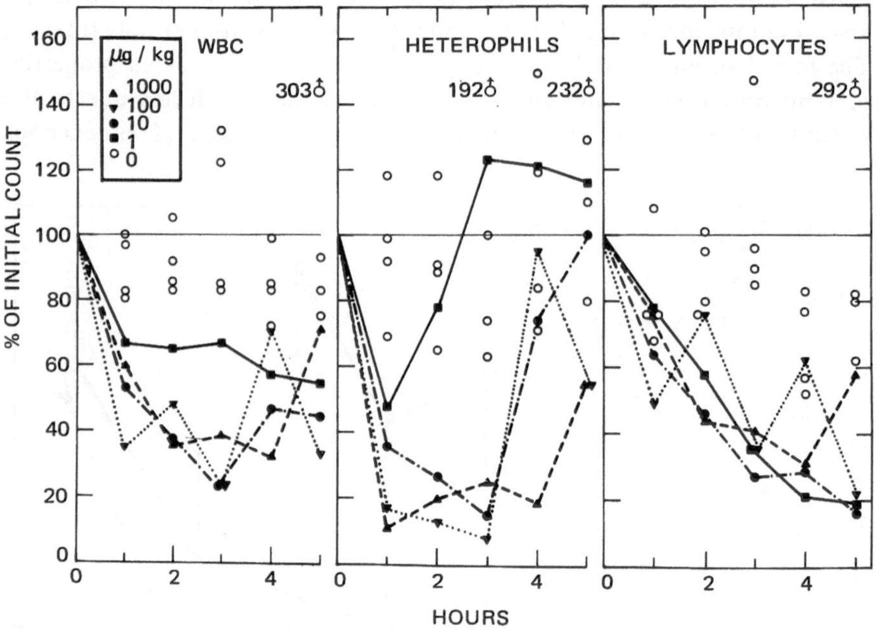

Fig. 3. Blood leucocyte changes in rabbits after different intravenous doses of poly (rI·rC). The symbols for 1 and 10 μg/kg are average values for groups of 4 animals; all other symbols are data for single animals. The controls received 5% dextrose.

Table 2. Toxicity of Intravenous Doses in Dogs

Group	Dose (mg/kg/day)	Number of daily doses	Mortality[a]	Time of death[b]
1	4.0	1	1/2	< 20 hours
	2.0	1	2/5	4 hours, < 20 hours
	1.0	1	1/1	2 hours[c]
	1.0	10[d]	1/4	5 days[c]
	0.5	10	0/4	
2	2.0	10	2/4	10[c], 11[c]
	1.0	10	0/4	
	0.5	10	0/2	

[a] Surviving animals were observed for 4 weeks after injection.
[b] Timed from the first injection.
[c] Killed for histopathological study.
[d] Each injection was given for a maximum of 10 doses on successive days except weekends.

this Symposium Hamilton has reported that batch CW-7 was among the more lethal of his preparations in mice). All of Group 2 in Table 2 were injected with the same batch of poly (rI·rC) used in the above studies in mice, rats, and rabbits. Two of these were killed at 10 and 11 days after having received

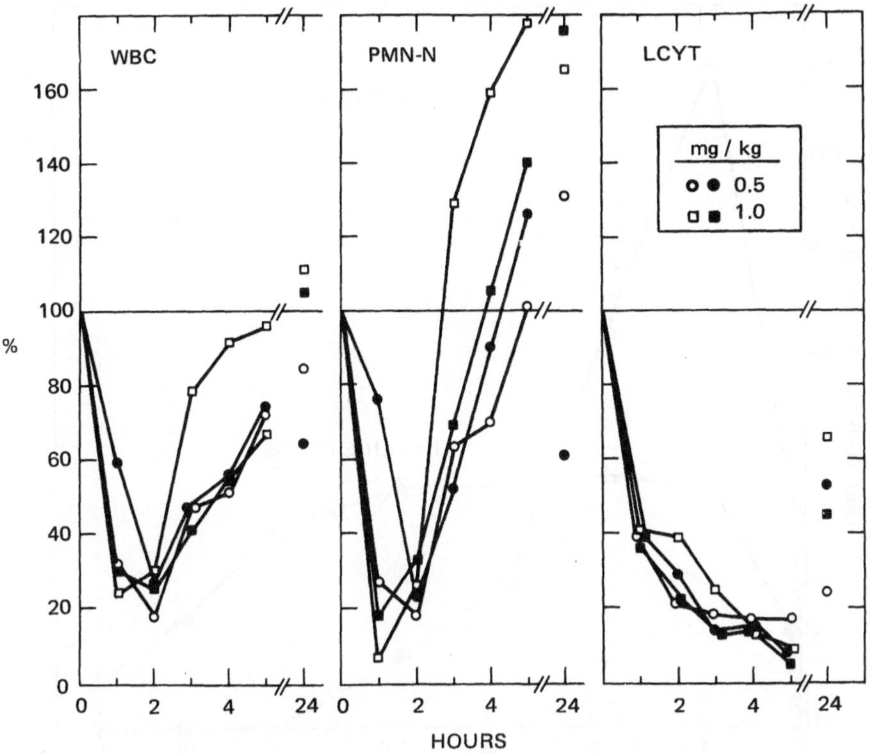

Fig. 4. Blood leucocyte changes in dogs after intravenous doses of poly (rI·rC). Each symbol is the result from a single animal. PMN-N, neutrophil granulocytes, and LCYT, lymphocytes.

8 injections of 2 mg/kg/day. The results listed in Table 2 indicate that doses of poly (rI·rC) greater than 0.5 mg/kg/day have significant lethality in dogs.

Most of the dogs given doses of 1 mg/kg or greater had signs of acute intoxication which were delayed in onset until 1 to 2 hours after injection. These included, to varying extent in different animals, emesis, salivation, tremor or shivering, injection of conjunctiva, dyspnea, loose stool or diarrhea, tenesmus, depressed behavior, and hypoactivity. In most of the affected animals the acute signs abated within 5 to 6 hours after injection. Anal temperatures were recorded hourly after injection of the first dose in each of the Group 2 dogs. Eight of the 10 had significant temperature elevations, ranging from 0.7 to 1.9° C, which were maximal at 3 hours. Leucopenia, as shown in Figure 4, was another acute response involving an earlier, transient decrease in neutrophils and a later, more slowly developing

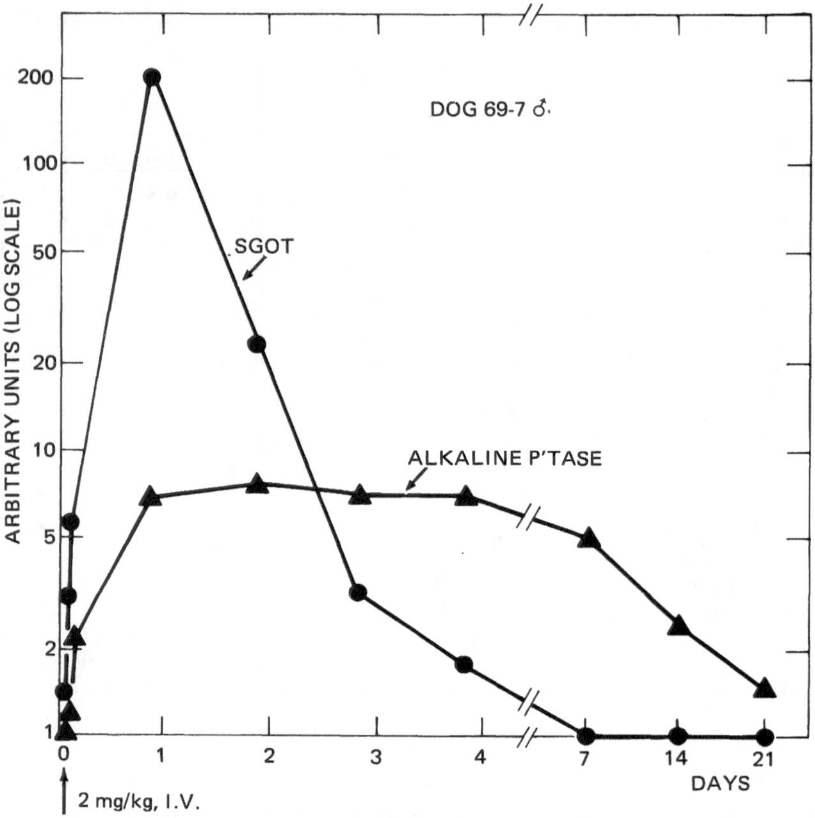

Fig. 5. Glutamic-oxaloacetic transaminase (SGOT) and alkaline phosphatase in the serum of a dog after a single intravenous injection of poly (rI·rC). The results are given as multiples of preinjection values.

fall in lymphocytes. Eosinophilic granulocytes also decreased in each of the animals of Figure 4; in 3 eosinophil counts were still less than 10% of initial values at 24 hours after injection. Showers of circulating normoblasts were not seen such as those observed within 1 hour after injection of endotoxin in this species (Weil and Spink, 1957).

In a minority of the dogs of Table 2 the course of acute intoxication was more fulminating than that described above. Two died unobserved between 4 and 20 hours after receiving a single dose of 2 or 4 mg/kg. Another was moribund at 3 hours and dead at 4 hours after an injection of 2 mg/kg. A fourth dog was killed when comatose at 2 hours after a single dose of 1 mg/kg. Its liver was dark purple and swollen with blood. Microscopic examination revealed extensive necrosis, involving most hepatocytes, and

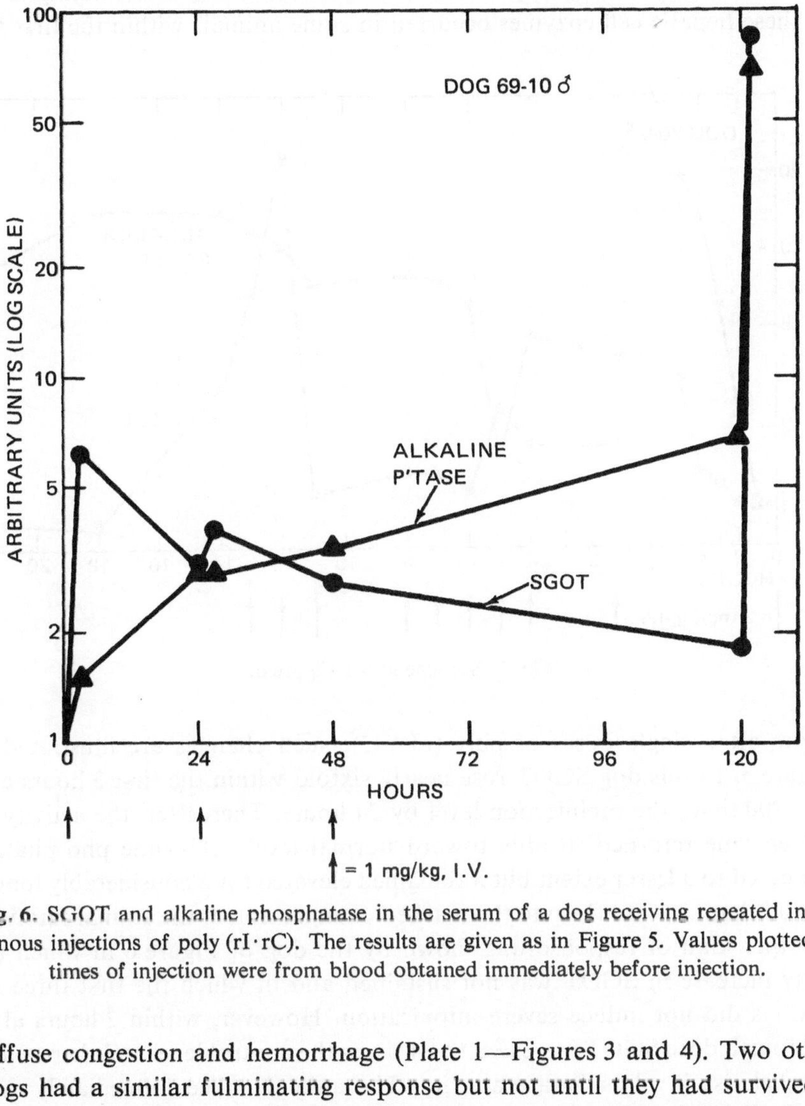

Fig. 6. SGOT and alkaline phosphatase in the serum of a dog receiving repeated intravenous injections of poly (rI·rC). The results are given as in Figure 5. Values plotted at times of injection were from blood obtained immediately before injection.

diffuse congestion and hemorrhage (Plate 1—Figures 3 and 4). Two other dogs had a similar fulminating response but not until they had survived a number of previously repeated, daily injections. One tolerated 3 daily doses of 1 mg/kg/day and then became moribund at 2 hours after the fourth dose (the dog killed at 5 days in Group 1, Table 2); the other received 7 doses of 2 mg/kg/day and became comatose at 3 hours after the eigth dose (killed at

11 days in Group 2). The livers of both animals had severe necrosis like that shown in Plate 1—Figure 4.

In order to follow the course and extent of hepatic injury the sera of most dogs were analyzed for changes in glutamic-oxaloacetic transaminase (SGOT) and alkaline phosphatase. Rapid increases in the circulating levels of these hepatic cell enzymes occurred in some animals within the first few

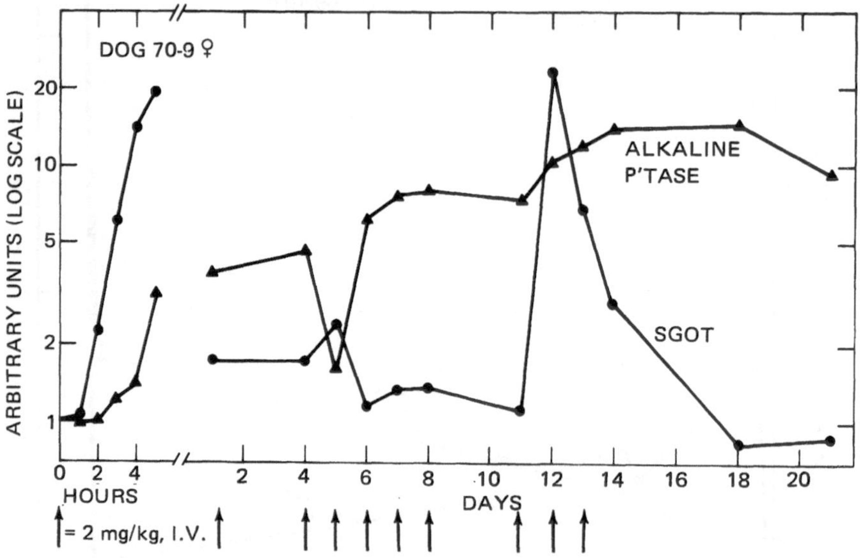

Fig. 7. See legend for Figure 6.

hours after single doses of poly (rI·rC); such changes are illustrated in Figure 5. In this dog SGOT rose nearly sixfold within the first 3 hours and was 200 times the preinjection level by 24 hours. Thereafter, the activity of the enzyme returned steadily toward normal levels. Alkaline phosphatase increased to a lesser extent but it remained elevated for a considerably longer period than SGOT. Despite the marked changes noted, the animal survived. Another kind of response was shown by the dog of Figure 6 in which the early increase in SGOT was not sustained and in which the first three injections did not induce severe intoxication. However, within 2 hours after the fourth dose SGOT increased to 84 times the control level and there was a parallel rise in phosphatase activity. The animal, killed immediately after the 2-hour blood sample had been drawn, was 1 of the 4 dogs described above in which severe hepatic necrosis was observed. Finally, Figure 7 presents the results obtained in a dog that survived 10 successive injections. In this example SGOT activity peaked significantly after the first, fourth, and eighth dose while the elevation of alkaline phosphatase was more sustained. In serum obtained at 4 weeks after the last injection alkaline phosphatase activity was again at control level.

Serum enzyme analyses were carried out in 20 of the dogs listed in Table 2, in 10 of Group 1 and in all of Group 2. Sixteen had significant increases in serum alkaline phosphatase, that is, to greater than twice

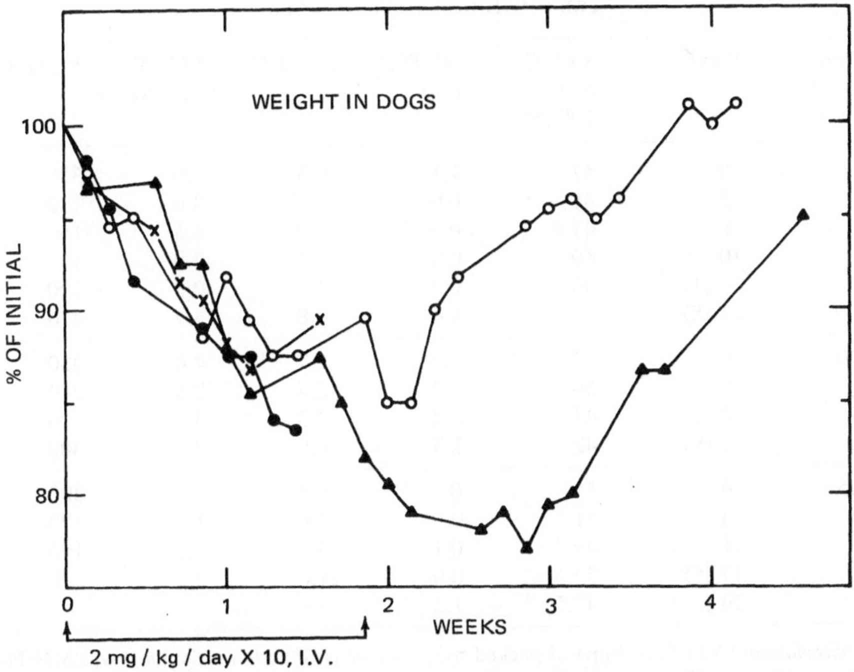

Fig. 8. Body weight in dogs receiving repeated intravenous injections of poly (rI·rC). Each symbol is the result from a single dog. Two of the dogs were killed at 10 and 11 days, respectively, for histopathological study.

preinjection levels; 13 of the same dogs had one or more peaks in SGOT like those illustrated in Figure 5, 6, and 7. The affected animals had received 5 different preparations of poly (rI·rC) in doses ranging from 0.5 to 2 mg/kg.

All dogs of Group 2, Table 2, were examined for evidence of intoxication at sites other than the liver. Half of those receiving 0.5 or 1 mg/kg/day and all of those given 2 mg/kg/day lost more than 5% of initial body weight. As shown in Figure 8 weight decreased steadily during treatment but began to recover within 1 week after the final injection. During the periods of loss the animals ate poorly or not at all. Pancytopenia was seen in the blood of each of the 10 animals. As in the examples of Table 3 the effects on circulating blood cells were moderate except for reticulocytopenia. Blood samples, analyzed at intervals like those listed in Table 3, had no conspicuous changes in blood glucose, in nonprotein nitrogen, sodium, potassium, and chloride in serum, and in one-stage prothrombin times in plasma. Sulfobromophthalein retention times remained at preinjection levels in the dogs treated with the lower doses; they did, however, increase significantly in two

Table 3. Blood Cells in Dogs[a]

Dose (mg/kg/day)	Day[b]	VPRC (ml/100 ml)	RETIC (%)	PMN-N	LCYT (1000/CMM)	PLTLT
2.0	0	47	1.1	9.3	3.0	320
	3	46.5	0.0	2.3	3.0	130
	8	41.5	0.0	5.0	1.5	150
	10	40	0.0	2.5	2.8	160
	14 (1)	36	0.2	11.3	2.3	190
	21 (8)	36	6.4	9.8	3.7	420
1.0	0	55	1.7	8.7	4.4	330
	3	50	0.2	2.4	2.5	240
	14 (1)	43	0.2	4.3	3.5	220
	21 (8)	42	2.3	6.9	2.9	300
0.5	0	51	0.8	9.4	3.0	300
	4	51	0.0	2.6	1.5	190
	11	49	0.1	3.2	1.8	190
	17 (6)	37.5	0.0	11.0	2.8	310
	24 (13)	42.5	1.3	...	...	...

Abbreviations: VPRC, volume of packed red blood cells; RETIC, reticulocytes; PMN-N, neutrophils; LCYT, lymphocytes; PLTLT, platelets.

[a] Dogs from Group 2 of Table 2.

[b] Timed from first injection; numbers in parentheses are days after last injection.

receiving 2 mg/kg/day in association with elevated activity of SGOT and serum alkaline phosphatase. One of these two animals also became jaundiced.

Two of the Group 2 dogs were killed for histopathological study (see Figure 8). One has been described above among the animals that had acute hepatic necrosis. In the same dog the epithelium of the small and large intestine was damaged with congestion, hemorrhage, and scattered necrotic cells at the surface and karyorrhexis and cells with swollen, pale nuclei in crypts. A few necrotic cells were also present in the crypts and surface epithelium of the stomach. Small numbers of karyorrhectic cells were seen in thymus and lymph nodes. The red pulp of the spleen was congested and had few hematopoietic elements. The liver of the second dog had scattered foci of congestion containing a few necrotic hepatic cells. All other hepatocytes were uniformly enlarged and vacuolated. Moderate amounts of lipid were present particularly in portal regions. Kupffer cells were prominent with ingested debris and eosinophilic cytoplasm. The cortical sinusoids of the adrenal glands had pyknotic nuclei and karyorrhexis. The remaining abdominal and thoracic viscera were normal in both dogs.

Comment. One of the remarkable features of the toxic effects of poly (rI·rC) is the precipitous and selective nature of the pathologic changes which are induced within a few hours after the administration of lethal doses. In mice and rats extensive destruction takes place in the surface epithelium of the small intestine, and degeneration is prominent in all lymphoid organs. In dogs hepatic cell necrosis is widespread. Because of the nonuniformity in the sites of major damage in experimental species, it is difficult to predict the lesions likely to be encountered in clinical trials. However, it seems prudent to anticipate that, regardless of location, lesions will arise precipitously when toxic, near-lethal doses are given.

The present studies have also shown a disparity in the potency of poly (rI·rC) among the four species tested. These may be arranged in the following order of decreasing susceptibility: rabbit > dog > mouse > rat. The sensitivity of the rabbit and rat differ by at least two orders of magnitude.

Rabbits are highly responsive to pyrogenic effects; doses as low as 1 µg/kg induce significant fevers. In this species poly (rI·rC) has a profound effect on circulating leucocytes. Leucopenia, involving an initial, transient, decrease in granulocytes and a more steadily developing depression in lymphocytes, is evident within 1 hour after injection. Acute hyperpyrexia and leucopenia are also seen in dogs. Poly (rI·rC) resembles, but is not identical with, bacterial endotoxin in effects on circulating leucocytes and body temperature. It seems reasonable to suppose that the fevers induced by poly (rI·rC) are mediated through disturbances of granulocytes like those caused by endotoxins.

References

Bennett, I. L. and Cluff, L. E. (1957). Bacterial pyrogens. Pharmacol. Rev. **9:** 427–475.

Leonard, B. J., Eccleston, E., and Jones, D. (1969). Toxicity of interferon inducers of the double-stranded RNA type. Nature **224:** 1023–1024.

Lindsay, H. L., Trown, P. W., Brandt, J., and Forbes, M. (1969). Pyrogenicity of poly I·poly C in rabbits. Nature **223:** 717–718.

Snell, E. S. and Atkins, E. (1969). The mechanisms of fever. *In* The Biological Basis of Medicine. **2**: 397–419. Ed. by E. E. Brittar. Academic Press, New York.

Weil, M. H. and Spink, W. W. (1957). A comparison of shock due to endotoxin with anaphylactic shock. J. Lab. Clin. Med. **50:** 501–515.

DISCUSSION

DR. W. BRAUN: Before our enthusiasm for the potential utilization of the stimulatory effects of polynucleotides is dampened by observations such as those recorded in the foregoing, I think it is worthwhile to reiterate the following points. First of all, toxic and immunogenic potentials of these polynucleotides differ greatly among different species. Furthermore, we have already heard that an alteration of the size and the nature of the molecule may provide an opportunity to lessen that aspect of the trigger mechanism which is associated with cell damage. Even though I concur fully with the current belief that interferon stimulation and cytotoxicity are causally related, it may well be possible to achieve the necessary signal without too much damage. I think that the potentiation of the stimulatory effects of poly A:U by substances such as theophylline or other stabilizers of cAMP (see paper by Braun *et al.*) might be very interesting to test in the interferon system to determine whether it might be possible, with the aid of such magnifiers of responses, to arrive at levels of interferon-inducing nucleotides that have minimal cell-damaging and toxic effects.

I think the data that were reported here by Dr. Jasmin from Dr. Matthe's group regarding success in the remission of leukemia in 4 patients with minimal doses of poly I:C, which did not produce any toxic effects, give cause for encouragement.

Finally, I think that we must keep in mind that we are dealing with multifaceted phenomena. We are dealing both with the capacity to induce antiviral effects as well as the capacity to stimulate classical immune responses. It certainly has been shown conclusively now that the latter, the stimulation of classical immune responses, can be achieved by polynucleotides that do not have toxic effects like the ones on which Drs. Philips and Stinebring have concentrated. I want to reiterate that there is no evidence for poly AU having any of the undesirable effects with which poly IC has now become notoriously connected.

DR. F. S. PHILIPS: There is only one comment I would have, Dr. Braun, to what you say. I am not yet aware of detailed studies of the toxicology and pharmacology of poly AU. The data that have been presented suggest that there is a wide difference in activity between poly IC and poly AU. Perhaps your optimistic feeling is entirely justified, but I think that until enough poly AU is made available for a detailed study of toxicology, we ought to hold off a minute before we assume that it is entirely without some potential danger to the cell.

DR. W. BRAUN: That, of course, is true. The data, although they are limited in comparison to those with poly IC, show that you cannot kill, at least mice, with any amount of poly AU. You do not get pyrogenic or any other easily detectable toxic effects in rabbits. We already have strong evidence that poly AU, for reasons that I discussed earlier, has entirely different pharmacological effects from poly IC.

Therefore, I think it will become increasingly important that people test the pharmacological effects of poly AU, and use poly AU rather than poly IC in systems in which they wish to stimulate classical immune responses rather than interferon.

DR. R. M. MORRELL: I have a question that is related to the pharmacology, particularly with the dog. The pyrimidine in the polynucleotide, cytosine, a DNA inhibitor, has been used as an antileukemic agent and has been studied in several species of animals. It has been found that the dog fails to respond with the usual hematologic repression in the absence of huge doses, mainly because it fails to oxidize the phosphoryl compound. I think, in view of Dr. Braun's concept of the mechanism of action of the compound, that this is an aspect that must be investigated in the future in relationship to the pharmacologic aspects of the polynucleotides. I wonder if, in reference to the dog work, there has been an opportunity to look at thyroid function and whether possibly there is a thyrotoxicosis, or thyrotoxis-like state, in which there is an increase in phosphorylization of the polynucleotides in the dog.

DR. F. S. PHILIPS: No, we have done no work on that. Perhaps others have. We are dealing with an entirely different kind of reaction. The cytosine, as you know, selectively inhibits DNA synthesis in many species. The rabbit cells are destroyed by this and other kinds of chemotherapy almost exclusively in the cell renewal system. You will find cell damage wherever there are mitotic cells, including damage to bone marrow and lymphoid germinal centers. The response of these centers to poly IC is not very great, and you certainly don't see any type of profound hepatic necrosis with this compound as you see with selective inhibitors of DNA synthesis.

DR. W. REGELSON: In our work with synthetic polysulfonates we were very much interested in the pattern of toxicity, in view of what has been reported many years ago by Lumier in regard to colloidal plasmic shock. The characteristics of blood were altered, resulting in a pathophysiology comparable to what one sees with the polynucleotides and with pyran copolymer. In fact, when one uses very large doses of pyran, one gets a hemolytic anemia very similar to what one sees when one perfuses blood through plastic extracorporeal circulation, implying that there is some mechanical effect of these macromolecules on cell surfaces, at least on the red cell surface. This, of course, can lead to changes in protein distribution and in the stability of blood elements. Therefore, I think some aspect of what one sees clinically, at least in regard to acute changes in leukocytes and adherence to blood vessels, is very similar to what one sees when one gives a massive intravenous dose, for example, of fibrinogen. In a sense, it may represent a syndrome reflecting distributions of molecular weight.

In regard to Dr. Stinebring's somewhat pessimistic view, while I agree with his dialectic materialism relating to these drugs, as a cancer chemotherapist I have to be optimistic. We can separate patterns of toxicity with the synthetic polymers that we are using, the polycarboxylates. We have, for example, one fraction of divinyl ether maleic anhydride obtained from George Butler that does not have the pathotoxicities reflected in blood metabolism and yet does have some of the other effects. So one might be able, by playing around with these molecules, to achieve patterns of specificity.

Incidentally, like Dr. Philips, we found that there is enhanced susceptibility to staphylococci following the administration of either pyran or poly IC. It also has been reported that poly IC may disrupt blood brain barriers. This is something else to consider.

DR. W. R. STINEBRING: Dr. Regelson, did you study these relatively nontoxic materials for interferon-inducing capacity?

DR. W. REGELSON: We have not looked at that yet.

DR. F. S. PHILIPS: May I make a comment and add something to what Dr. Regelson has said? First of all, he said that, being a cancer chemotherapist, he has to be hopeful. I would like to point out something. Dr. Hamilton discussed a whole group of cancer therapeutic agents from the standpoint of their comparative effects against the RES. You can take several views of these substances. You can say they are some of the world's best mouse and rat poisons, but, nevertheless, each and every one of them has had use against tumor in man of a certain kind. So you can either take a very dismal view of poly IC because of what I and others have said in our formal presentations, or you can say that there is a good possibility that you may find one or another kind of tumor that has the type of acute responses that we have seen in some of the normal situations.

I should add one other thought. I don't think we know what the acute response of normal tissue is going to be in man. There is such a vast difference between the mouse and the rat on one hand, and the dog on the other hand, that I think we have to keep our fingers crossed as to what the limiting cytotoxic response of normal tissues is going to be in patients. But I think we can be pretty sure that when it occurs, it is going to be a rather massive one, because it has been massive, fulminating, and rapidly developing in those species already studied.

DR. M. R. HILLEMAN: Dr. Stinebring did a very fine job of summarizing contemporary anxieties concerning poly IC application. There is, however, a sequence of investigations that still must be carried out, and these data that are not presently available will hopefully permit a meaningful judgment. The first problem, as we view it, is to determine the relevance of interferon induction by poly IC to prevention and/or treatment of viral disease in man and domestic animals. This is done very cautiously under conditions that will not present a significant hazard to the recipient. One cannot conclude a drug is bad based on a simple listing of unrefined adverse reaction data recorded to date. One must consider five principal factors in making a judgment, namely, source and quality of the drug, animal species used, dose given, route of administration, and frequency of administration. It is not difficult to recognize, as an example, that poly IC (sometimes of undefined quality or composition) has been given (often in fairly large amount) into animals (for example, the dog) that are exquisitely sensitive to it by a most unfavorable route (intravenous). Giving the substance by the respiratory route, which requires a smaller dose, causes essentially no adverse effect. Perhaps, even more important, poly IC has been relatively nontoxic in man even when given in large and repeated doses by intravenous route. In perspective, it seems likely that if the criteria for contemporary criticism of poly IC had been applied to penicillin (based on guinea pig studies), or to aspirin, we would hardly have these substances for use today. The second problem, as we see it, is to improve the molecular constitution of the substance so as to dissociate its adverse effects from its beneficial effects. The name of the game is toxicity:activity ratio and this is the key point in essentially all pharmaceutical development.

The development of poly IC for application to human or animal medicine is in a very early stage of development and it is far too early to make significant or meaningful judgments. We simply do not have the information. My own guess is that the acute toxicity findings that have been reported to date are relatively unimportant. The more significant findings may relate to more subtle effects, such as reported by Dr. Talal, which are slower to develop and which relate to immunologic potentiation. In relation to this, it is perhaps worthy of note that man and animal must be continuously assaulted with double-stranded RNA's of viral origin resulting from repeat viral infections. Perhaps some of the far-reaching questions that need to be answered have already been put to test in nature.

DR. A. G. JOHNSON: There are now available exquisitely sensitive assays for circulating endotoxin in the blood. I am wondering if anybody has looked for circulating endotoxin in the blood of animals receiving poly IC, on the hypothesis that one function of poly IC might be to liberate these kinds of materials.

DR. W. R. STINEBRING: I don't know that it was done, but I certainly agree with you, it should be done.

DR. F. S. PHILIPS: I think Dr. Johnson brought up a very interesting point on which we have no information, but we do have a thought that intrigues us. We don't know enough of the vast field of immunology and endotoxicology to know whether the answers are available. But I would like to point out that dogs responding to anaphylactic shock show acute necrosis, with a time course very like that seen in dogs responding with acute lethal hepatic response to poly IC. As near as we can gather from the literature, although pathotoxicity is seen with endotoxin in the dog, it is nowhere near as massive. What is seen is scattered foci of hepatic necrosis developing 12 or 14 hours later. We have wondered whether there is a possibility that many of the phenomena that we have seen with poly IC are somehow related to a pre-existing sensitization to gram negative products. This is sheer speculation, but it seems worthy of following up.

DR. W. BRAUN: While it is possible that poly IC might in some way liberate endotoxins, I don't think it is a necessary assumption. We know that the effects of endotoxin, including the effects due to pre-existing sensitization, include membrane damage in the same way that poly IC seems to damage membranes. Therefore, I don't see any necessity for assuming that poly IC has to liberate endotoxin in order to produce the effects; both poly IC and endotoxins are polyanions and action membranes. Therefore, they may produce the same triggering without our having to assume that one makes available the other.

DR. W. R. STINEBRING: We are struck by the toxicity in the hypersensitive phenomena. Of course, as you well know, it has been postulated by various people from *in vitro* data that there is a release of cytotoxic factors from cells affected by cytotoxic agents. I am sure the same thing is also true in hyperreactivity. I must point out, however, that we were able to differentiate between endotoxin reactivity and delayed hypersensitivity in the tuberculin type hypersensitivity. This could not be done on the basis of gross pathology but by the findings that we saw with the help of other techniques.

I am sure there are many interrelationships. After all, we are dealing with cells of the RES system. As has been pointed out by numerous people, there is active material that could influence other cells in this system, or other cells beyond the system. I think your point is certainly well taken. It might be endotoxin, but it could also be other endogenous cell substances that are obviously very important.

Part V

Polynucleotide-Associated Events in Immunocompetent Cells

CELL TO CELL COOPERATION IN THE IMMUNE RESPONSE: A ROLE FOR MACROPHAGE RNA-ANTIGEN COMPLEXES?

GEORGES E. ROELANTS

National Institute for Medical Research, Mill Hill, London NW 7

There is accumulating evidence that the induction of antibody formation to most antigens requires, under physiological conditions, a cooperation among several cells: a phagocytic cell, predominantly of the macrophage type and two cells of the lymphoid series, an "antigen reactive cell" playing a role in antigen recognition but unable to secrete antibody, and an antibody-forming cell precursor.

The role of each type of cell and the nature of their interaction is not clear at present. In the framework of this Symposium on the Biological Effects of Polynucleotides I will discuss one of the proposed mechanisms for cell cooperation.

The "Macrophage RNA-Antigen Complex" Hypothesis

It has been postulated that, after phagocytosis by macrophages, antigens or fragments of antigens (antigenic determinants) become associated with RNA and that antigen-RNA complexes are transferred to the antibody-forming cell precursor to trigger cell proliferation and specific antibody production.

This concept was first based on the finding that crude filtrates of peritoneal exudate (P.E.) cells exposed to T2 bacteriophages, when incubated with lymph node cells *in vitro*, induced in some cases the production of anti-T2 antibody and that this phenomenon was inhibited by ribonuclease treatment of the extract (Fishman, 1961). Subsequent experiments showed that similar results could be obtained using RNA preparations from P.E. cells instead of whole filtrates (Fishman and Adler, 1963). The presence of antigen was later demonstrated in similar RNA preparations using the same antigen, T2 phage (Friedman *et al.*, 1965) or Maia squinada hemocyanin (Askonas and Rhodes, 1965). When RNA was extracted immediately after the addition of ^{131}I-hemocyanin to P.E. cells, that is, before any "processing" of the antigen could take place, an immunogenic RNA containing ^{131}I labelled material could also be obtained (Askonas and Rhodes, 1965). The amount of antigen calculated to be present on the immunogenic RNA preparation on the basis of radioactivity was about 20 to 30 times less than the minimum amount of free hemocyanin eliciting a response in mice (Askonas and Rhodes, 1965). Thus, the association with RNA seemed to enhance the immunogenicity of the antigen.

Gottlieb *et al.* (1967) after incubation of P.E. cells with T2 bacteriophages assayed activity of RNA fractions on splenic fragments. Low levels of T2 inhibition could be induced by the 28S fraction from sucrose density gradients; the activity persisted after degradation of the RNA to 4–6 S fragments. Incorporation of ^{3}H-uridine showed that only 4–5 S RNA had been newly synthesized during the time of incubation of P.E. cells with T2 phage.

By DNA hybridization experiments, it was shown that RNA from P.E. cells not exposed to T2 phage (Gottlieb *et al.*, 1967) or exposed to R17 phage (Gottlieb, 1968a) competed for annealing with that of cells exposed to T2. Those results, like those of Askonas and Rhodes, indicate that activity resides in a pre-existing RNA fraction rather than in RNA synthesized following exposure to antigen. When examined in cesium sulfate equilibrium gradients (Gottlieb *et al.*, 1967), the bulk of the P.E. cells RNA showed a density of 1.665. A minor component, consisting of about 5% of the total RNA with a density of 1.588, accounted for the total immunogenic activity of the preparation, and the location in sucrose gradients was first reported to be only in the 28 S region (Gottlieb, 1968a) but also appeared in 4–6 S RNA (Gottlieb, 1968b). This minor band, as well as the immunogenic activity, disappeared after treatment of the RNA preparation with pronase (Gottlieb *et al.*, 1967; Gottlieb, 1968a). It could be directly iodinated with ^{125}I (Gottlieb, 1968a), and also contained ^{125}I label after incubation of the P.E. cells with iodinated phages (Gottlieb, 1968b). Its formation was not impaired by actinomycin D (Gottlieb, 1968b). When RNA was extracted from P.E. cells that had not been incubated with antigen, the minor band also appeared and contained protein (Gottlieb *et al.*, 1967). After incubation with antigen it contained antigenic fragments of low molecular weight (Gottlieb, 1969).

From these observations it was concluded that the "minor band" consisted of an antigen-RNA complex which "may be the essential means by which the information eliciting specific antibody production is processed, even though the RNA itself is not specific" (Gottlieb *et al.*, 1967).

Yet, even though the enhanced immunogenicity of some antigens complexed to macrophage RNA had been established and the complex had been partially characterized, there was no evidence that antigen-RNA complexes were indeed formed within viable cells and transferred to lymphocytes. Thus, our interest was to examine some parameters of antigen-RNA association to try to determine if this was a physiological phenomenon or rather an interesting artifact. The work I am going to describe now was done in collaboration with Dr. J. W. Goodman.

Association with RNA is not Related to Immunogenicity

We argued that if the macrophage was the cell responsible for "antigen recognition" and if the formation of macrophage RNA-antigen complexes

Table 1. The Association of Various Molecules with RNA from Peritoneal Exudate Cells

Molecule	Total dose range		No. of experi-ments	Efficiency of binding[a]
	μg	μc		
^{3}H-Poly-γ-D-glutamic acid[b]	40–550	(2.04–28.05)	44[c]	$1.7–10 \times 10^{-4}$
^{3}H-Poly-α-D-glutamic acid	110–550	16.17–80.85	4	$1.8–4.0 \times 10^{-4}$
^{14}COOH-dextran	572–11,450	0.49–9.96	20	$1.3–4.9 \times 10^{-4}$
^{3}H-Dextran	260	24.96	2	Below detection
^{125}I-Myeloma protein	40	1.62	2	4.7×10^{-5}
^{125}I-(TGAL)	4	8.72	4[d]	5.0×10^{-5}
^{14}C-Testosterone	10	1.75	2	Below detection
^{3}H-Estradiol	10[e]	14.68	2	Below detection
^{3}H-Cortisone	10[e]	9.54	2	Below detection
^{14}C-DL-Glutamic acid	75	3.42	2	4.0×10^{-5}
^{14}C-L-Glutamic acid	75	111.22	2	4.0×10^{-5}

[a] Micrograms of "antigen" complexed per microgram of RNA per microgram of total antigen (intracellular or in homogenate). Corrected for controls (antigen and purified RNA mixed in phosphate buffer and subjected to the same extraction procedure).

[b] No difference was found between the alum-precipitated and the methylated bovine serum albumin-complexed forms (5).

[c] Some experiments were done with mouse and guinea pig P.E. cells.

[d] These experiments were done with P.E. cells from CBA and C57 mice.

[e] 1 part of radioactive compound was diluted with 100 parts of cold hormone on a weight basis.

(Roelants and Goodman, 1969; reprinted with permission from the *Journal of Experimental Medicine*.)

had a role in immune induction, there would be some degree of differentiation between molecules on the basis of immunogenicity in the binding process. Different compounds including natural and synthetic polypeptides, a protein, polysaccharides, amino acids, and steroid hormones were assayed for their capacity to form complexes with RNA from P.E. cells (Table 1) (Roelants and Goodman, 1968, 1969).

There was no correlation between RNA association and immunogenicity but rather between binding and the charge of the antigen. All the molecules bearing negative charges (poly-glutamic acid, IgG, carboxylated dextran, poly (tyr, glu)-poly-DL-ala-polylys, glutamic acid) showed some association in contrast to uncharged molecules (dextran, testosterone, estradiol, and cortisone) regardless of their size, complexity, and immunogenicity. The RNA could not be saturated by increasing the antigen-RNA ratio within practical limits, suggesting that the RNA involved was not specific. Moreover, the synthetic polypeptide poly (tyr, glu)-poly-DL-ala-polylys (TGAL) was binding to the same extent to P.E. cells RNA from mice of strain C57 which are good responders to this polypeptide and from strain CBA which are poor responders (McDevitt and Sela, 1965; Roelants and Goodman, 1969).

Association with RNA is not an Enzyme-Dependent Reaction

Using the same number of P.E. cells from the same rabbit in order to minimize trivial differences, the kinetics of antigen-RNA association were studied by varying the incubation time from 1 minute to 48 hours and the temperature from 2 to 37 °C. No significant differences were found in antigen-RNA ratios or the efficiencies of binding under these conditions. The same findings were also obtained whether cell homogenates were used instead of living cells and when these homogenates were subjected to treatment with pronase or dithiothreitol, to a temperature of 80 °C for 5 minute, or when they were incubated with antigen in the presence of 0.5% sodium dodecyl sulfate.

These results (Roelants and Goodman, 1969) suggested that the formation of RNA-antigen complexes was not an enzyme-mediated reaction and hence rendered unlikely covalent binding between RNA and antigen.

Association with RNA Does not Require Newly Synthesized RNA

In experiments in which nuclei and mitochondria were removed from P.E. cell homogenates and the remaining fraction incubated with antigen, the efficiency of binding was the same as when unfractionated homogenates were used.

When transcription was blocked by actinomycin D, as shown by the lack of ^{14}C-uridine incorporation into RNA, again no impairment of binding was observed.

These results (Roelants and Goodman, 1969) confirmed that there is no need for RNA synthesis after the introduction of antigen to form complexes, as had already been reported earlier (Askonas and Rhodes, 1965; Gottlieb *et al.*, 1967; Raska and Cohen, 1968). They also indicate that association of antigen to RNA is merely a passive phenomenon. The confirmation of this point and of the noncovalent nature of the bond was provided using purified RNA (Roelants and Goodman, 1969). An association, qualitatively and quantitatively identical to that found *in vivo*, could take place *in vitro* provided that Mg^{++} ions were present.

Formation of Antigen Complexes with RNA from P.E., HeLa, and E. coli Cells

The synthetic polypeptide TGAL was incubated with homogenates of rabbit, guinea pig, C57 and CBA mice P.E., HeLa, or *E. coli* cells. The RNA was extracted with cold phenol (Roelants and Goodman, 1968) or hot phenol (Scherrer and Darnell, 1962) and submitted to zonal sedimentation in cesium sulfate (Roelants and Goodman, 1968).

In all gradients (Figure 1) a major band of RNA was observed at a density of 1.68. After hot phenol treatment this was the only band seen in the gradient. After cold phenol extraction one or two "minor bands" were found at densities ranging from 1.56 to 1.63. These results are in contrast with reports of Gottlieb (Gottlieb *et al.*, 1967; Gottlieb, 1968a, 1968b, 1969) who described a minor band with a constant density of 1.588 whether using T2 phage, a synthetic copolymer glu-ala-tyr, or no antigen at all, although quite surprisingly, the density of the "major band" consisting of pure RNA varied between 1.620 and 1.676.

When ^{125}I labeled TGAL was used, radiolabel was associated with the RNA extracted by cold phenol and in cesium sulfate appeared concentrated in the "minor band" and free at the top of the gradient (Figure 1). When hot phenol was used no radiolabel was detected in the RNA preparation. When no antigen was used the lower density band was also present after the cold phenol extraction but absent when hot phenol was used. The important point was that the same amount of ^{125}I-TGAL was bound by RNA from P.E., HeLa, or *E. coli* cells and in each case the complexes appeared as a minor band of lower density in cesium sulfate. The immunogenicity and adjuvanticity of the RNA bands isolated from several cell species are under investigation. These findings clearly show that the formation of antigen-RNA complexes depends on the method used for RNA preparation and is

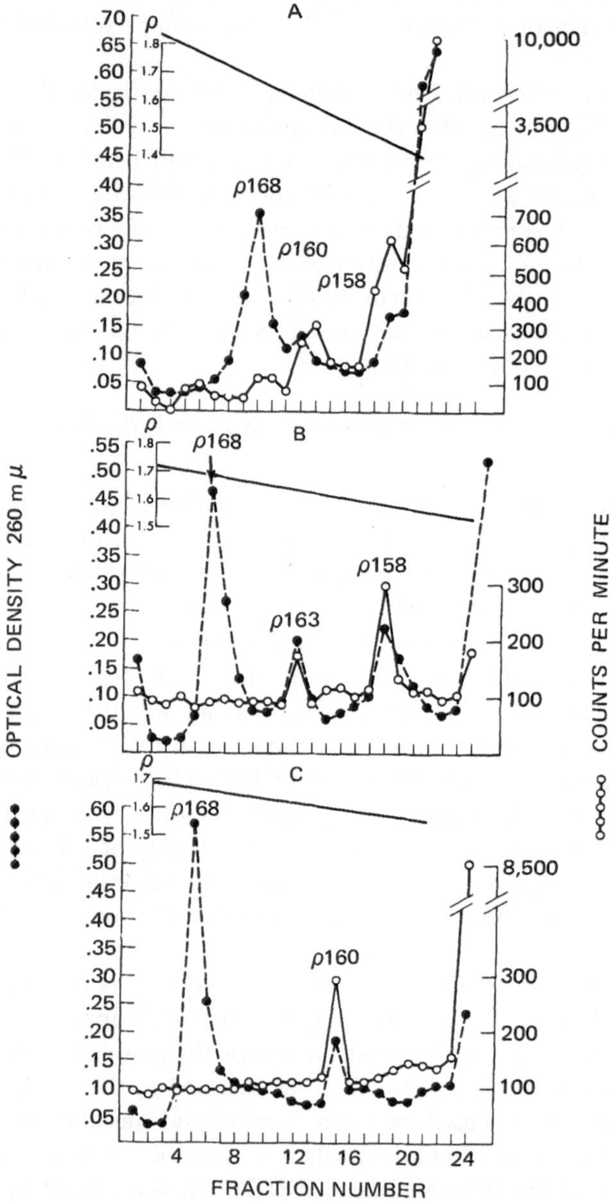

Fig. 1. Equilibrium sedimentation in cesium sulfate of RNA extracted by cold phenol from homogenates of (A) P.E. cells; (B) HeLa cells; (C) *E. coli* incubated with ^{125}I-(TGAL) (Roelants and Goodman, unpublished observation).

not specific for macrophages. Moreover, protein-RNA appears as a minor band also in the absence of antigen.

A Physiologically Significant Role for Macrophage RNA-Antigen Complexes in Immune Induction?

Several facets of antigen-RNA association have been examined: The formation of complexes was unrelated to the immunopotency of the antigen, was not an enzyme-dependent reaction, did not involve covalent binding, did not require RNA synthesis following introduction of the antigen, did not involve antigen specific-RNA, depended on the method used for RNA preparation, and was not specific for macrophages.

Specificity was absent at every level examined: the antigen, the RNA, the association mechanism, and the cell involved. In our opinion, this makes an important physiological role for those complexes very unlikely. The association between antigen and RNA within viable cells has never been demonstrated and the RNA extraction procedure involves disrupting the organization of the cell: The most direct explanation for the appearance of these "antigen-RNA complexes" is that they represent artifacts of preparation.

There is strong additional evidence against a major role in immune induction for small antigen fragments such as present on P.E. cells ribonucleoprotein complexes. (1) There is an inverse correlation between the electrical charge of the antigen and that of the antibody it elicits. In the case of antihapten antibody this correlation depends on the net overall charge of the antigen rather than the charge within the limited area around the haptenic determinant (reviewed in Sela, 1969). (2) Antigenic determinants are mostly conformation-dependent (reviewed in Goodman, 1969; Sela, 1969). (3) The response to a hapten is highly carrier specific (Mitchison, 1967) and requires the recognition of determinants on the carrier in addition to the haptenic determinant itself (Rajewsky *et al.*, 1969). Thus, the recognition of most antigenic determinants and the triggering of the immune response takes place while the immunogenic molecule is still intact and not after breakdown.

The enhanced potency of some antigens when complexed to RNA is not so surprising. It is well known that aggregating antigen or making it insoluble (for example, by alum precipitation, by coating on bentonite, to give but two of a variety of methods) intensifies the antibody response. On the other hand, the adjuvant action of polynucleotides was studied by several speakers of this Symposium.

All this makes us look for another mechanism for macrophage-lymphocyte cooperation. The association of undegraded antigen with macrophage cell membrane and the increased immunogenicity of macrophage-bound antigen may indicate that the role of the macrophage is not to process the antigen but to expose it in a more effective way to lymphoid cells (Askonas *et al.*, 1968; Unanue and Askonas, 1968a 1968b; Unanue *et al.*, 1969; Unanue and Cerottini, 1970) although this hypothesis is unable to account for all the observations (Askonas and Jaroskova, 1970).

The role of phagocytes and the mechanism of action of polynucleotides and other adjuvants in immune induction remain challenging questions.

ACKNOWLEDGEMENT

I wish to thank Dr. B. A. Askonas for helpful discussions in the preparation of this manuscript.

References

Askonas, B. A. and Rhodes, J. M. (1965). Immunogenicity of antigen containing RNA preparation from macrophages. Nature **205:** 470–474.

——— Auzins, I., and Unanue, E. R. (1968). Role of macrophages in the immune response. Bull. Soc. Chim. Biol. **50:** 1113–1128.

——— and Jaroskova (1970). Antigen in macrophages and antibody induction. In press.

Beiser, S. M., Erlanger, B. F., Agate, F. J., Jr., and Lieberman, S. (1959). Antigenicity of steroid-protein conjugates. Science **129:** 564–565.

Fishman, M. (1961). Antibody formation in vitro. J. Exp. Med. **114:** 837–856.

——— and Adler, F. L. (1963). Antibody formation initiated *in vitro*. II. Antibody synthesis in x-irradiated recipients of diffusion chambers containing nucleic acids derived from macrophages incubated with antigen. J. Exp. Med. **117:** 595–602.

Friedman, H. P., Stavitsky, A. B., and Solomon, J. M. (1965). Induction *in vitro* of antibodies to phage T2: Antigen in the RNA extract employed. Science **149:** 1106–1107.

Goodman, J. W. (1969). Immunochemical specificity: Recent conceptual advances. Immunochemistry **6:** 139–149.

Gottlieb, A. A., Glisin, V. R., and Doty, P. (1967). Studies on macrophage ribonucleic acid involved in antibody production. Proc. Natl. Acad. Sci. **57:** 1849–1856.

Gottlieb, A. A. (1968a). Antigens, RNA's and macrophages. R.E.S.J. Reticuloendothel. Soc. **5:** 270–281.

——— (1968b). The antigen-RNA complex of macrophages. *In* Nucleic Acids in Immunology. 471–486. Ed. by O. J. Plescia and W. Braun. Springer-Verlag, New York.

——— (1969). Studies on the binding of soluble antigens to a unique ribonucleoprotein fraction of macrophage cells. Biochemistry **8:** 2111–2116.

McDevitt, H. O. and Sela, M. (1965). Genetic control of the antibody response. I. Demonstration of determinant specific differences in response to synthetic polypeptide antigens in two strains of inbred mice. J. Exp. Med. **122:** 517–531.

Mitchison, N. A. (1967). Antigen recognition responsible for the induction in vitro of the secondary response. *In* Cold Spring Harbor Symposium on Quantitative Biology. **13:** 431–439.

Rajewsky, K., Schirrmacher, V., Nase, S., and Jerne, N. K. (1969). The requirement of more than one antigenic determinant for immunogenicity. J. Exp. Med. **129:** 1131–1143.

Raska, K. and Cohen, E. P. (1968). RNA in mouse cells exposed to different antigens. Nature **217:** 720–723.

Roelants, G. E. and Goodman, J. W. (1968). Immunochemical studies on the poly-γ-D-glutamyl capsule of *Bacillus anthracis*. IV. The association with peritoneal exudate cell ribonucleic acid on the polypeptide in immunogenic and nonimmunogenic forms. Biochemistry **7:** 1432–1440.

——— (1969). The chemical nature of macrophage RNA-antigen complexes and their relevance to immune induction. J. Exp. Med. **130:** 557–574.

Scherrer, K. and Darnell, J. E. (1962). Sedimentation characteristics of rapidly labeled RNA from HeLa cells. Biochem. Biophys. Res. Com. **7:** 486–490.

Sela, M. (1969). Antigenicity: Some molecular aspects. Science **166:** 1365–1374.

Unanue, E. R. and Askonas, B. A. (1968a). Persistence of immunogenicity of antigen after uptake by macrophages. J. Exp. Med. **127:** 915–926.

——— (1968b). The immune response of mice to antigen in macrophages. Immunology **15:** 287–296.

Unanue, E. R., Cerottini, J.-C., and Bedford, M. (1969). Persistence of antigen on the surface of macrophages. Nature **222:** 1193–1195.

Unanue, E. R. and Cerottini, J.-C. (1970). The immunogenicity of antigen bound to the plasma membrane of macrophages. J. Exp. Med. **131:** 711–725.

ANTIGEN CAPTURE BY A UNIQUE RIBONUCLEOPROTEIN OF MACROPHAGE CELLS*

A. ARTHUR GOTTLIEB

Institute of Microbiology
Rutgers University, The State University of New Jersey
New Brunswick, New Jersey

The association of antigens with polynucleotides has been a subject of intense interest in the past years. In particular, several reports have appeared which suggest that the association of antigen(s) with ribonucleic acids *in vitro* leads to an augmentation of the immunogenic properties of the substances so complexed (Fishman *et al.* 1961, 1963; Askonas and Rhodes, 1965). Polynucleotides have also been implicated as nonspecific stimulators of the immune response (Braun and Nakano, 1967).

It is important to recognize that although such artifical interactions lead to enhanced immunogenicity of certain antigens, this fact by itself does not imply that such a mechanism necessarily operates *in vivo*. Much interest has been generated by the observations of Fishman and co-workers which indicated that antigens taken up by peritoneal cells were associated with ribonucleic acids and that these bulk RNA preparations had the capacity to stimulate specific antibody formation against the antigen to which the peritoneal cells had been exposed (Fishman *et al.*, 1961, 1963). Recent work by Roelants and Goodman (1969) dramatically demonstrates that many substances carrying negative charges at physiological *p*H do associate with RNA in the presence of magnesium ion. It is argued by these investigators, on the basis of their observations, that all associations of RNA with anti-

* This work was supported by NIH grants AI-8307 and AI-9850. A portion of this work was carried out in the Department of Medicine, Harvard Medical School under a Career Development Award (1-K3-GM-28302-02).

genic substances are physiologically irrelevant. It is the purpose of this report to disprove this conclusion.

About 3 years ago, we demonstrated that bulk RNA extracted from macrophages could be fractionated into two components by banding the RNA preparation in a cesium sulfate density gradient at equilibrium (Gottlieb *et al.*, 1967). The most prominent RNA band observed in the banding profile had a density of 1.68 g/cc, consistent with other density measurements of mammalian RNA in this kind of system, while the minor RNA component had a banding density of 1.58 g/cc and had the unusual property of carrying all of the immunogenic activity of the bulk RNA prepared from T2 phage-infected peritoneal macrophages. A similar concentration of sheep red blood cell antigens into the minor RNA component was independently shown by Bishop and co-workers (1967). This minor light-density component was subsequently shown to constitute 5% of the total cellular RNA and could be found only in peritoneal exudate cells (Gottlieb, 1968). In nonimmunized peritoneal cells, the band was also present but lacked immunologic activity. We have designated this light density component as macrophage RNP. Our studies have previously estimated that the molecular weight of this RNP complex is 12,000 and that the complex contains about 30% protein. These values place an upper limit of 3600 daltons on the size of the antigenic determinant which can be accommodated by the complex (Gottlieb and Straus, 1969). Necessarily, those antigenic determinants that are larger than 36 amino acids would not be found in the RNP complex. In the T2 bacteriophage system, it has been clearly shown that the fragments of T2 antigens associated with RNP specifically recognize neutralizing antibody against T2 phage (Gottlieb, 1969).

If macrophages are exposed to large concentrations of various soluble antigens (1 mg/ml or higher), gross association of these antigens with the bulk RNA of the macrophage is noted (Roelants and Goodman, 1968). In our hands, at very low concentrations (1–3 μg/ml), antigen is selectively taken up into the RNP complex (Gottlieb, 1969). As the dose of antigen increases, large amounts of antigen begin to be associated with bulk macrophage RNA, while saturation of the RNP complex is reached. In order to show this, it was necessary to develop systems in which antigens linked to RNP could be selectively separated from antigens nonspecifically associated with bulk RNA. We have successfully employed two continuous polyacrylamide gel electrophoretic systems for this purpose: one at *p*H 7.6, the other at *p*H 3.8. In the neutral system, RNP has a mobility of 1.20 with respect to *E. coli* sRNA and the major band of RNA from cesium sulfate gradients, while synthetic antigens such as GAT and GLT are well retarded (Gottlieb and Straus, 1969). In the acidic system, RNP migrates behind sRNA with a relative mobility of 0.75 (Figure 1). The latter system has the special advantage that the glutamic acid residues of GAT and GLT are all in the neutral form at *p*H 3.8. Consequently, these polymers do not migrate in this system. Using these two gel electrophoresis systems, it is possible to score specifically for the appearance of these labeled antigens in RNP. Thus, one can clearly discriminate between antigens bound to RNP and antigenic fragments which contaminate the bulk RNA preparation. With this in mind, we now present the result of studies dealing with the interaction of several synthetic antigens with RNP.

The first question we need to answer is simply whether there is any discrimination exhibited by the macrophage in binding fragments of

Table 1. Antibody Response of Rats to D- and L-$Glu^{60}Ala^{30}Tyr^{10}$

Animal no.	Immunized with	Antibody titer to cells coated with	
		L-GAT	D-GAT
1	100 μg L-GAT	1:128	0
2	100 μg L-GAT	1:32	0
3	100 μg L-GAT	1:64	0
4	100 μg L-GAT	1:16	0
5	100 μg L-GAT	1:32	0
1	100 μg D-GAT	0	0
2	100 μg D-GAT	0	0
3	100 μg D-GAT	0	0
4	100 μg D-GAT	0	0
5	100 μg D-GAT	0	0

Five rats were given 100 μg of L-$Glu^{60}Ala^{30}Tyr^{10}$ in complete Freund's adjuvant, in the footpads. A similar set of 5 rats received D-$Glu^{60}Ala^{30}Tyr^{10}$. Sera were recovered 2 weeks post-immunization and assay for anti-GAT antibodies carried out by the technique of tanned red cell hemagglutination. (These assays were performed by Dr. Sidney Leskowitz.)

antigen(s) to RNP. We consider the following system, namely, the synthetic polymers, L-Glu^{60} Ala^{30} Tyr^{10} and D-Glu^{60} Ala^{30} Tyr^{10}. The L-form is a good antigen and is capable of eliciting precipitating antibody in the rat (Table 1), while the D-form is at best a poor antigen in this species. L-GAT and D-GAT were iodinated to equivalent specific activities and were given to comparable numbers of macrophages; the RNA was extracted and run on the polyacrylamide gel systems. The amount of label associated with the RNP moiety was determined in multiple populations of macrophages using several concentrations of L- and D-GAT. The degree of labeling of the RNP moiety under these conditions is shown in Figure 2. It is clearly seen that it is possible to saturate the RNP complex with either optical form of the GAT polymer. Moreover, the saturation level for D-GAT is very much lower than that for L-GAT. This indicates that the association of GAT with RNP is not simply an ionic type of interaction, since such an interaction could not discriminate between optical forms of GAT.

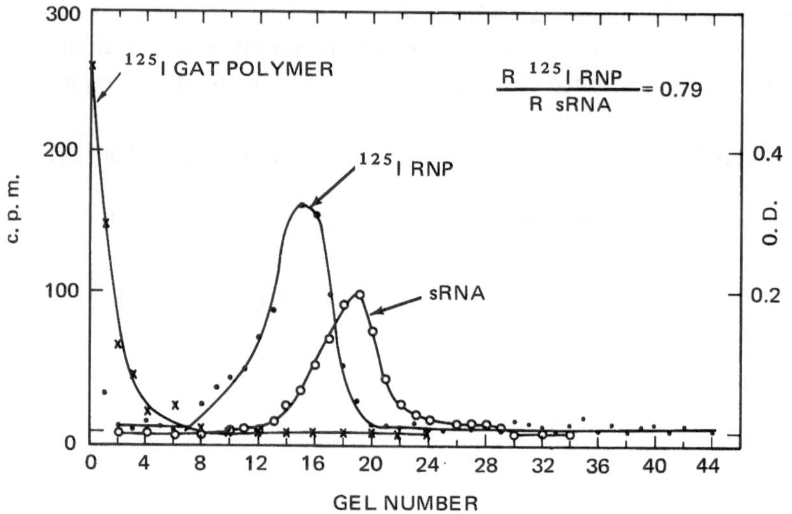

Fig. 1. *Polyacrylamide gel electrophoresis profile of* ^{125}I *RNP at pH 3.8.* Purified ^{125}I labeled RNP was subjected to electrophoresis in a 5% gel at *p*H 3.8 in the presence of 100 µg of *E. coli* sRNA. ^{125}I labeled GAT (m.w. = 50,000) was run in paralled on an identical gel. The composite result is displayed here. Migration is from left to right toward the anodic pole.

A similar experiment can be performed with another synthetic polypeptide, L-GLT. It can be again noted in Figure. 2 that saturation of RNP for L-GLT occurs and that this level differs from that of either L- or D-GAT. This indicates that the binding of different antigens to RNP exhibits considerable variation for different antigens and again supports the idea that this binding is not a matter of nonspecific ionic association between polymer and polynucleotides.

A second question is what effect does exposure of macrophage populations to one antigen have on the uptake of a second antigen by RNP. To answer this, macrophages were exposed to selected polymers [L-GLT, D-GAT, and keyhole limpet hemocyanin (KLH)] for 2 hours prior to addition of saturating levels of labeled L-GAT for another 2 hours. Bulk RNA was recovered and subjected to electrophoresis in the polyacrylamide gel system. The results of these studies are compiled in Figure 3 in which the relative specific activity of fragments of ^{125}IL-GAT in RNP are scored as a function of pretreatment of the macrophages with the various copolymers. It is clear that pretreatment of the cells with cold L-GAT results in the

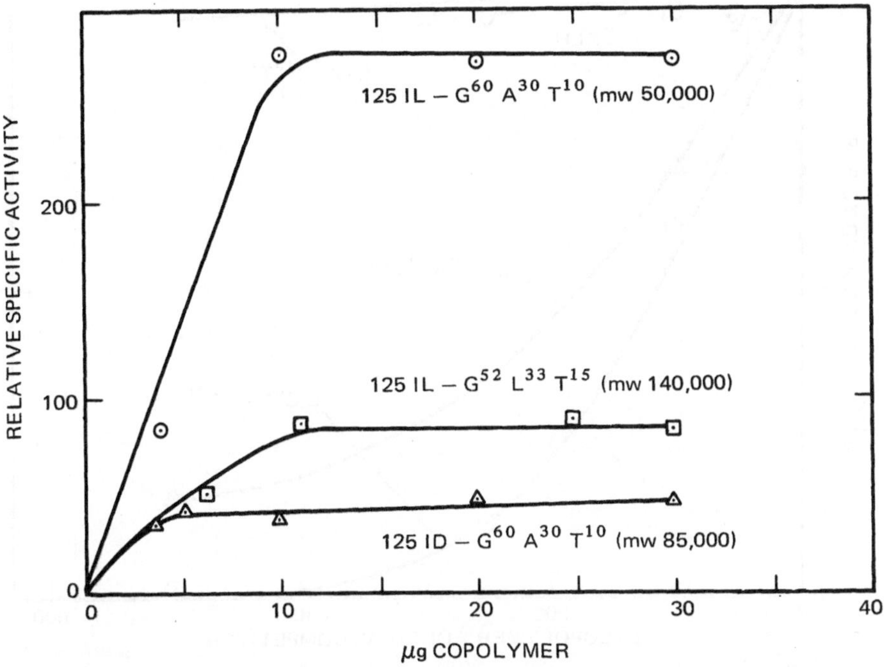

Fig. 2. *Saturation profiles of copolymer fragments on RNP.* Sets of 2×10^8 peritoneal exudate cells were incubated for 2 hours at 37° in 4.0 ml of MEM Eagle's Medium for suspension in the presence of varying amounts the copolymers indicated. RNA was recovered by conventional techniques and electrophoresed on polyacrylamide gels (*p*H 3.8) in paralled with RNP standards. The counts associated with RNP were determined. The relative specific activity was determined from the equation shown in Table 2, assuming that 5% of the total cellular RNA was RNP. The relative specific activity is displayed here as a function of copolymer dosage.

complete suppression of uptake of fragments of ^{125}IL-GAT into RNP. Rather surprisingly, the D-GAT copolymer is nearly as effective in this respect, despite the marked difference noted in the amount of D- and L-GAT bound to RNP at saturation. This indicates that there may be sufficient catabolism of the D-polymer by the macrophage to provide adequate numbers of fragments to suppress the binding of the L-form. This must mean that there is an additional mechanism in the macrophage that limits the amount of the D-form of GAT which binds to RNP. It is also of interest to note that the antigens KLH and L-GLT have little effect on the binding of L-GAT to RNP. Moreover, no dissociation of any of these labeled

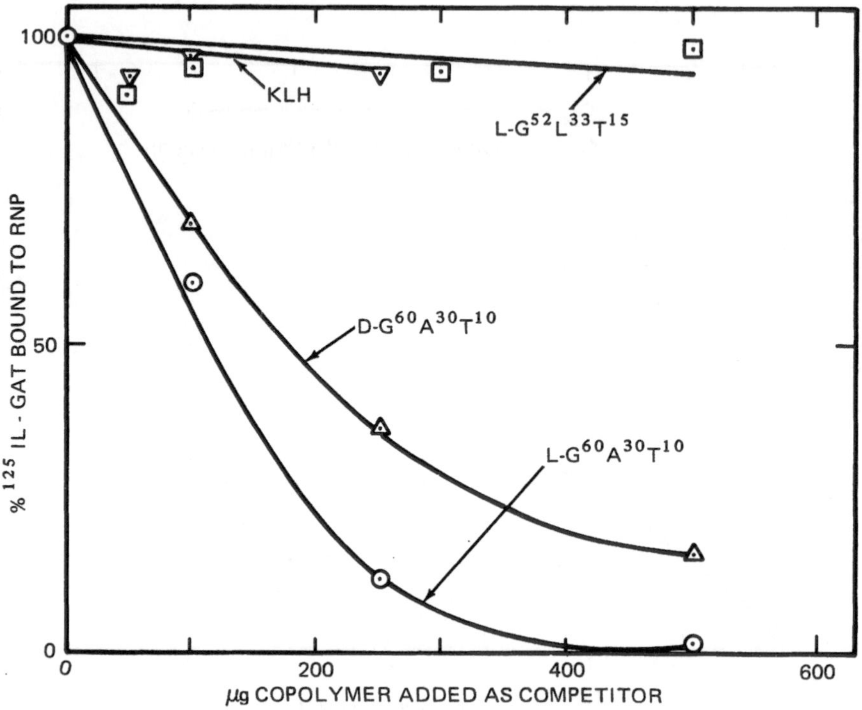

Fig. 3. *Competitive interactions with* L-GAT. Sets of 2×10^8 peritoneal exudate cells were incubated for 2 hours at 37 °C with varying dosages of L-GAT, D-GAT, L-GLT, and keyhole limpet hemocyanin (KLH) prior to addition of 15 μg of ^{125}I L-GAT to the cultures. Incubation was continued for another 2 hours, the RNA recovered and electrophoresed on polyacrylamide gels. The labeling of RNP was computed as described in Figure 2 and the degree of binding (expressed as a percentage of the noncompeted control) displayed as a function of the dosage of copolymer added as competitor.

antigens to RNP was seen if 0.01 *M* versene was included in the gel buffers. This strongly suggests that the linkage of copolymers to RNP does not involve magnesium ion. The suppression of the binding of L-GAT fragments to RNP by the D-GAT polymer can also be explained by interaction of the bulk D-GAT polymer with a site in the macrophage (possibly an enzyme) through which these fragments must pass en route to binding to RNP.

It is of interest to note that a homogenate of macrophages is nearly as efficient in linking L-GAT to RNP as is the whole cell. Homogenates of myeloma tumor cells which contain large quantities of RNA do not incorporate significant amounts of L-GAT into RNP-like species (Table 2). This again supports the idea that we are not dealing with nonspecific ionic interactions.

Table 2. Comparative Effectiveness of Linkage of GAT to RNP by Whole Macrophages, Macrophage Homogenate, and Myeloma Homogenate

	Relative specific activity[a]
2×10^8 whole cells + 10 µg ^{125}I L-GAT	278.6
TKM homogenate of 2×10^8 cells + 10 µg ^{125}I L-GAT	260.4
TKM homogenate of 2×10^8 myeloma cells + 10 µg ^{125}I L-GAT	6.8

[a] Relative specific activity $= \dfrac{\text{cpm/µg RNP}}{\text{cpm copolymer added/total no. of polymer residues}}$

2×10^8 peritoneal cells were homogenized in 3.0 ml of 0.01 *M* Tris (*p*H 7.25); 0.025 *M* KCl; 0.040 *M* $MgCl_2$ (TKM buffer) using a Dounce Homogenizer. A homogenate of 2×10^8 myeloma tumor cells was produced in a similar manner. 10 µg of ^{125}I L-GAT was added to these homogenates as well as to a population of 2×10^8 peritoneal cells in MEM Eagles for suspension medium. Incubation was carried out for 2 hours at 37 °C following which RNA was prepared and electrophoresed on polyacrylamide gels. The specific activity of RNP was computed using the equation indicated.

These observations taken in concert indicate that the association of the negatively charged synthetic antigens with RNP is not simply due to ionic interactions or to magnesium bridges between negative groups on the antigen and phosphate groups of the RNA. Such ionic mechanisms could not distinguish between D- and L-forms of GAT copolymer. This conclusion is supported by the fact that the related synthetic antigen L-GLT does not interfere with the binding of L-GAT to RNP. Certainly there are many regions of the GLT molecule that closely resemble those of GAT and yet there is no competition between these forms. Clearly, simple ionic interactions of the polymers with RNA could not account for these observations. Instead, it appears that there are complex mechanisms within the macrophage that regulate the uptake of different antigens into the RNP molecule. Although the nature of these mechanisms is still unclear, our view would be that single antigenic determinants bind to separate RNP molecules to form an RNP complex which is immunogenic for that antigen. The RNA carrier molecule appears to be the same for each antigenic determinant. We must bear in mind that there are some antigens that require complete preservation of native tertiary structure for immunogenicity. Such antigenic determinants would not survive breakdown by macrophages and would not be found in the RNP complex. However, local regions of tertiary structure at least as large as 3600 daltons can be preserved, as shown by studies described previously (Gottlieb, 1969b).

More recent studies indicate that most of the radioactivity recoverable in the macrophage after the administration of ^{125}I L-GAT is, by the criteria of chromatography on Bio-Gel columns, between 1600–3600 daltons in molecular weight. It is of interest to note that the estimated maximal size of an antigenic determinant in the RNP complex is 3600 daltons.

The mode of operation of this complex in the regulation of the immune response *in vivo* is a subject of great interest to us. As of now, we have very little information about this, but we can suggest two general mechanisms that might be involved. Since macrophages do not make antibody, it is possible that RNP operates at the level of the lymphocyte either to derepress an operon(s) for synthesis of antibody messenger RNA, or by some other control function at the level of the lymphocyte nucleus. Possibly, as Braun has suggested (1969), there are two steps involved in induction of the immune response:

(a) degradation and "processing" of antigen, possibly by linkage of the antigenic determinant to RNP, and

(b) a second contact of the lymphocyte with antigen which serves to convert the memory cell to a proliferating antibody-producing cell.

A second role for RNP might be in the control of antigen dosage to the lymphocyte, possibly by preventing the presentation of large quantities of native antigen to the lymphocyte.

Finally, we must comment on the question of binding of antigen(s) to the surface of macrophages. Although it is clear that many antigens are found on the surface of macrophages, it is by no means certain that this surface-bound antigen rather than antigen linked to RNP is the moiety which interacts with the lymphocyte *in vivo*. It may be that surface-bound antigen has no more relevance to antibody induction than do gross aggregates of antigen with the bulk RNA of macrophages.

ACKNOWLEDGMENT

The author wishes to express his appreciation to Dr. Sidney Leskowitz for helpful discussion of several aspects of this study, as well as generous supplies of L- and D-GAT.

References

Askonas, B. A. and Rhodes, J. M. (1965). Immunogenicity of antigen-containing RNA preparations from Macrophages. Nature **205:** 470–474.

Bishop, D. C., Pisciotta, A. V., and Abramoff, P. (1967). Synthesis of normal and "immunogenic" RNA in peritoneal macrophage cells. J. Immunology **99:** 751.

Braun, W. (1969). *In* Immunological Tolerance. 189–195. Academic Press, New York.

——— and Nakano, M. (1967). Antibody formation: Stimulation by polyadenylic and polycytidylic acids. Science **157:** 819.

Fishman, M. (1961). Antibody formation *in vitro*. J. Expt. Med. **114:** 837–856.

——— and Adler, F. L. (1965). Antibody formation initiated *in vitro* II. J. Exp. Med. **117:** 595–602.

Gottlieb, A. A. (1968). The antigen-RNA complex of macrophages. *In* Nucleic Acids in Immunology. Springer-Verlag, New York.

——— (1969a). Studies on the binding of soluble antigens to a unique ribonucleoprotein fraction of macrophage cells. Biochemistry **8:** 2111–2116.

——— (1969b). Macrophage ribonucleoprotein: Nature of the antigenic fragment. Science **165:** 592–594.

——— Glisin, V. R., and Doty, P. (1967). Studies on macrophage RNA involved in antibody production. Proc. Natl. Acad. Sci. **57:** 1849.

——— and Straus, D. S. (1969). Physical studies on the light density ribonucleoprotein of macrophage cells. J. Biol. Chem. **244:** 3324.

Roelants, G. E. and Goodman, J. W. (1968). Immunochemical studies on the poly γ-D-glutamyl capsule of *Bacillus anthracis* IV. Biochemistry **7:** 1432–1440.

——— (1969). The chemical nature of macrophage RNA-antigen complexes and their relevance to immune induction. J. Exp. Med. **130:** 557.

OBSERVATIONS ON THE REGULATORY EFFECTS OF THE TRANSFER RNA MINOR BASE, N^6-Δ^2-(ISOPENTENYL)ADENOSINE, ON HUMAN LYMPHOCYTES

ROBERT C. GALLO * and JACQUELINE WHANG-PENG **

Human Tumor Cell Biology Branch, National Cancer Institute
National Institutes of Health, Bethesda, Maryland

Before beginning the topic of this paper, we want to present some introductory comments on the background that led us into studies on transfer RNA and unique components of tRNA, such as isopentenyladenosine. Our laboratory has been interested in the biochemical mechanisms involved in the conversion of normal human leukocytes to leukemic cells. From recent autoradiographic and cell kinetic studies (Astaldi and Mauri, 1953; Gavasto *et al.*, 1964; Craddock and Nakai, 1962), it has become increasingly evident that human leukemia, in particular, acute leukemia, may be regarded as a disorder of the normal process of cellular differentiation rather than primarily a proliferative disorder. The evidence for this is that when leukemic blast cells of the bone marrow are compared to the blast cells of normal bone marrow, rates of DNA synthesis and mitosis are quite comparable (Astaldi and Mauri, 1953; Gavasto *et al.*, 1964; Craddock and Nakai, 1962). In fact, if anything, the normal "blast" cells may proliferate at a rate faster than the leukemic cell. How then do leukemic cells accumulate? In the normal bone marrow, 5% or less of the cells are primitive cells or so-called "blast" cells. There is a continuous process of maturation, during which the primitive cells are converted to mature leukocytes. During the progression to the granulocyte (the fully mature leukocyte), two major morphological changes occur. (1) Chromatin is converted from the template-

* Head, Section on Cellular Control Mechanisms.
** Senior Investigator.

active, loosely arranged, so-called euchromatin to the pyknotic, darkly staining, and template-inactive heterochromatin. (2) Accompanying the change in the nucleus, the cytoplasm originally staining blue progressively accumulates round, enzyme-filled structures, the lysosomes.

The structural changes in the nucleus and cytoplasm are associated with correspondingly significant changes in function. The nucleus in the blast cell and in early differentiation is genetically active, that is, acts as template for DNA and RNA synthesis, while in the fully differentiated cell the DNA does not replicate, little or no RNA is made, and the cell does not divide. The cytoplasmic functional changes are numerous. Most striking is the change in the pattern of cytoplasmic proteins. "Immature cell proteins" are predominant in the immature cells, that is, the enzymes involved in DNA and RNA synthesis and in the pathways leading to their synthesis (Gallo, 1970; Smith and Baker, 1959, 1960; Silber *et al.*, 1963; Rabinowitz, 1966). Although many of these enzymes are present in the more differentiated cells, their activities are markedly reduced. Conversely, mature leukocytes contain a class of proteins present in very small amounts (or not at all) in the immature cells. These are the lysosomal enzymes, the hydrolytic enzymes that act to degrade ingested foreign material, that is, to carry out the major function (phagocytosis) of the mature granulocyte (Valentine, 1960; Xefteris *et al.*, 1961), and miscellaneous enzymes, including enzymes involved in thymidine metabolism (Gallo, 1970; Gallo and Perrt, 1968), and in carbohydrate metabolism (Rabinowitz, 1966), and which we would like to refer to as "mature cell protein".

In the acute leukemias this process is impaired. The human acute leukemic cells contain enzymes in quantity and type that appear to parallel that of normal *immature* cells (Gallo, 1970; Rabinowitz, 1966). They retain the capacity to replicate DNA and to proliferate. This failure to differentiate usually leads to a slow but progressive increase in the pool of these cells and the clinical disorder we see as acute leukemia.

To account for the difference in quantity in a large number of enzymes between the normal *immature* cell (or the leukemic cell) versus the fully differentiated normal leukocyte, we have felt that control of protein synthesis at the translational level, specifically through tRNA, would be one likely mechanism. Changes in tRNA during differentiation could theoretically alter the type and particularly the quantity of many proteins present in the cell. A failure to develop this change or the synthesis of nonfunctional or misfunctional tRNA's might result in the inability to synthesize the full complement of mature cell protein. One such protein(s) might be repressor protein. In turn, deficient quantity of repressor would lead to the continued ability of the cell to synthesize DNA and to divide, for example, the

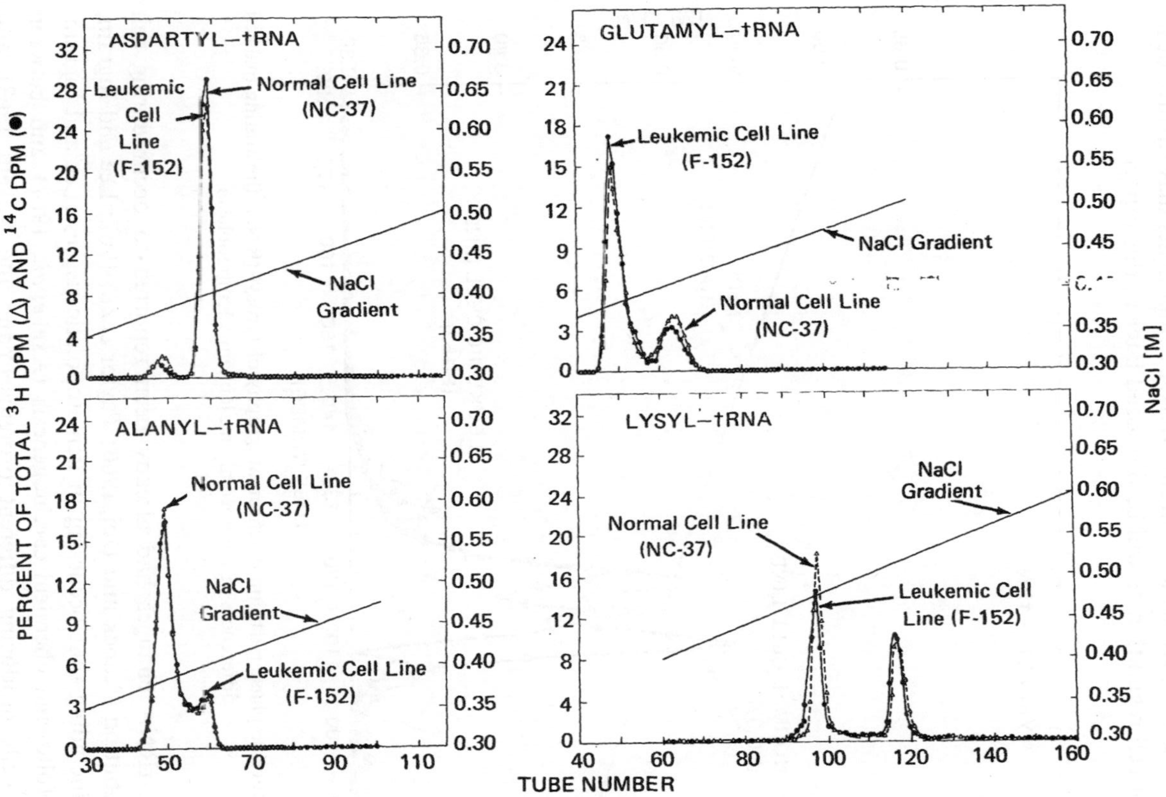

Fig. 1. Reverse phase partition co-chromatography (freon column) of 4 aminoacyl-tRNAs with no significant differences between normal human lymphoblasts (NC-37) and leukemic human lymphoblasts (F-152). Columns were run at 23 °C. Flow rates were 1.5 ml/min with a 2 L linear NaCl gradient (0.30–0.65 *M*) containing 0.01 *M* Na-acetate, *p*H 4.5; 0.01 *M* $MgCl_2$; and 0.001 *M* EDTA.

undifferentiated cell. In fact, there is already some evidence that normal mature leukocytes contain repressor proteins which are absent or deficient in leukemic cells (Paran *et al.*, 1969). In addition to their central role in protein synthesis, tRNA's or specific tRNA components may involve other levels of cell regulation as will be discussed later in this report.

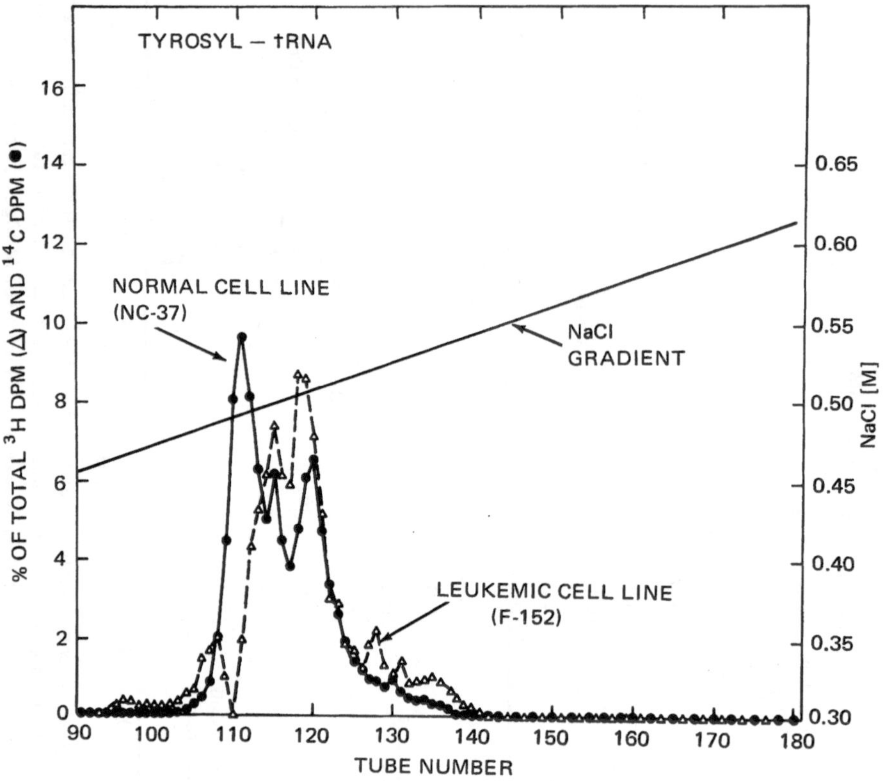

Fig. 2. Reverse phase partition co-chromatography of tyrosyl-tRNA, illustrating marked differences in the normal and leukemic lymphoblasts.

Since tRNA changes had already been reported to occur during cell differentiation (Kanek and Doi, 1966; Taylor *et al.*, 1967; Lee and Ingram, 1967), including that occurring in hematopoietic tissues (Lee and Ingram, 1967), following chemical carcinogenesis (Axel *et al.*, 1967) and between similar cells producing different proteins (Yang and Novelli, 1968), and since techniques are available for careful analysis of tRNA (Weiss and Kelmers, 1967), we have initiated a systematic investigation of the tRNA's in leukocyte differentiation and following neoplastic transformation (Gallo,

1969; Gallo and Pestka, 1970). Recently we completed a chromatographic analysis of the tRNA's for all 20 amino acids in normal and leukemic human lymphoblasts (Gallo and Pestka, 1970). Although these cells were identical by light and electron microscopy, had identical generation times, and were grown under identical conditions, a few significant differences

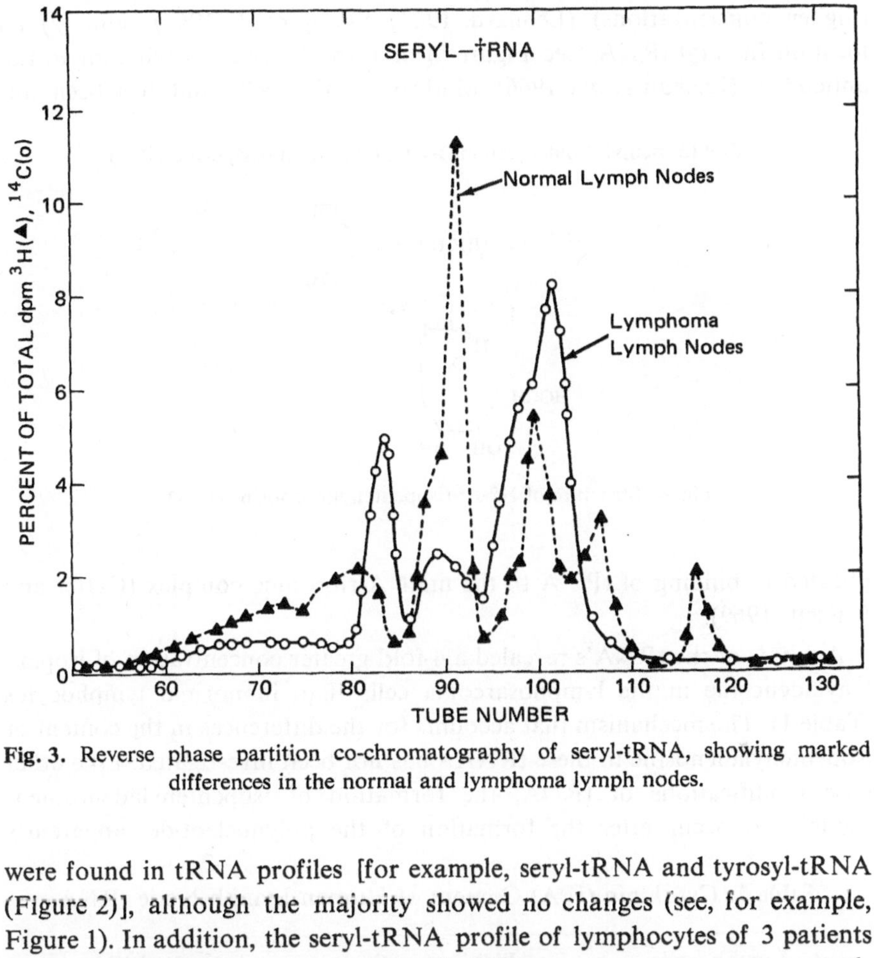

Fig. 3. Reverse phase partition co-chromatography of seryl-tRNA, showing marked differences in the normal and lymphoma lymph nodes.

were found in tRNA profiles [for example, seryl-tRNA and tyrosyl-tRNA (Figure 2)], although the majority showed no changes (see, for example, Figure 1). In addition, the seryl-tRNA profile of lymphocytes of 3 patients with lymphosarcoma were markedly different from normal controls (Figure 3). In contrast, the seryl-tRNA profile of a patient with Burkitt's lymphoma was identical to the normal (Goldstein, Skoog, and Gallo, unpublished results).

A tRNA difference that can account for differences in tRNA elution profiles between tumor and normal cells is variation in minor base components, for example, through methylation or other alkyl side group modification that result in "minor base" formation. Both tyrosyl-tRNA and seryl-tRNA have been shown to contain the minor base, N^6-Δ^2-(isopentenyl)adenosine (Figure 4 and 5; Zachan *et al.*, 1966; Madison *et al.*, 1967). Of the tRNA minor bases, isopentenyladenosine is of unusual interest because: (1) it is a known potent plant cytokinin, that is, it can stimulate plant cell division (lower concentrations) or differentiation (higher concentrations) (Leonard, 1964; Skoog *et al.*, 1967); and (2) its location in seryl-tRNA (see Figure 5) or tyrosyl-tRNA is adjacent to the anticodon (Zachan *et al.*, 1966; Madison *et al.*, 1967) and has been im-

6-N-(3-methyl-2-butenylamino)-9-β-D ribofuranosylpurine (IPA)

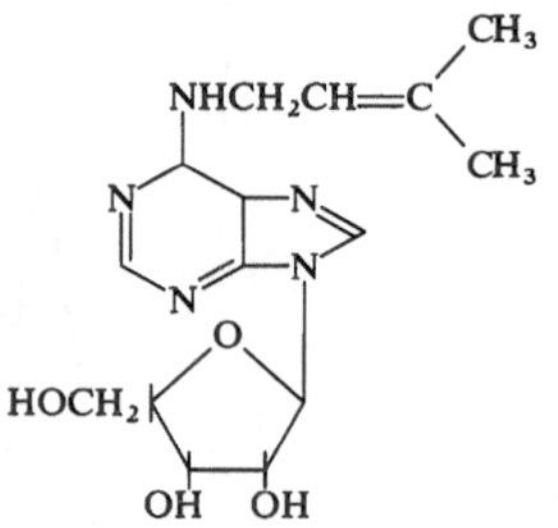

Fig. 4. Structure of N^6-Δ^2-(isopentenyl)adenosine (IPA).

plicated in binding of tRNA to the mRNA-ribosome complex (Gefter and Russell, 1969).

Analysis of the tRNA's revealed a 4-fold greater concentration of isopentenyladenosine in the lymphosarcoma cells than normal lymphocytes (Table 1). The mechanism that accounts for the differences in the content of isopentenyladenosine in these tRNA's has not been investigated. Like other base modifications of tRNA, the formation of isopentenyladenosine is thought to occur after the formation of the polynucleotide, apparently

Table 1. Cytokinin (IPA) Content of Human Lymph Node tRNA

Lymph node	Maximum stimulus to growth of tobacco callus[a]	Estimated IPA (μg/mg tRNA)
Normal	2-fold	0.075
Lymphosarcoma	8-fold	0.30

[a] tRNA concentrations tested varied over a range of 0.32–2.6 mg/l. The assays were performed by Dr. F. Skoog, University of Wisconsin.

derived from mevalonic acid (Peterkofsky, 1968; Fittler *et al.*, 1968). Fittler, Kline, and Hall have found an enzyme that catalyzes the insertion of the isopentenyl side group onto the adenosine moiety of tRNA (Peterkofsky, 1968). Among other possibilities, the greater amount of isopentenyladenosine

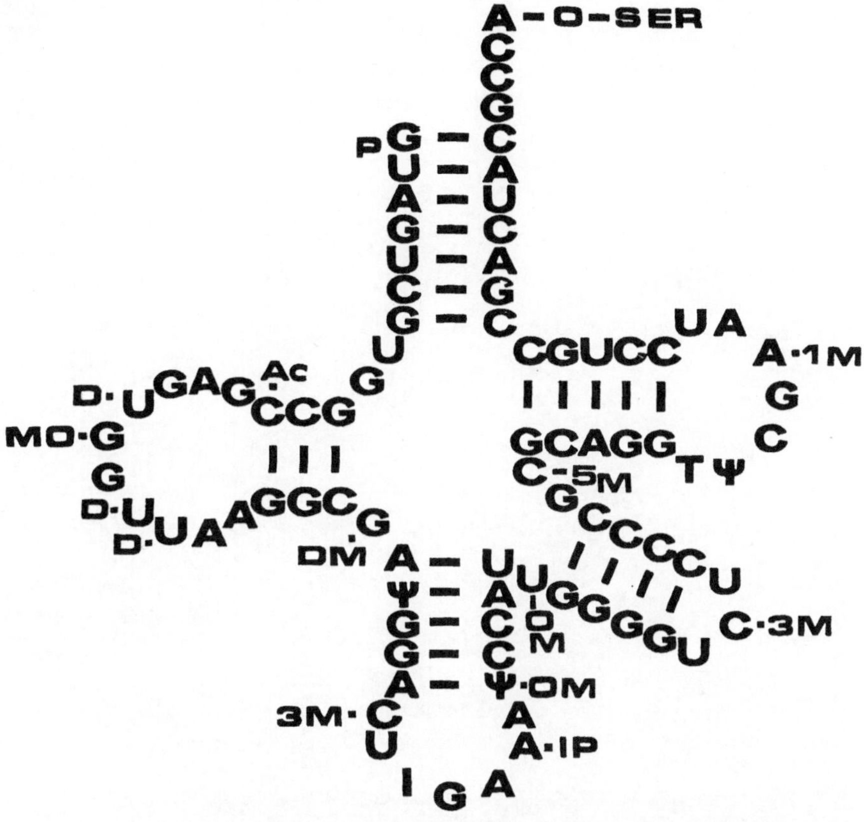

Fig. 5. Cloverleaf model of rat liver seryl-tRNA. Note isopentenyladenosine adjacent to the 3′ end of the IGA anticodon. The figure was taken from Staehelin *et al.*, Nature **219:** 1364 (1968) and reproduced here with the permission of the editors and the authors.

in the lymphoma tRNA could be due to: (1) a higher proportion of tRNA's containing isopentenyladenosine per total tRNA; (2) greater pools of isopentenyl precursors; (3) increased activity of the Fittler-Kline-Hall enzyme; and (4) the presence of an inhibitor of the cytokinin assay in the normal tRNA preparations.

A study of the biological effects of isopentenyladenosine and of tRNA's containing isopentenyladenosine has been initiated. We have shown that tRNA's can be taken up by mammalian cells (Figure 6; Herrera *et al.*, 1969; Herrera *et al.*, submitted for publication).

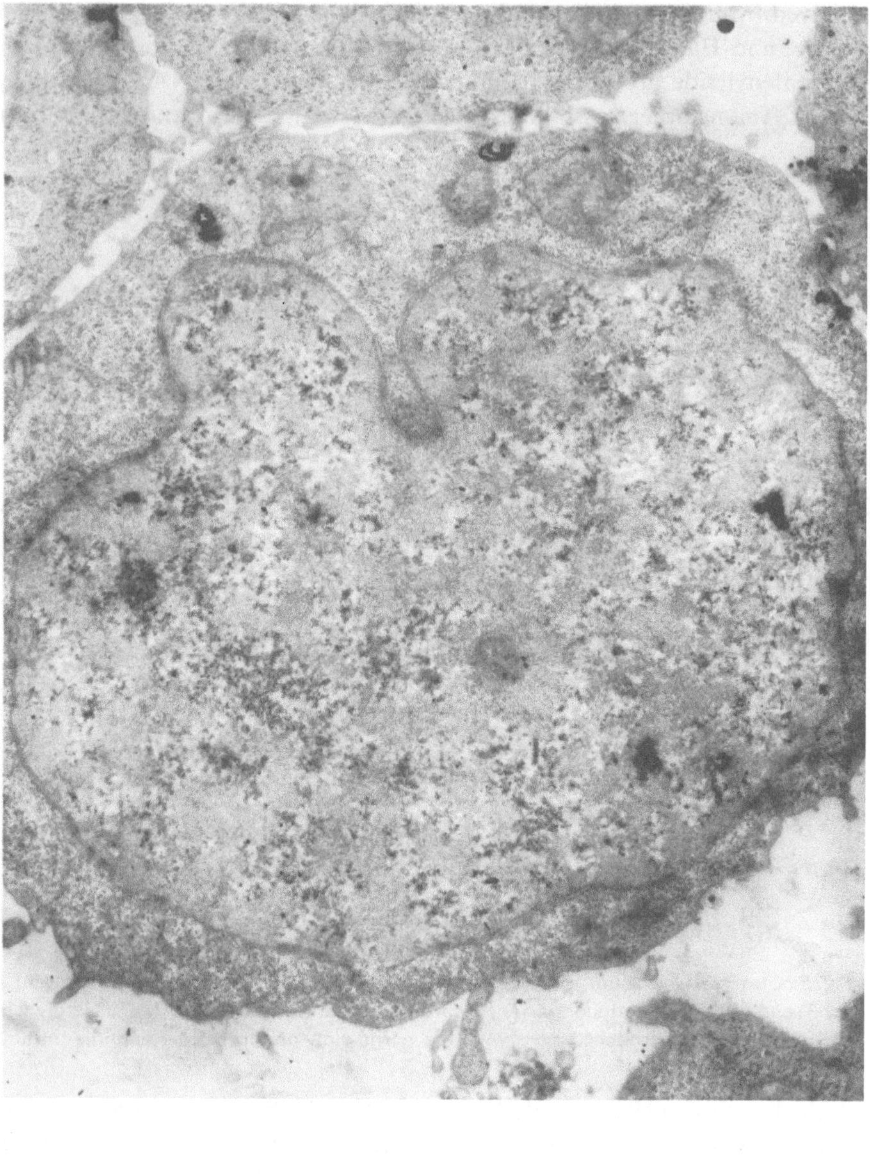

Fig. 6. Electron microscopic radioautograph of an L1210 mouse leukemic cell previously incubated with *E. coli* [^{14}C]-tRNA. The intracellular grains are *E. coli* tRNA, as subsequently verified by reisolation of the tRNA and distinguishing *E. coli* tRNA from L1210 tRNA by carrying out reactions with purified *E. coli* aminoacyl-tRNA synthetases and L1210 tRNA methylases.

A more detailed analysis of the effects of isopentenyladenosine itself was undertaken and forms the major topic of this paper. The cells utilized in this study were short-term cultures of human lymphocytes isolated from the peripheral blood of healthy donors, purified by nylon column chromatography (Greenwalt *et al.*, 1962), and partially synchronized with the mitogen,

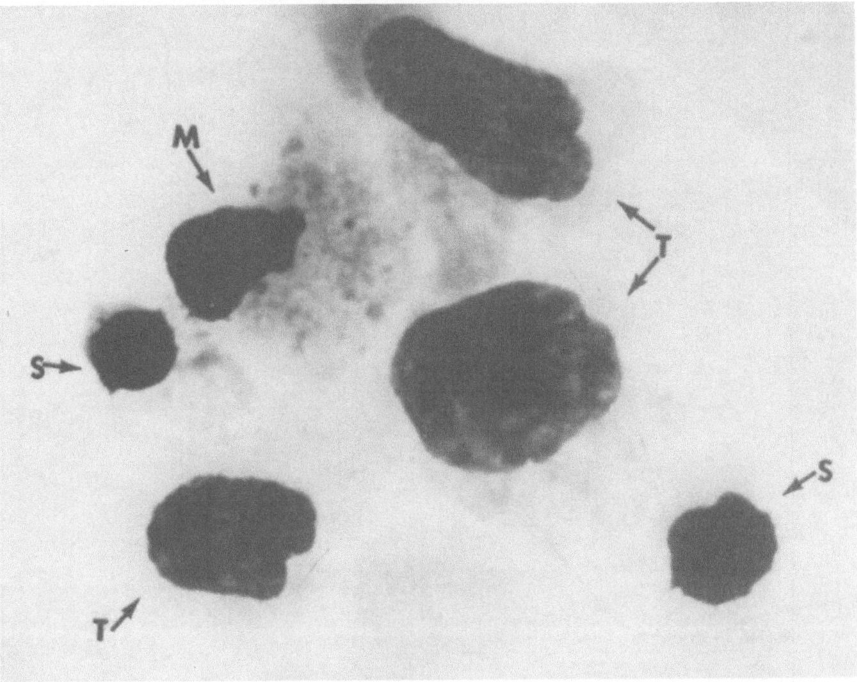

Fig. 7. Light microscopy photograph of leukocytes incubated with phytohemmagglutinin illustrating the spectrum of changes found in stimulated lymphocytes. S, small lymphocyte (morphology of a non-PHA stimulated cell); T, morphologically transformed lymphocyte (blast cell); M, macrophage. Taken from Lewis *et al.*, Amer. J. Obstet. Gynec. **96:** 287 (1966), and reproduced here with the permission of the editors and authors.

phytohemagglutinin (PHA). Transformation of small lymphocytes to blast-like cells by PHA is of interest as a model for immunological questions as well as for studies on the mechanism(s) of conversion of a "G_1" cell to a cell "in cycle" (Figure 7). The major events following addition of PHA to fresh lymphocytes appear to be an initial interaction at the cell membrane followed by gene derepression, macromolecular synthesis, morphological transformation, and mitosis (Figure 8).

Isopentenyladenosine (10^{-5}–10^{-6}M) added early in the cell cycle is a potent inhibitor of human lymphocyte DNA synthesis, transformation, and mitosis induced by PHA (Figure 9 and 10; Gallo *et al.*, 1969a, 1969b). Lower concentrations (10^{-7}–10^{-6}M) added late in the cell cycle enhance DNA synthesis and transformation of human lymphocytes (Figure 9 and 10; Gallo *et al.*, 1969a, 1969b). These observations are in some respects anal-

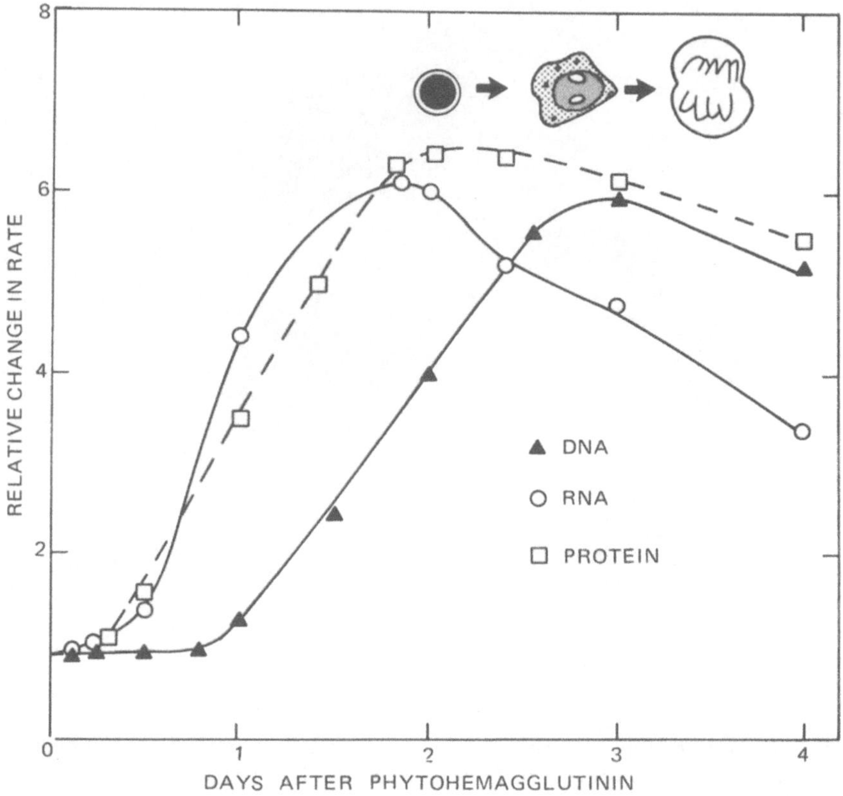

Fig. 8. Relative change in rates of DNA, RNA, and protein synthesis in PHA stimulated lymphocytes. Lymphocytes were incubated with PHA, pulsed for 4 hours with labeled precursors at varying intervals, and harvested at the indicated times. The radioactivity in the isolated DNA, RNA, and protein was measured and the change in specific activity (cpm/mg macromolecule) determined. Top of figure schematically illustrates the transformation and mitosis of stimulated cells.

ogous to cytokinin activity in plants, that is, lower concentrations increase mitosis while higher concentrations result in differentiation. The inhibitory effect was also found in rat spleen lymphocytes (Hacker, 1969). The stimulatory effect of isopentenyl-adenosine may be directly on DNA synthesis since no stimulation occurs when the agent is added prior to the cells entering S phase of the cell cycle. The same magnitude of stimulation

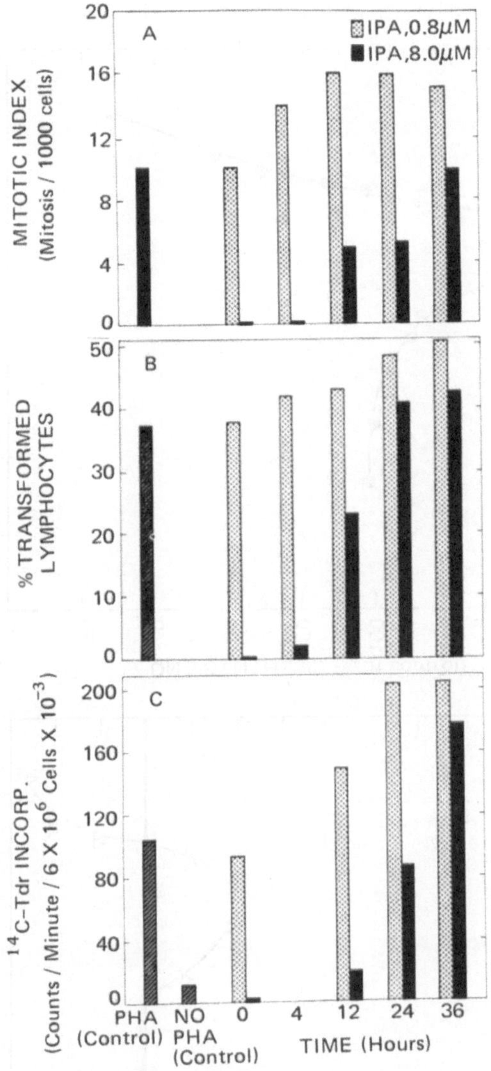

Fig. 9. Lymphocyte DNA synthesis (c), transformation (b), and mitosis (a) following stimulation with PHA: time course of inhibitory and stimulatory effects of isopentenyladenosine. PHA was added at the start of incubation, and the cells were harvested at 48 hours. Isopentenyladenosine (8 μM, solid bars or 0.8 μM, stippled bars) was added at the indicated times after PHA. The striped bars are control sample (with and without PHA) receiving no isopentenyladenosine. For measurements of DNA synthesis thymidine-2-C^{14} was added at 24 hours. Note the inhibitory effects of 8 μM IPA added early in the course of incubation which disappear when added later in the cycle. Note stimulatory effects of 0.8 μM added late in incubation. The figure was reproduced from Gallo *et al.*, Science **165:** 400 (1969), with the permission of the editors.

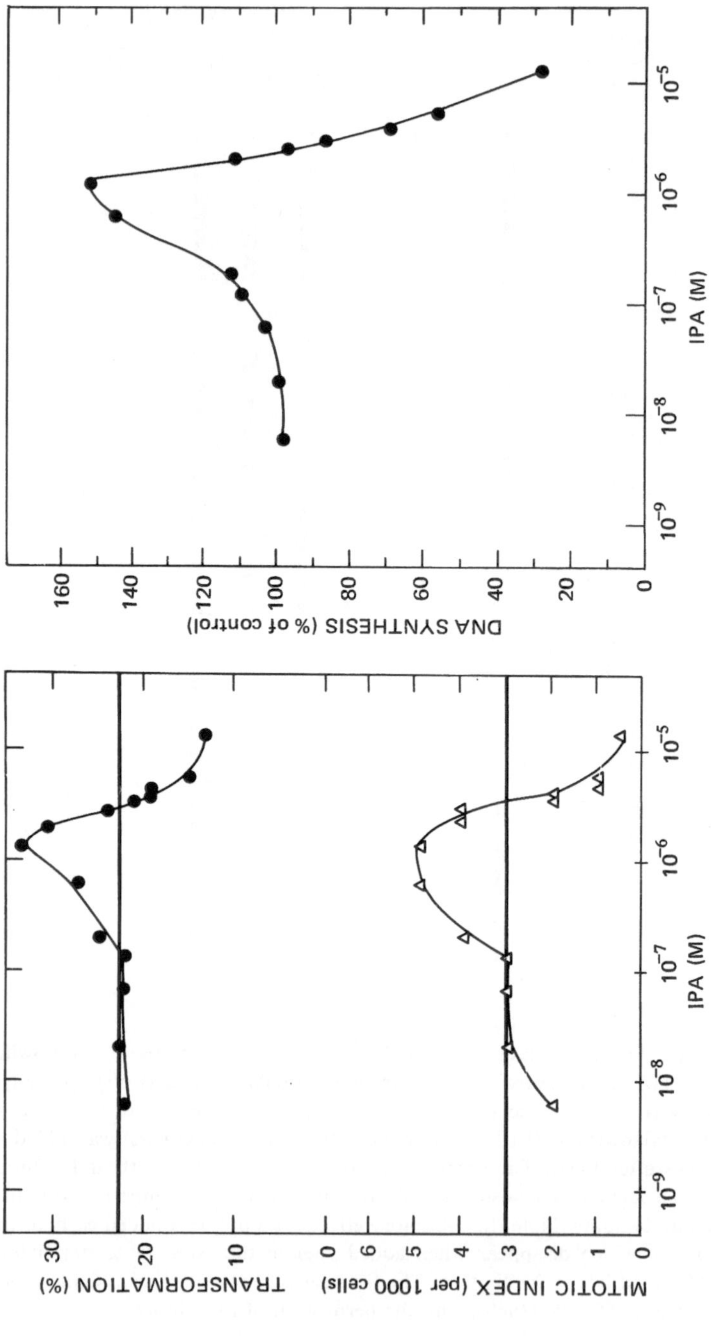

Fig. 10. Effects of varying concentrations of isopentenyladenosine on lymphocyte DNA synthesis (lower panel), transformation, and mitosis (upper panel). Isopentenyladenosine was added 24 hours after PHA. Cells were harvested at 48 hours after PHA. The figure was reproduced from Gallo *et al.*, Science **165:** 400 (1969), with the permission of the editors.

was observed at all concentrations of PHA (Burroughs Wellcome) from 0.25 mg to 1.0 mg per culture bottle (6 ml). Above this concentration, PHA itself causes cell death. Further analysis of the changes produced by isopentenyladenosine will be limited to its inhibitory effect.

Analysis of the Inhibitory Effects of Isopentenyladenosine on Transformation and Mitosis of the PHA Stimulated Human Lymphocyte.

(1) *The primary mechanism for the inhibitory effects of isopentenyl adenosine is not on DNA or protein synthesis.* Although DNA synthesis is markedly inhibited under the conditions described above, it was clear that this could

Table 2. Effect of IPA on Binding of Seryl-tRNA to Ribosome-Messenger RNA Complex[a]

System	Seryl-tRNA rabbit liver		Seryl-tRNA human liver		Seryl-tRNA *E. coli*	
	cpm	μμmoles bound	cpm	μμmoles bound	cpm	μμmoles bound
Complete	806	3.98	182	0.90	472	2.33
Complete-poly UC	118	0.58	64	0.32	151	0.75
Complete-poly UC + IPA 1×10^{-6} M	123	0.61	66	0.33	159	0.79
Complete-poly UC + IPA 5×10^{-4} M	116	0.57	67	0.33	150	0.74
Complete + IPA 1×10^{-6} M	829	4.09	176	0.87	497	2.45
Complete + IPA 1×10^{-5} M	807	3.98	170	0.84	475	2.35
Complete + IPA 1×10^{-4} M	759	3.75	185	0.91	468	2.31
Complete + IPA 5×10^{-4} M	785	3.88	187	0.92	423	2.09

[a] The reaction mixture contained a final volume of 0.05 ml: 0.02 *M* MgAc; 0.05 *M* KAc; 0.05 Tris-Ac, *p*H 7.2; 3.05 A_{260} units of washed *E. coli* ribosomes; 0.21 A_{260} units of poly UC; 0.50 A_{260} units of seryl-tRNA; rabbit seryl-tRNA 2506 cpm (12.38 μμm); human seryl-tRNA (2.54 μμm); and *E. coli* seryl-tRNA (8.03 μμm). The specific activity (cpm/A_{260}) of each seryl-tRNA was as follows: rabbit seryl-tRNA, 8130; human seryl-tRNA, 2962; and *E. coli* seryl-tRNA, 3500. Reaction mixtures were incubated for 20 minutes at 24° and then poured into a Millipore filter individually. Taken from Nirenberg and Leder (1964). We are very grateful to Dr. S. Pestka, National Cancer Institute, NIH, for his suggestions on the binding and for generously providing ribosomes and *E. coli* seryl-tRNA.

not be the primary site of action. This was true for a number of reasons but most clearly from the fact that the *onset* of DNA synthesis in these cells (under our conditions) is not until 24–30 hours after the addition of PHA and start of incubation. The inhibitory effects of isopentenyladenosine on DNA synthesis, transformation, or mitosis did not occur when added at

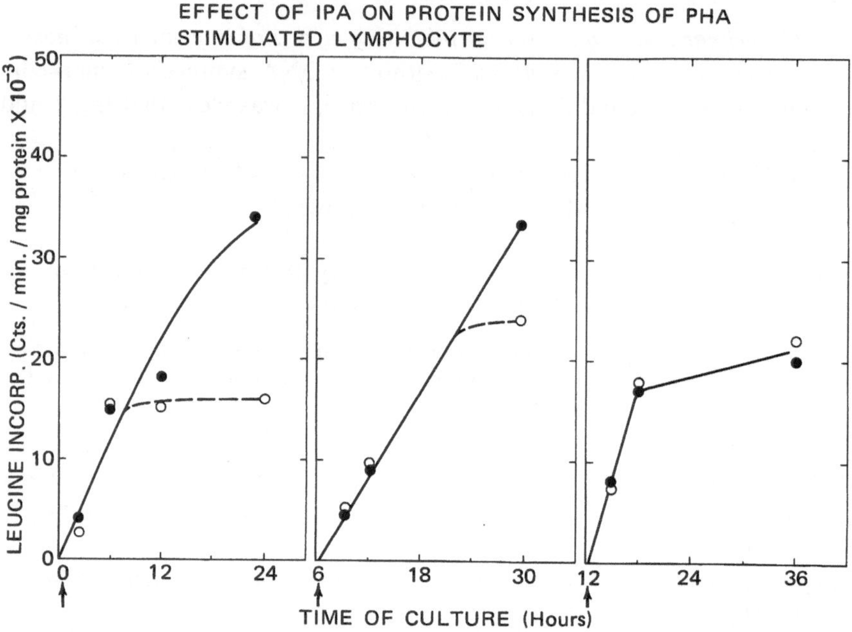

Fig. 11. Effect of isopentenyladenosine on protein synthesis of PHA stimulated lymphocyte. PHA was added at the start of incubation. Isopentenyladenosine was added at either the start of incubation (left panel), after 6 hours (middle panel), or after 12 hours (right panel). In each case [^{14}C]-leucine (0.5 μc/ml) was added with the isopentenyladenosine. Samples were harvested at the indicated times and protein synthesis determined by measuring the incorporation of the leucine into hot TCA insoluble material. •—•, no isopentenyladenosine; o—o, 3 μM isopentenyladenosine.

that period or thereafter. Inhibition only occurs when the IPA is added just prior to PHA, with the PHA at the start of incubation, or within the first 10–20 hours of culture, that is, when the cells are in G_0 or G_1. The effect is not one of general toxicity since there were no chromosomal abberations, changes in trypan blue staining, or visible structure changes by light or electron microscopic examination.

Since isopentenyladenosine is located adjacent to the anticodon of some tRNA's and is thought to be involved in binding of these tRNA's to the mRNA-ribosome complex, we felt that the most likely mechanism of action of exogenously added isopentenyladenosine was inhibition of protein synthesis by competing with a ribosomal site for the isopentenyl moiety of tRNA. However, this was not verified. As shown in Table 2, there was no inhibition of seryl-tRNA binding by isopentenyladenosine. Nonetheless, as

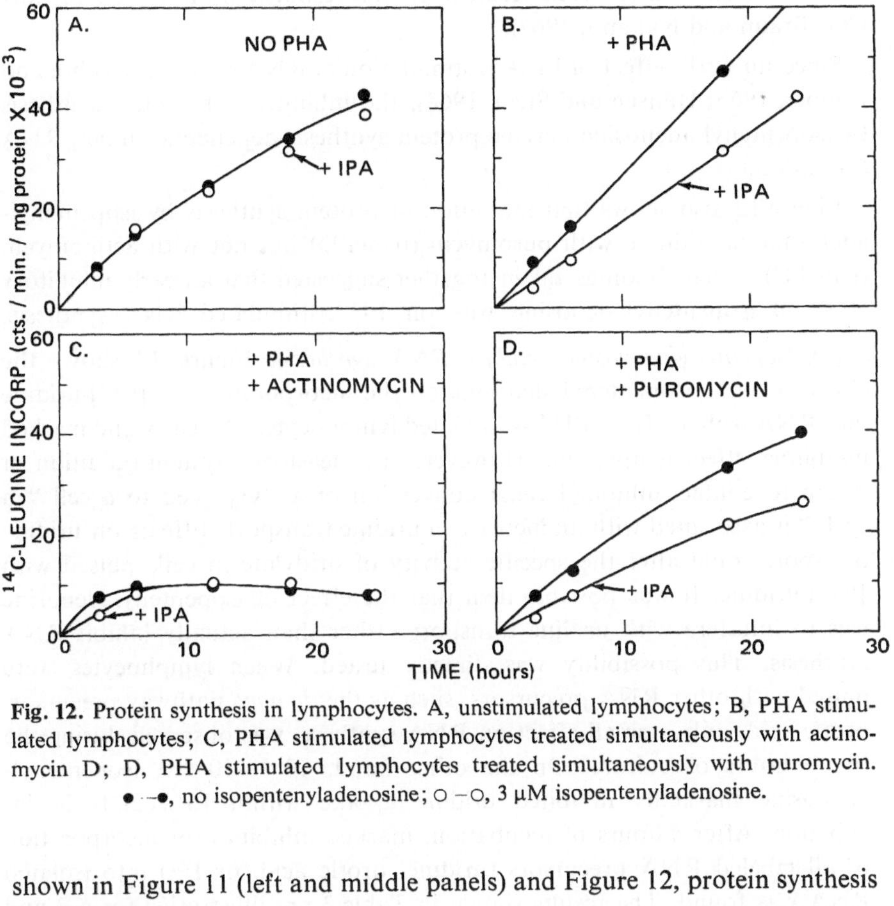

Fig. 12. Protein synthesis in lymphocytes. A, unstimulated lymphocytes; B, PHA stimulated lymphocytes; C, PHA stimulated lymphocytes treated simultaneously with actinomycin D; D, PHA stimulated lymphocytes treated simultaneously with puromycin. •–•, no isopentenyladenosine; ○–○, 3 μM isopentenyladenosine.

shown in Figure 11 (left and middle panels) and Figure 12, protein synthesis is inhibited.

It is also evident (Figure 11, right panel) that when the isopentenyl adenosine is added after 12 hours of incubation there is no inhibition of protein synthesis. These findings indicate that inhibition of protein synthesis is not the primary effect. This is also demonstrated by the findings discussed below. Figure 12A shows that isopentenyladenosine has no effect on the *basal* level of protein synthesis, that is, human lymphocytes incubated in the absence of PHA. However, the increase in protein synthesis brought about by PHA is completely inhibited. This is shown in Figure 12B, where in the presence of 3 μM isopentenyladenosine the rate of protein synthesis is reduced back to the level of the non-PHA stimulated cells. These findings are reminiscent of the report by Braun and Nakano in mouse spleen cells. They showed that isopentenyladenosine had no effect on the normal response of these cells to immunization with sheep red cells. However, it markedly inhibited the enhancement of this response by poly AU or poly CG (Braun and Nakano, 1967).

Since an early effect of PHA is stimulation of RNA synthesis (Rubin and Cooper, 1965; Hansen and Stein, 1968), the inhibition of protein synthesis by isopentenyl-adenosine may be protein synthesis dependent on new RNA formation.

Figure 12 also shows that inhibition of protein synthesis by isopentenyladenosine is additive with puromycin (panel D) but not with actinomycin (panel C). These findings taken together suggested that an early inhibitory effect of isopentenyladenosine was on PHA-stimulated RNA synthesis.

(2) *Isopentenyladenosine inhibits RNA synthesis.* Figure 13 shows the effect of 3 μM isopentenyladenosine on the incorporation of [^{14}C]-uridine into RNA isolated from PHA-stimulated lymphocytes. An early and marked inhibitory effect is apparent. However, in at least one system (addition of serum to contact-inhibited cells) conversion of a "G_0" cell to a cell "in cycle" is associated with an increase in uridine transport. Effects on uridine transport could alter the specific activity of uridylate in cells pulsed with [^{14}C]-uridine. It was possible then that the effect of isopentenyladenosine was to interfere with uridine transport rather than actually inhibit RNA synthesis. This possibility was directly tested. When lymphocytes were pulsed with other RNA precursors, such as the *de novo* pathway precursor, orotic acid, only a small effect on RNA synthesis was observed during the first 4 hours of culture. On the other hand, 10^{-5}—10^{-6} M isopentenyladenosine markedly inhibited uridine uptake within 15 seconds of incubation. After 4 hours of incubation, marked inhibition of incorporation of all labeled RNA precursors (uridine, orotic acid, or P^{32}) into isolated RNA was found. The results shown in Table 3 are illustrative for a 2 and 16-hour incubation. Thus, isopentenyladenosine does inhibit RNA synthesis, but the magnitude of the effect is not as pronounced early in the course of incubation as suggested by the data in Figure 13. With uridine as the labeled precursor then, the effects are apparently complicated by an additional inhibition of uridine uptake.

(3) *Inhibition of RNA synthesis is not through direct inhibition of uridylate formation.* PHA has been reported to induce uridine kinase activity, and the

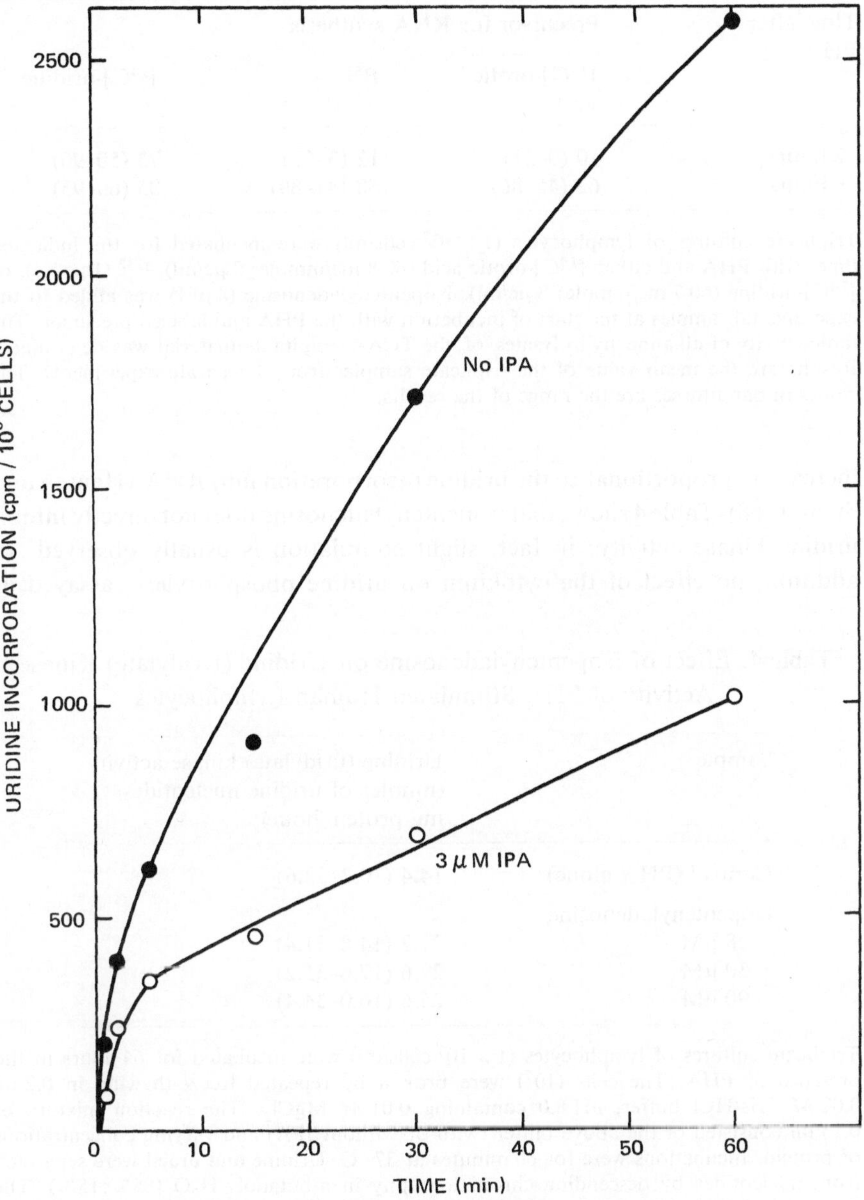

Fig. 13. Lymphocyte RNA synthesis in the first hour of stimulation with PHA.

Table 3. Percent Inhibition by 8 μM Isopentenyladenosine of *de novo* and Salvage Pathways for RNA Synthesis of the PHA Stimulated Human Lymphocyte

Time after PHA	Precursor for RNA synthesis		
	[^{14}C]-orotic acid	P^{32}	[^{14}C]-uridine
2 hours	10 (0–21)	12 (3–22)	73 (50–90)
16 hours	62 (45–86)	58 (41–89)	85 (60–95)

Triplicate cultures of lymphocytes (1×10^7 cells/ml) were incubated for the indicated times with PHA and either [^{14}C]-orotic acid (60.8 mc/mmole; 3 μc/ml), P^{32} (1 μc/ml), or [^{14}C]-uridine (60.7 mc/mmole; 3 μc/ml). Isopentenyladenosine (8 μM) was added to the experimental samples at the start of incubation with the PHA and labeled precursor. The radioactivity of alkaline hydrolysates of the TCA-precipitable material was determined. Results are the mean value of the triplicate samples from 3 separate experiments. The values in parentheses are the range of the results.

increase is proportional to the uridine incorporation into RNA (Hansen and Stein, 1968). Table 4 shows that isopentenyladenosine does not directly inhibit uridine kinase activity; in fact, slight stimulation is usually observed. In addition, no effect of the cytokinin on uridine phosphorylase, assayed as

Table 4. Effect of Isopentenyladenosine on Uridine (Uridylate) Kinase Activity of PHA-Stimulated Human Lymphocytes

Sample	Uridine (uridylate) kinase activity (nmoles of uridine nucleotides/ mg protein/hour)
Control (PHA alone)	14.4 (10.2–23.6)
Isopentenyladenosine	
8 μM	21.2 (14.8–31.4)
30 μM	27.6 (17.6–35.2)
90 μM	25.6 (16.0–34.4)

Triplicate cultures of lymphocytes (1×10^7 cells/ml) were incubated for 64 hours in the presence of PHA. The cells (10^7) were broken by repeated freeze-thawing in 0.2 ml 0.02 *M* Tris-HCl buffer, *p*H 8.0 containing 0.01 *M* $MgCl_2$. The reaction mixture of 0.15 ml consisted of the above buffer (with or without IPA) and varying concentrations of protein. Incubations were for 60 minutes at 37 °C. Uridine and uracil were separated from nucleotides by descending chromatography in *n*-butanol: H_2O (85%:15%). The procedure was similar to that reported by Hausen and Stein (1968). The values are the mean of the triplicates from two experiments. The values in parentheses are the range of the results.

described previously (Gallo and Perry, 1969), was found. On the other hand, we found that it does inhibit the *synthesis* of uridine kinase; that is, in control and in PHA-stimulated cells, overnight incubation in the presence of 8 μM isopentenyladenosine resulted in a marked reduction in total enzyme activity (Table 5). The assays were carried out after extensively washing the cells to remove the isopentenyladenosine. Although this reduction in the level of the enzyme might contribute in part to the diminished incorporation of uridine into RNA, it cannot account for the reduction in orotic acid incorporation nor for the reduced incorporation of all labeled precursors (orotic acid, P^{32}, and uridine) after intervals as short as 4–6 hours. We conclude that there is a true inhibition of RNA synthesis, a reduction in the *synthesis* of uridine kinase activity, and an apparent effect on uridine transport. The inhibition of uridine kinase biosynthesis must involve the mechanism of gene derepression for this enzyme, the same mechanism which is stimulated by PHA.

Table 5. Inhibition of the Biosynthesis of Uridine (Uridylate) Kinase by Isopentenyladenosine

Sample	Uridine (uridylate) kinase activity (nmoles of uridine nucleotides/mg protein/hour)
(a) Control (no PHA)	1.55
(b) + PHA	15.3
(c) PHA + isopentenyl-adenosine (8 μM)	0.28

Triplicate culture of lymphocytes were preincubated 16 hours, PHA was then added to flask (b), and PHA + 8 μM isopentenyladenosine to flask C. Cells were harvested 24 hours later, washed to remove the isopentenyladenosine, and assayed for the levels of uridine (uridylate) kinase as described in Table 4.

(4) *The inhibition of RNA synthesis is not due to an increase in RNA degradation.* Lymphocytes were incubated for 16 hours in the presence of PHA and [^{14}C]-uridine. The cells were washed with 20 ml of media, containing 1 mM cold uridine, 3 times, divided into 2 flasks containing the usual media but with 1 mM uridine and actinomycin (10 μg/ml). In addition, 1 flask contained 8 μM isopentenyladenosine. The cells were then incubated again, aliquots were taken at intervals, and the radioactivity in RNA determined. As shown in Figure 14, the rates of RNA degradation in the isopentenyladenosine-treated and control flasks were identical.

(5) *Inhibition of RNA synthesis includes all species separable by sucrose gradients.* The inhibitory effects of isopentenyladenosine on lymphocyte RNA synthesis were investigated in the presence of PHA (Figure 15A) and in the absence of PHA (Figure 15B). The RNA was isolated and

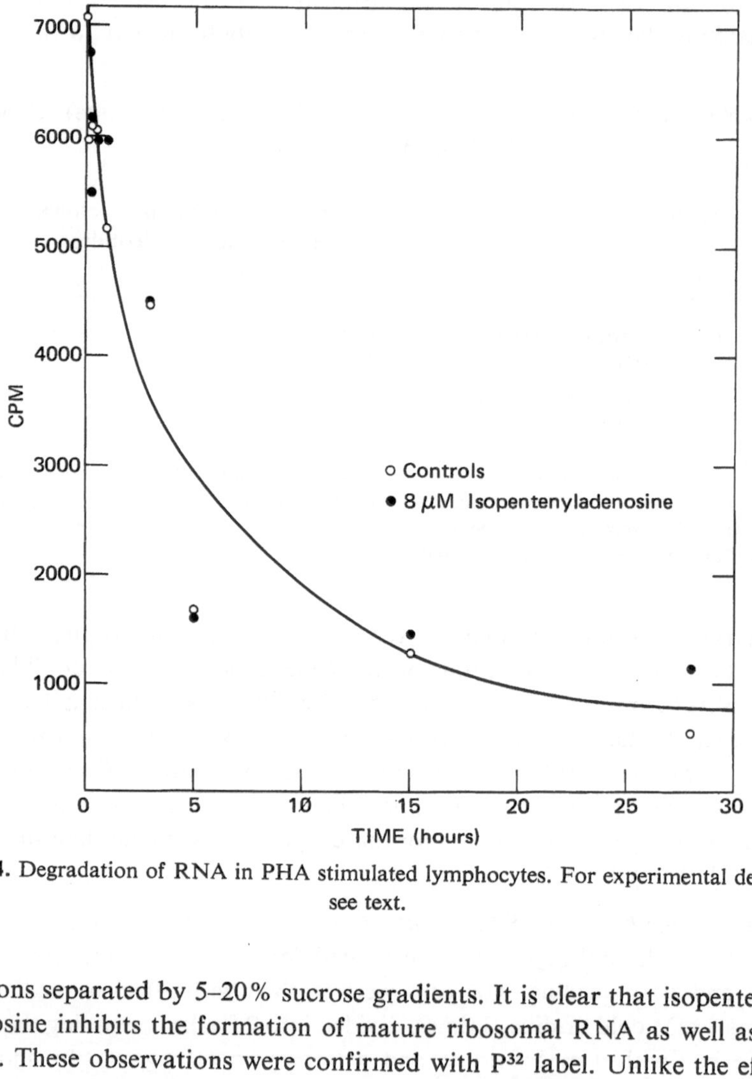

Fig. 14. Degradation of RNA in PHA stimulated lymphocytes. For experimental details, see text.

fractions separated by 5–20% sucrose gradients. It is clear that isopentenyladenosine inhibits the formation of mature ribosomal RNA as well as 4S RNA. These observations were confirmed with P^{32} label. Unlike the effect on protein synthesis, Figure 15B illustrates that "baseline" RNA synthesis (no PHA stimulation) is also inhibited.

(6) *Isopentenyladenosine does not significantly interfere with RNA maturation.* Sucrose gradient centrifugation of RNA labeled pulse with [^{3}H]-

uridine at 1, 2, 6, and 18 hours after PHA addition did not reveal interference with maturation of the newly synthesized RNA. Although [^{14}C]-methyl-methionine incorporation into RNA was inhibited by isopentenyladenosine, the inhibition was consistent with the diminished amount of newly synthesized RNA available for methylation. Wainfan and Borek

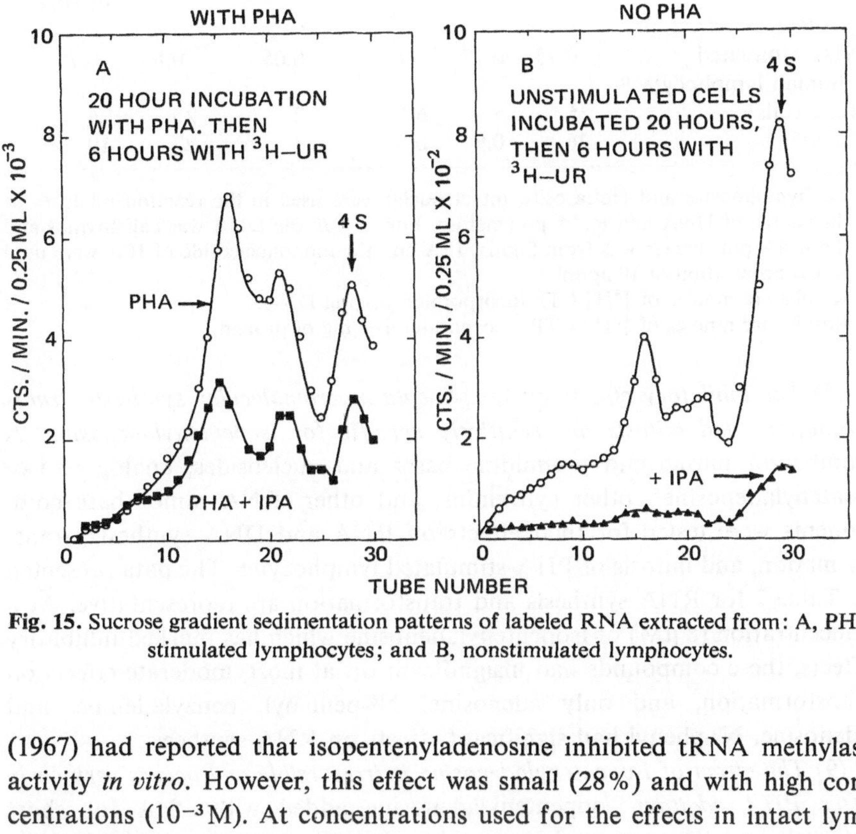

Fig. 15. Sucrose gradient sedimentation patterns of labeled RNA extracted from: A, PHA stimulated lymphocytes; and B, nonstimulated lymphocytes.

(1967) had reported that isopentenyladenosine inhibited tRNA methylase activity *in vitro*. However, this effect was small (28%) and with high concentrations (10^{-3} M). At concentrations used for the effects in intact lymphocytes (10^{-6}—10^{-5} M), we found no inhibition of tRNA methylase activities.

(7) *RNA polymerase activity is not altered.* Neither isopentenyl adenosine nor its mononucleotide had any effect on RNA polymerase activity of lymphocytes, HeLa cells, or *E. coli* (Table 6). The mononucleotide, which was generously provided by Dr. Nelson Leonard, Department of Chemistry, University of Illinois, was tested because of the possibility that the effect of isopentenyladenosine in whole cells was dependent on enzymatic conversion to the nucleotide, which might not take place under the conditions of the *in vitro* assay.

Table 6. Lack of Effect of Isopentenyladenosine on the RNA Polymerase Activity of PHA Stimulated Lymphocytes, HeLa Cells, and *E. coli*

Source	Reaction mixture[a]					
	Complete	2 °C	DNAase	Actinomycin	IPA	Mononucleotide of IPA
PHA stimulated human lymphocytes[b]	0.75	0.17	0.15	0.05	0.85	0.74
HeLa cells[b]	25	3	6	2	25	24
E. coli[c]	26	0.9	0.5	3	27	19

[a] For lymphocytes and HeLa cells, intact nuclei were used in the reaction mixtures as the source of DNA and RNA polymerase. For *E. coli*, the DNA was calf thymus and the RNA polymerase was from Sigma. IPA and the mononucleotide of IPA were used at a concentration of 10 μg/ml.
[b] Results are nmoles of [^{3}H]-CTP incorporated per mg DNA.
[c] Results are nmoles of [^{3}H]-CTP incorporated per mg of protein.

(8) *The inhibitory effects on lymphocyte macromolecular synthesis, transformation, and mitosis are relatively specific for isopentenyladenosine.* A number of purine and pyrimidine bases and nucleosides, analog of isopentenyladenosine[1], other cytokinins, and other tRNA minor base components were tested for their effects on RNA and DNA synthesis, transformation, and mitosis of PHA-stimulated lymphocytes. The data presented in Table 7 for RNA synthesis and transformation are representative. At a concentration (8 μM) of isopentenyladenosine which has marked inhibitory effects, these compounds had insignificant or, at most, moderate effects on transformation, and only adenosine, N^6-pentenyl, benzyladenine, and adenosine, N^6-phenyl had significant effects on RNA synthesis.

(9) *The effect of isopentenyladenosine is irreversible and occurs very early after PHA addition.* Isopentenyladenosine added with PHA for short intervals and then removed by washing the cells markedly diminished the response of the cells to subsequent addition of more PHA. The measurements made were RNA synthesis and morphological transformation (Table 8). These findings indicate that the effect of the cytokinin is partly irreversible. The data also suggest that the mechanism of inhibition is one of the very early events occurring after addition of PHA.

(10) *Isopentenyladenosine does not appear to compete with PHA for a lymphocyte membrane binding site.* The binding of PHA to lymphocyte cell

[1] We are very grateful to Dr. C. A. Nichol, Roswell Park Memorial Institute for the supply of many of the N^6 substituted adenosine derivatives.

Table 7. Effect of N^6-Δ^2-Isopentenyl)Adenosine and Other Purine or Pyrimidine Compounds on RNA Synthesis and Transformation of PHA-Stimulated Lymphocytes (48-hour harvest)

Purine or pyrimidine added	PHA	RNA synthesis[a] (% of control)	Trans-formation (% of control)
None	—	7	2
None	+	100	100
N^6-(Δ^2-isopentenyl)adenosine	+	5	0
Analogs of N^6-(Δ^2-isopentenyl)adenosine			
Benzyladenine	+	72	56
N^6-(Δ^2-isopentenyl)adenine	+	88	76
Adenosine, N-pentenyl	+	19	66
Adenosine, N-(2-ethoxyethyl)	+	96	84
Adenosine, N-phenyl	+	62	65
Other cytokinins			
6-furfurylaminopurine (kinetin)	+	98	104
6-(3-methyl-4-trans-hydroxy-2-	+	102	96
butenylamino)purine (zeatin)	+	102	96
Other purine and pyrimidine nucleosides and deoxynucleosides			
Inosine	+	105	102
Adenosine	+	100	96
Deoxyadenosine	+	97	100
Thymidine	+	86	80
Deoxycytidine	+	94	103
Other transfer RNA minor base components			
Pseudouridine	+	94	97

[a] Results are the mean of duplicates. The concentration of all compounds was 8 μM.

surface receptor sites appears to be the initial step in the induction of transformation and mitosis (Kornfield and Kornfield, 1969). It was of particular interest to determine if isopentenyladenosine interfered with this initial event. Lymphocytes were incubated either in the presence or in the absence of 8 μM isopentenyl-adenosine and with varying concentrations of PHA for 48 hours. The cells were then harvested and the percent of cells transformed was determined. As shown in Figure 16, increasing concentrations of PHA did not overcome the inhibition by isopentenyladenosine. Although these results are not conclusive, they indicate that the effect of isopentenyladenosine is not by competition with PHA for a lymphocyte membrane binding site.

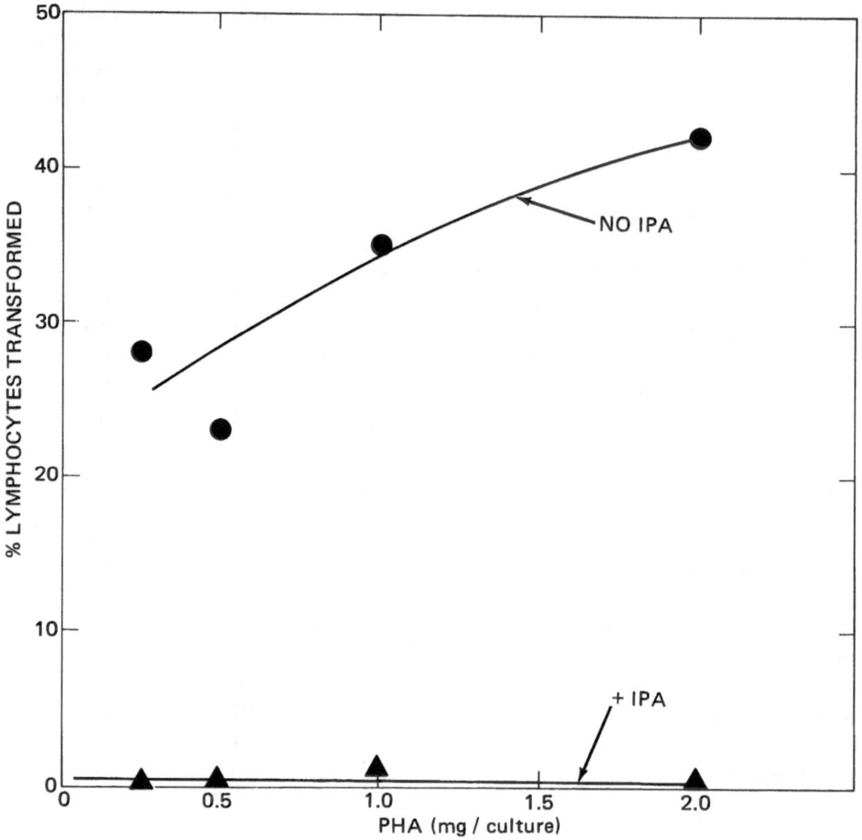

Fig. 16. Failure of increasing amounts of PHA to overcome the inhibitory effect of isopentenyladenosine (8 μM) on lymphocyte transformation. Cells were incubated in the presence of PHA or PHA plus isopentenyladenosine for 48 hours.

(11) *Isopentenyladenosine does not inactivate PHA.* PHA and isopentenyladenosine (8 μM) were incubated for 1 and 16 hours. The samples were then extensively dialyzed to remove the isopentenyladenosine. The transformation and mitogenic activity of the PHA in the dialysis bag was then tested and compared to control PHA (incubated without isopentenyladenosine and dialyzed). The results were identical. We conclude that the effect of isopentenyladenosine is prior to genome activation but after the interaction of PHA at the membrane.

Table 8. Evidence for Irreversibility of the Isopentenyladenosine Effect

Conditions[a]	RNA synthesis		Transformation (% of control)
	1 hour	20 hours	
A. PHA 1 hour, wash, add PHA and [^{14}C]-uridine pulse	1150	75,220	100
B. PHA + IPA 1 hour, wash, add PHA and [^{14}C]-uridine pulse	550	25,600	67
C. PHA 1 hour, wash, no additional PHA, [^{14}C]-uridine pulse	762	45,635	87

[a] Lymphocytes were incubated in three separate flasks: A, B, and C. Flasks A and C contained PHA, flasks B PHA and 8 μm IPA. At 1 hour the incubation was terminated, the cells were washed 3 times with 20 ml of media and pulsed with [^{14}C]-uridine. In addition, PHA was added again to flasks A and B. Aliquots were taken at intervals for measurement of RNA synthesis and at 48 hours for determination of transformation.

(12) *Inhibition of macromolecular synthesis and transformation are partially overcome with dibutyryl cyclic AMP.* As discussed above, prior to the activation of the genome and the resultant enhancement of macromolecular synthesis, PHA appears to bind to the lymphocyte membrane. Of great importance but least understood are the events that trigger genome activation after the interaction of PHA with the membrane sites. It is known that phosphorylation of histones occurs prior to the marked increase in RNA synthesis. What is not yet clear is the relationship of the histone changes to transcription, and what initially triggers the histone phosphorylation. The extraordinary information accumulating that implicates cyclic AMP as a second messenger in the action of some hormones led Smith, Steiner, and Parker to investigate cyclic AMP changes in lymphocytes. They have reported a 25–300% increase in cyclic AMP within 2–3 minutes of incubation with PHA, associated with an increase in the membrane enzyme adenyl cyclase. It was hypothesized that cyclic AMP may be the secondary messenger in the PHA activation of lymphocytes (Smith *et al.*, 1970).

We have found that the effects of isopentenyladenosine are partially reversed by dibutyryl cyclic AMP (Figure 17). Of interest are the results in the absence of isopentenyladenosine. Depending on concentration, dibutyryl cyclic AMP alone stimulates or inhibits PHA induced lymphocyte transformation. In Figure 18 the stimulatory → inhibitory concentration

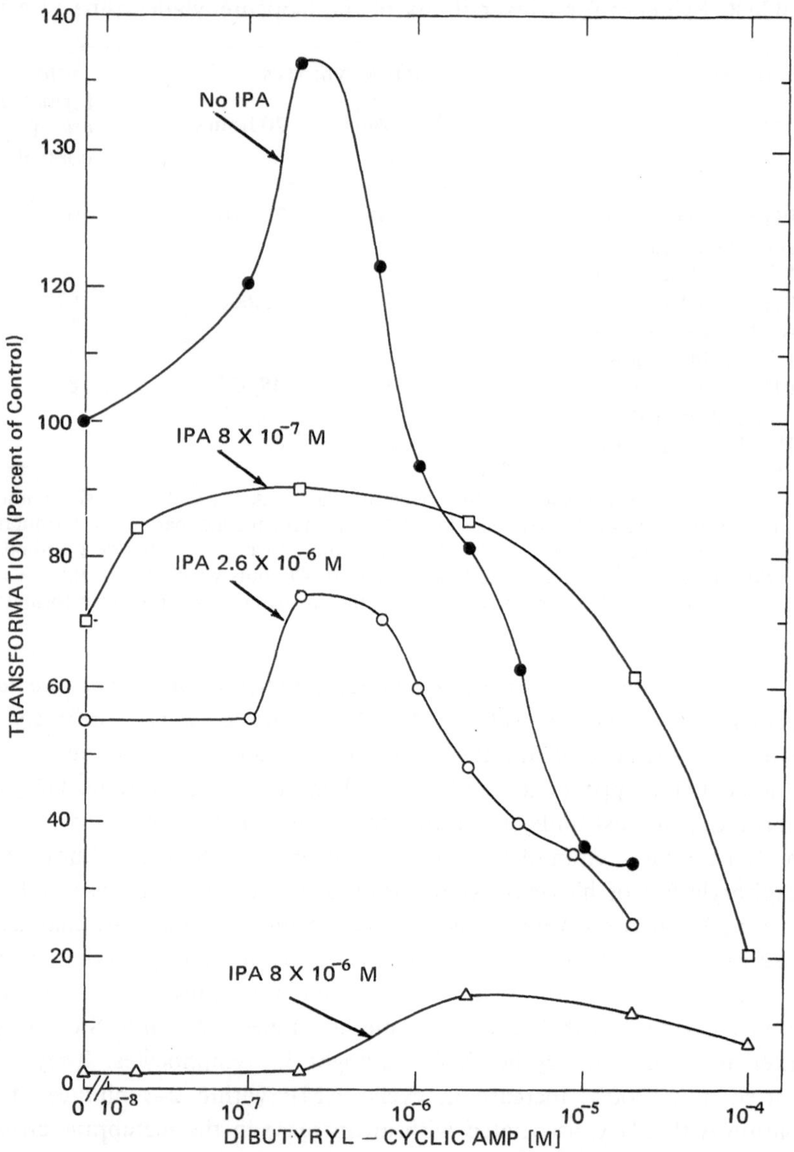

Fig. 17. The effect of dibutyryl-cyclic-AMP and isopentenyladenosine on PHA induced lymphocyte transformation. PHA, isopentenyladenosine, and dibutyryl-cyclic-AMP were added at the start of incubation and the cells harvested at 48 hours. Note that dibutyryl-cyclic-AMP partly overcomes the inhibitory effect of isopentenyladenosine and alone can enhance or inhibit transformation.

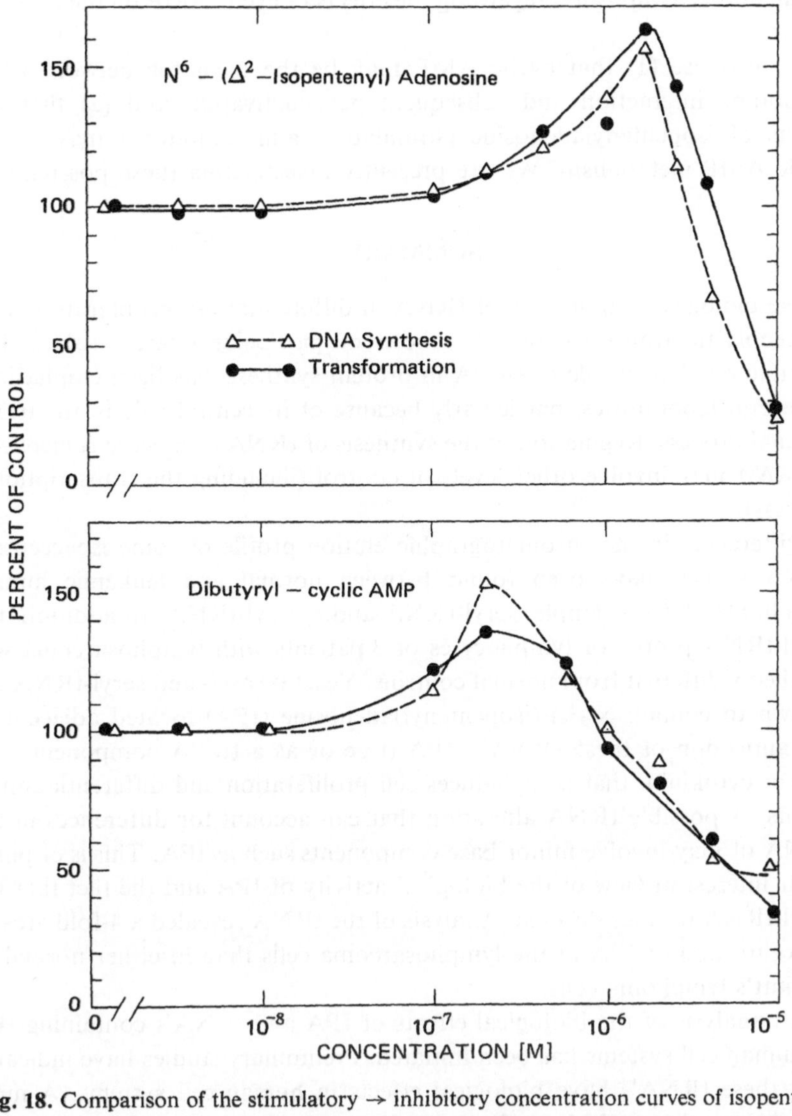

Fig. 18. Comparison of the stimulatory → inhibitory concentration curves of isopentenyladenosine (added 24 hours after PHA) and dibutyryl-cyclic-AMP (added with PHA) on lymphocyte transformation and DNA synthesis.

curves of isopentenyladenosine and dibutyryl cyclic AMP are compared. The curves (DNA synthesis and percent of lymphocytes transformed) are remarkably similar for isopentenyladenosine and dibutyryl cyclic AMP. The results showing that cyclic AMP may enhance transformation of these immunocompetent cells are strikingly similar to the results reported by

Braun in this Symposium regarding the effects of cyclic AMP on the immune responses.

We propose (1) that cyclic AMP may be the mediator between PHA membrane interaction and subsequent gene activation and (2) that the effects of isopentenyladenosine (stimulatory and inhibitory) may be on cyclic AMP metabolism. We are presently investigating these possibilities.

SUMMARY

The biological importance of tRNA in differentiation, proliferation, and in neoplastic transformation has been of particular interest to us. The possible regulatory role of tRNA in protein synthesis has been emphasized in several laboratories, particularly because of its central role in the translational process. Regulation of the synthesis of tRNA *or specific components of tRNA* may involve other levels of control (including the transcriptional process).

Differences in the chromatographic elution profile of some isoaccepting tRNA species have been found between normal and leukemic human lymphoblasts, for example, seryl-tRNA and tyrosyl-tRNA. In addition, the seryl-tRNA profile of lymphocytes of 3 patients with lymphosarcoma was markedly different from normal controls. Yeast tyrosyl- and seryl-tRNA are known to contain N^{6}-$^{2}\Delta$-(isopentenyl)adenosine (IPA) located adjacent to the anticodon of these tRNA's. IPA (free or as a tRNA component) is a potent cytokinin, that is, it induces cell proliferation and differentiation in plants. A possible tRNA alteration that can account for differences in the tRNA of may involve minor base components such as IPA. This is of particular interest in view of the biological activity of IPA and the fact that the seryl-tRNA profiles differed. Analysis of the tRNA revealed a 4-fold greater concentration of IPA in the lymphosarcoma cells than in either normal or Burkitt's lymphoma cells.

An analysis of the biological effects of IPA and tRNA's containing IPA in human cell systems has been initiated. Preliminary studies have indicated that these tRNA's have biological effects in human cell systems. A more detailed analysis of the effects of IPA on phytohemagglutinin (PHA) induced transformation and mitosis of human lymphocytes was undertaken. Transformation is preceded by a series of events involving an initial interaction of PHA at the cell membrane followed by gene derepression, morphological transformation, and mitosis.

We have found that IPA (10^{-5}—10^{-6}M) added early in the cell cycle is a potent inhibitor of PHA induced macromolecular synthesis and prevents subsequent transformation and mitosis. Lower concentrations of IPA

(10^{-7}—10^{-6}M) added late in the cell cycle enhance DNA synthesis and transformation.

Analysis of the inhibitory effects have shown that: (1) The effect is partially irreversible. (2) The effect is quite specific for IPA since IPA analogs or derivatives, some other cytokinins, and other tRNA minor base components have little or no effect. (3) Inhibition is not overcome with any purine or pyrimidine compounds. (4) There is little or no effect on the "basal" level of lymphocyte protein synthesis, but all protein synthesis induced by PHA and presumably dependent on new RNA synthesis is blocked. (5) Inhibition of protein synthesis is not by interference in the binding of seryl-tRNA to a ribosome-poly U, C complex. (6) Inhibition of protein synthesis is preceded by an inhibition of RNA synthesis. The inhibition of RNA synthesis involves all species separable by sucrose gradients. (7) Inhibitory concentrations of IPA have no direct effect on the activity of uridine kinase but inhibits the induction of uridine kinase by PHA. (8) IPA has no effect on RNA polymerase nor on ribonuclease activities, and only a slight inhibitory effect on RNA methylases. (9) The inhibition by IPA is not due to inactivation of PHA nor to competition with PHA for a membrane binding site. (10) The inhibition is partially overcome with cyclic AMP.

Although the precise biochemical mechanism for the effect of IPA is not known, we conclude that it is after the binding of PHA to the membrane and prior to the subsequent gene activation.

It has recently been reported that PHA induces adenyl cyclase in lymphocytes within a few minutes of incubation. We have found that dibutyryl-cyclic-AMP, like IPA, promoted DNA synthesis and transformation of lymphocytes at low concentrations while higher concentrations were inhibitory. It is possible that a cyclic AMP mechanism is the "mediator" between the PHA membrane effect and subsequent gene activation. We hypothesize that the mechanism of IPA may involve cyclic AMP metabolism which may account for the remarkable similarity in the concentration curves (stimulation→inhibition) of IPA and cyclic AMP.

ACKNOWLEDGEMENT

It is a pleasure to acknowledge our gratitude to Dr. T. R. Breitman, Laboratory of Physiology, National Cancer Institute, National Institutes of Health for his continued interest in this work, for his frequent and stimulating good counsel, and for reviewing the manuscript. We are also grateful to Turid Knutsen, Carla Davis and Alva Russell for excellent technical assistance.

References

Astaldi, G. and Mauri, C. (1953). Rev. Belge. Pathol. Med. Exptl. **23:** 69.

Axel, R., Weinstein, J. B., and Farber, E. (1967). Patterns of transfer RNA in normal rat liver and during hepatic carcinogenesis. Proc. Natl. Acad. Sci. U.S. **58:** 1255

Braun, W. and Nakano, M. (1967). Antibody formation: Stimulation by polyadenylic and polycytidylic acids. Science **157:** 819.

Craddock, C. G. and Nakai, G. S. (1962). Leukemic cell proliferation as determined by *in vitro* deoxyribonucleic acid synthesis. J. Clin. Invest. **41:** 360.

Fittler, F., Kline, L. K., and Hall, R. H. (1968). Biosynthesis of N^6-(Δ^2-isopentenyl)adenosine. The precursor relationship of acetate and mevalonate to the Δ^2-isopentenyl group of the transfer ribonucleic acid of microorganisms. Biochemistry **7:** 940.

——— (1968). N^6-(Δ^2-isopentenyl)adenosine: Biosynthesis *in vitro* by an enzyme extract from yeast and rat liver. Biochem. Biophys. Res. Commun. **31:** 571.

Gallo, R. C. (1969). Transfer RNA's in human leukemia. J. Cell. Physiol. **74:** 149.

——— (In press a). Physiology of human leukemic leukocytes. II. Biochemical considerations. *In* Regulation of Hematopoiesis. Ed. by A. Gordon. Appleton-Century-Crofts, New York.

——— and Pestka, S. (1970). J. Molec. Biol. **52:** 195.

——— and Perry, S. (1968). Enzyme abnormality in human leukaemia. Nature **218:** 465.

——— (1969). The enzymatic mechanisms for deoxythymidine synthesis in human leukocytes. IV. Comparisons between normal and leukemic leukocytes. J. Clin. Invest. **48:** 105.

Gallo, R. C., Whang-Peng, J., and Perry, S. (1969a). Isopentenyladenosine stimulates and inhibits mitosis of human lymphocytes treated with phytohemagglutinin. Proc. Am. Assoc. for Cancer Res. **10:** 28.

——— (1969b). Isopentenyladenosine stimulates and inhibits mitosis of human lymphocytes treated with phytohemagglutinin. Science **165:** 400.

Gavosto, F., Pileri, A., Bachi, C., and Pegoràro, L. (1964). Proliferation and maturation defect in acute leukaemia cells. Nature **203:** 92.

Gefter, M. L. and Russell, R. L. (1969). Role of modifications in tyrosine transfer RNA: A modified base affecting ribosome binding. J. Molec. Biol. **39:** 145.

Greenwalt, T. J., Gajewsik, M., and McKenna, J. L. (1962). A new method for preparing buffy coat-poor blood. Transfusion **2:** 221.

Hacker, B. and Feldbush, T. L. (1969). N^6-(Δ^2-isopentenyl)adenosine. Effects upon nucleic acid synthesis in lymphocytes *in vitro* and the development of immunologic hypersensitivity *in vivo*. Biochem. Pharm. **18:** 847.

Hausen, P. and Stein, H. (1968). On the synthesis of RNA in lymphocytes stimulated by phytohemagglutinin. I. Induction of uridine-kinase and the conversion of uridine to UTP. European J. Biochem. **4:** 401.

Herrera, F., Adamson, R. H., and Gallo, R. C. (1969). Uptake of transfer RNA by leukemic cells. Blood (abstr.) **34:** 826.

Herrera, F., Adamson, R. H., and Gallo, R. C. (1970). Uptake of transfer RNA by leukemia cells. Proc. Natl. Acad. Sci.

Kaneko, I. and Doi, R. H. (1966). Alteration of valyl-sRNA during sporulation of *Bacillus subtilis*. Proc. Natl. Acad. Sci. U.S. **55:** 564.

Kornfield, S. and Kornfield, R. (1969). Solubilization and partial characterization of a phytohemagglutinin receptor site from human erythrocytes. Proc. Natl. Acad. Sci. U.S. **63:** 1439.

Lee, J. C. and Ingram, V. M. (1967). Erythrocyte transfer RNA: Change during chick development. Science **158:** 1330.

Leonard, N. J. (1964). Trans. Morris County Res. Counc. **1:** 11.

Madison, J. T., Everett, G. A., and Kung, H. (1967). Oligonucleotides from yeast tyrosine transfer ribonucleic acid. J. Biol. Chem. **242:** 1318.

Paran, M., Ichikawa, Y., and Sachs, L. (1969). Feedback inhibition of the development of macrophage and granulocyte colonies. II. Inhibition by granulocytes. Proc. Natl. Acad. Sci. U.S. **62:** 81.

Peterkofsky, A. (1968). The incorporation of mevalonic acid into the N^6-(Δ^2-isopentenyl)adenosine of transfer ribonucleic acid in Lactobacillus acidophilus. Biochemistry **7:** 472.

Rabinowitz, Y. (1966). DNA polymerase and carbohydrate metabolizing enzyme content of normal and leukemic glass column separated leukocytes. Blood **27:** 470.

Rubin, A. D. and Cooper, H. L. (1965). Evolving patterns of RNA metabolism during transition from resting state to active growth in lymphocytes stimulated by phytohemagglutinin. Proc. Natl. Acad. Sci. U.S. **54:** 469.

Silber, R., Gabrio, B. W., and Huennekens, F. M. (1963). Studies on normal and leukemic leukocytes. VI. Thymidylate synthetase and deoxycytidylate deaminase. J. Clin. Invest. **42:** 1963.

Skoog, F., Hamzi, H. Q., Szweykowska, A. M., Leonard, N. J., Carraway, K. L., Fujii, T., Helgeson, J. P., and Leoppky, R. N. (1967). Cytokinins: Structure-activity relations. Phytochemistry **6:** 1169.

Smith, J. W., Steiner, A. L., and Parker, C. W. (1970). Early effects of phytohemagglutinin (PHA) on lymphocyte cyclic AMP levels. Fed. Proc. **29:** 369.

Smith, L. H., Jr. and Baker, F. A. (1959). Pyrimidine metabolism in man. I. The biosynthesis of orotic acid. J. Clin. Invest. **38:** 798.

——— (1960). Pyrimidine metabolism in man. III. Studies on leukocytes and erythrocytes in pernicious anemia. J. Clin. Invest. **39:** 15.

Stent, G. S. (1964). The operon: On its third anniversary. Science **144:** 816.

Taylor, M. W., Granger, G. A., Buck, C. A., and Holland, J. J. (1967). Similarities and differences among specific tRNA's in mammalian tissues. Proc. Natl. Acad. Sci. U.S. **57:** 1712.

Valentine, W. N. (1960). The metabolism of the leukemic leukocyte. Am. J. Med. **28:** 699.

Wainfan, E. and Borek, E. (1967). Differential inhibitors of tRNA methylases Molec. Pharm. **3:** 595.

Weiss, J. F. and Kelmers, A. D. (1967). A new chromatographic system for increased resolution of transfer ribonucleic acids. Biochemistry **6**: 2507.

Xefteris, E., Mitus, W. J., Mednicoff, I. B., and Dameshek, W. (1961). Leukocytic alkaline phosphatase in busulfan induced remissions of chronic granulocytic leukemia. Blood **18**: 202.

Yang, W. K. and Novelli, G. D. (1968). Isoaccepting tRNA's in mouse plasma cell tumors that synthesize different myeloma protein. Biochem. Biophys. Res. Commun. **31**: 534.

Zachau, H., Dutting, D., and Feldman, H. (1966). Nucleotide sequences of two serine-specific transfer ribonucleic acids. Angew. Chem. **78**: 392.

DISCUSSION

DR. W. BRAUN: I would like to ask Dr. Roelants and Dr. Gottlieb whether, assuming that the macrophage population contains primary antigen handling cells that are meaningfully involved in immune responses, there is not the possibility that both of them are right. We are dealing with extremely heterogeneous cell populations, and different cells in the macrophage populations probably have different functions. Therefore, is it conceivable that you may have both a nonspecific as well as specific formation of complexes?

DR. A. A. GOTTLIEB: I think the issue Dr. Braun raises is an important one, particularly with respect to the question whether these studies on peritoneal cells have anything to do with what goes on in the lymph nodes. We quite honestly don't know. My view on this is that I don't have any fundamental fault with the data that Dr. Roelants has presented here or in the past with respect to gross association of antigens and RNA. My only quarrel is that we are talking about specific binding with very specific species. On the question of whether this goes on in lymph nodes, my view is that I would like to believe it. I have no reason not to believe it.

DR. W. BRAUN: May I raise another question? Dr. Roelants refers to the importance of the conformation and the charge of the antigen in evoking the specificity of the antibody response. As we have discussed here, it has become quite clear that antigen is required at least twice in the activation of antibody formation. Antigen is not only required at the level of activation of the stem cell, but also at the level of driving the precursor cell to antibody formation. Therefore, is it not conceivable that the importance of the charge and conformation expresses itself at that latter antigen-dependent stage rather than at the early one?

DR. G. E. ROELANTS: Antigen bound to RNA? What could it do there?

DR. W. BRAUN: I am not referring to antigen bound to RNA. I am just trying to point out that the criticism that has been leveled against the small piece of antigenic residue, combined with the RNA of the macrophages being insufficient for controlling responses based on conformation, may not necessarily be valid if the whole molecule expresses its influence at the final antigen-dependent stage rather than at the initial one.

DR. G. E. ROELANTS: Yes, maybe in some systems. But it seems to become clear from the work on thymus-bone-marrow interactions that what is happening is a cooperation between the thymocytes, that is, the thymus-dependent cells, and the bone-marrow-derived cells, and antigen is acting to increase the interaction between the two, and in many cases that is the only thing you are required to have to obtain an immune response.

DR. W. BRAUN: That does not exclude the possible involvement of a primary antigen handling cell, preceding these final events, and the question I am raising is whether the initial event may involve a form of the antigen that differs from the form of the antigen that is involved in the second event in which antigen interacts with bone-marrow-derived and thymus-dependent cells.

DR. G. E. ROELANTS: I don't know.

DR. W. BRAUN: There is no experimental evidence available, I know, but I am trying to point out that since we now know that antigen is required twice in antibody formation, this suggests the possibility that the form of the antigen required may not be identical in both these events.

DR. G. E. ROELANTS: Maybe yes.

DR. A. A. GOTTLIEB: I also think that while the issue of thymus-bone-marrow interaction is quite widely accepted, this does not preclude interactions with the macrophage for some antigens. I think this is an entirely separate issue. We don't really know which antigens require such interactions; there seem to be some which do not, and we don't yet know the requirement of the macrophage for immunogenicity. However, we may recall the experiments of Mosier, which made it quite clear that the macrophages are required for the initiation of responses to sheep red blood cells. Maybe more complicated antigens, such as cells with surface antigens, require processing whereas soluble antigens may not; we just don't know.

DR. W. BRAUN: I think we all would agree with this. Furthermore, it has become evident that the macrophage does not play a role in secondary responses, and so it seems to me that in many of the recent experiments that try to exclude the participation of primary antigen-handling cells that belong to the macrophage population, it is not quite clear whether one is really dealing with a primary or secondary response.

DR. S. HECHT: Dr. Gallo, while I find it somewhat disappointing that cytokinins other than the isopentenyl adenosine have greatly diminished effects in your system, I know that the ribosides that you tested are all compounds that are also much less active as cytokinins. Have you tested any of the other three cytokinins that are known to be components of tRNA, or any of the potent synthetic cytokinins?

DR. R. C. GALLO: The answer is no.

DR. L. D'ANTONIA: In the light of the adjuvant effects discussed here today and yesterday with poly AU and poly IC, has the binding affinity of various antigens to poly AU and poly IC been determined? Would such complexes, if they do form, facilitate the processing of the antigen, possibly through the macrophage system, or even permit the antigen to bypass the macrophage?

DR. W. BRAUN: This has not been studied, but in view of the variety of materials that can be stimulated by poly AU or poly IC, I have much doubt that it involves a direct interaction between the polynucleotides and the antigen.

DR. J. W. GOODMAN: Dr. Gottlieb, you mentioned that in your RNP, derived from T2-exposed macrophages, you did have antigenic determinants that reacted with anti-T2 antibody. It is a common finding that antibody will react with fragments of antigen, but there is usually several orders of magnitude difference in the affinity of the reaction with fragments compared to the intact antigen. I wonder if you have any data on the relative binding of your anti-T2 antibody with the RNP fragment and with the intact T2.

DR. A. A. GOTTLIEB: I think that is an important point. The difficulty in giving you a direct answer is the fact that we don't have a weight basis for the amount of T2 antigen in RNP. We can only score for the ability of that preparation to act as an immunogen. However, I will point out that the amount of T2 involved, the amount of the T2 fragment, is really very small.

DR. J. W. GOODMAN: If I can ask you one more question: I wonder if you have tried, since we feel that charge is a very important feature in the formation of these complexes, any completely uncharged antigen.

DR. A. A. GOTTLIEB: We have tried some uncharged molecules and we didn't find very much binding to RNP. We would like to look at TGAL but we have not done it.

DR. D. C. BISHOP: I want to make a comment, particularly to Dr. Roelants, to indicate that the isolation of RNP molecules from macrophages has really nothing to do with phenol extraction. When we take macrophages and incubate them with iodine-labeled synthetic copolymers, particularly L-GAT, we find, in homogenates of macrophages, RNP molecules that are primarily in the supernatant, in the cytoplasmic fraction. About 85% is in the macrophage cytoplasm and 15–20% is associated with the ribosomes of macrophages. So we do not produce our effects by using phenol.

DR. G. E. ROELANTS: But aren't you using phenol?

DR. D. C. BISHOP: In many of the studies phenol was used to purify the RNA. But I am talking about the preparation of simple homogenates which show that there is a nucleoprotein in the cytoplasm of macrophages and this material possess the density of the RNP that we have defined and described previously. The remainder is bound to macrophage ribosomes, and is released from these ribosomes by methods that are frequently used to remove charged molecules from the ribosomes, such as washing the ribosomes with ammonium chloride.

INDEX